Digital Audio Technology

Digital Audio Technology

A guide to CD, MiniDisc, SACD, DVD(A), MP3 and DAT

Fourth edition

Edited by

Jan Maes and Marc Vercammen

Sony Service Centre (Europe)

Previous editions edited by

Luc Baert, Luc Theunissen, Guido Vergult, Jan Maes and Jan Arts

Sony Service Centre (Europe)

Focal Press

OXFORD   AUCKLAND   BOSTON   JOHANNESBURG   MELBOURNE   NEW DELHI

Focal Press
An imprint of Butterworth-Heinemann
Linacre House, Jordan Hill, Oxford OX2 8DP
225 Wildwood Avenue, Woburn, MA 01801-2041
A division of Reed Educational and Professional Publishing Ltd

A member of the Reed Elsevier plc group

First published as *Digital Audio and Compact Disc Technology* 1988
Second edition 1992
Third edition 1995
Reprinted 1995, 1998
Fourth edition 2001

British Library Cataloguing in Publication Data
A catalogue record for this book is available from the British Library

Library of Congress Cataloguing in Publication Data
A catalogue record for this book is available from the Library of Congress

ISBN 0 240 51654 0

For information on all Focal Press publications visit our website at
www.focalpress.com

Printed and bound in Great Britain

PLANT A TREE

British Trust for
Conservation Volunteers

FOR EVERY TITLE THAT WE PUBLISH, BUTTERWORTH-HEINEMANN
WILL PAY FOR BTCV TO PLANT AND CARE FOR A TREE.

Contents

Preface

The past century has witnessed a number of inventions and developments which have made music regularly accessible to more people than ever before. Not the least of these were the inventions of the conventional analog phonograph and the development of broadcast radio. Both have undergone successive changes or improvements, from the 78 rpm disc to the 33⅓ rpm disc, and from the AM system to the FM stereo system. These improvements resulted from demands for better and better quality.

More than 20 years ago, another change took place which now enables us to achieve the highest possible audio fidelity yet – the introduction of digital technology, specifically the compact disc. Research and development efforts, concentrated on consumer products, have made the extraordinary advantages of digital audio systems easily accessible at home. The last decade has witnessed an exponential growth of digital media, disc based as well as network based. To address these new media, the latest edition of this book includes the newest developments; as a result, the title has been changed to visualize this evolution. Sony is proud to have been one of the forerunners in this field, co-inventor of the compact disc digital audio system and inventor of the MiniDisc, which has led to an entirely new level of quality music.

Sony Service Centre (Europe) NV

A short history of audio technology

Early years: from phonograph to stereo recording

The evolution of recording and reproduction of audio signals started in 1877, with the invention of the phonograph by T. A. Edison. Since then, research and efforts to improve techniques have been determined by the ultimate aim of recording and reproducing an audio signal faithfully, i.e., without introducing distortion or noise of any form.

With the introduction of the gramophone, a disc phonograph, in 1893 by P. Berliner, the original form of our present record was born. This model could produce a much better sound and could also be reproduced easily.

Around 1925 electric recording was started, but an acoustic method was still mainly used in the sound reproduction system: where the sound was generated by a membrane and a horn, mechanically coupled to the needle in the groove in playback. When recording, the sound picked up was transformed through a horn and membrane into a vibration and coupled directly to a needle which cut the groove onto the disc.

Figure I shows Edison's original phonograph, patented in 1877, which consisted of a piece of tin foil wrapped around a rotating cylinder.

Vibration of his voice spoken into a recording horn (as shown) caused the stylus to cut grooves into a tin foil. The first sound recording made was Edison reciting 'Mary had a little lamb' (Edison National History Site).

Figure I Edison's phonograph.

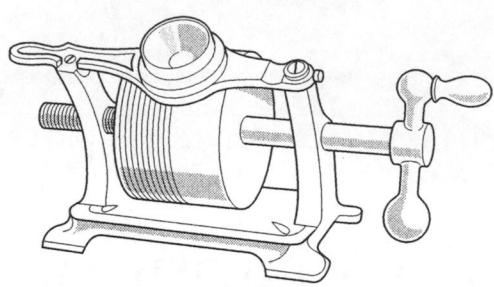

Figure II shows the Berliner gramophone, manufactured by US Gramophone Company, Washington, DC. It was hand-powered and required an operator to crank the handle up to a speed of 70 revolutions per minute (rpm) to get a satisfactory playback (Smithsonian Institution).

Figure II Berliner gramophone.

Further developments, such as the electric crystal pick-up and, in the 1930s, broadcast AM radio stations, made the SP (standard playing 78 rpm record) popular. Popularity increased with the development, in 1948 by CBS, of the 33⅓ rpm long-playing record (LP), with about 25 minutes of playing time on each side. Shortly after this, the EP (extended play) 45 rpm record was introduced by RCA with an improvement in record sound quality. At the same time, the lightweight pick-up cartridge, with only a few grams of stylus pressure, was developed by companies like General Electric and Pickering.

The true start of progress towards the ultimate aim of faithful recording and reproduction of audio signals was the introduction of stereo records in 1956. This began a race between manufacturers to produce a stereo reproduction tape recorder, originally for industrial master use. However, the race led to a simplification of techniques which, in turn, led to the development of equipment for domestic use.

Broadcast radio began its move from AM to FM, with consequent improvement of sound quality, and in the early 1960s stereo FM broadcasting became a reality. In the same period, the compact cassette recorder, which would eventually conquer the world, was developed by Philips.

Developments in analog reproduction techniques

The three basic media available in the early 1960s – tape, record and FM broadcast – were all analog media. Developments since then include the following.

Developments in turntables

There has been remarkable progress since the stereo record appeared. Cartridges, which operate with stylus pressure of as little as 1 gram, were developed and tonearms which could trace the sound groove perfectly with this 1-gram pressure were also made. The hysteresis synchronous motor and DC servo motor

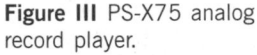

Figure III PS-X75 analog record player.

were developed for quieter, regular rotation and elimination of rumble. High-quality heavyweight model turntables, various turntable platters and insulators were developed to prevent unwanted vibrations from reaching the stylus. With the introduction of electronic technology, full automation was performed. The direct drive system with the electronically controlled servo motor, the BSL motor (brushless and slotless linear motor) and the quartz-locked DC servo motor were finally adopted together with the linear tracking arm and electronically controlled tonearms (biotracer). So, enormous progress was achieved since the beginning of the gramophone: in the acoustic recording period, disc capacity was 2 minutes on each side at 78 rpm, and the frequency range was 200 IIz–3 kHz, with a dynamic range of 18 dB. At its latest stage of development, the LP record frequency range is 30 Hz–15 kHz, with a dynamic range of 65 dB in stereo.

Developments in tape recorders

In the 1960s and 1970s, the open reel tape recorder was the instrument used both for record production and for broadcast, so efforts were constantly made to improve the performance and quality of the signal. Particular attention was paid to the recording and reproduction heads, recording tape as well as tape path drive mechanism with, ultimately, a wow and flutter of only 0.02% wrms at 38 cm s^{-1}, and of 0.04% wrms at 19 cm s^{-1}. Also, the introduction of compression/expansion systems such as Dolby, dBx, etc. improved the available signal-to-noise ratios.

Professional open reel tape recorders were too bulky and too expensive for general consumer use, however, but since its invention in 1963 the compact cassette recorder began to make it possible for millions of people to enjoy recording and playing back music with reasonable tone quality and easy operation. The impact of the compact cassette was enormous, and tape recorders for recording and playing back these cassettes became quite indispensable for music lovers, and for those who use the cassette recorders for a myriad of purposes such as taking notes for study, recording speeches, dictation, for 'talking letters' and for hundreds of other applications.

Inevitably, the same improvements used in open reel tape recorders eventually found their way into compact cassette recorders.

Jo Coleman

Information Update Service

Butterworth-Heinemann

FREEPOST SCE 5435

Oxford

Oxon

OX2 8BR

UK

Keep up-to-date with the latest books in your field.

Visit our website and register now for our FREE e-mail update service, or join our mailing list and enter our monthly prize draw to win £100 worth of books. Just complete the form below and return it to us now! (FREEPOST if you are based in the UK)

www.bh.com

Please Complete In Block Capitals

Title of book you have purchased:..

....

Subject area of interest:..

Name:...

Job title:...

Business sector (if relevant)..

Street:..

Town:.. County..

Country:... Postcode:...

Email:...

Telephone:..

How would you prefer to be contacted: Post ☐   e-mail ☐   Both ☐

Signature:.. Date:..

☐ Please arrange for me to be kept informed of other books and information services on this and related subjects (✔ box if not required). This information is being collected on behalf of Reed Elsevier plc group and may be used to supply information about products by companies within the group.

FOR OFFICE USE ONLY

Butterworth-Heinemann,
a division of Reed Educational
& Professional Publishing Limited.
Registered office: 25 Victoria Street,
London SW1H 0EX.
Registered in England 3099304.
VAT number GB: 663 3472 30.

BUTTERWORTH
HEINEMANN

A member of the Reed Elsevier plc group

Figure IV TC-766-2 analog domestic reel-to-reel tape recorder.

Limitations of analog audio recording

Despite the spectacular evolution of techniques and the improvements in equipment, by the end of the 1970s the industry had almost reached the level above which few further improvements could be performed without increasing dramatically the price of the equipment. This was because quality, dynamic range and distortion (in its broadest sense) are all determined by the characteristics of the medium used (record, tape, broadcast) and by the processing equipment. Analog reproduction techniques had just about reached the limits of their characteristics.

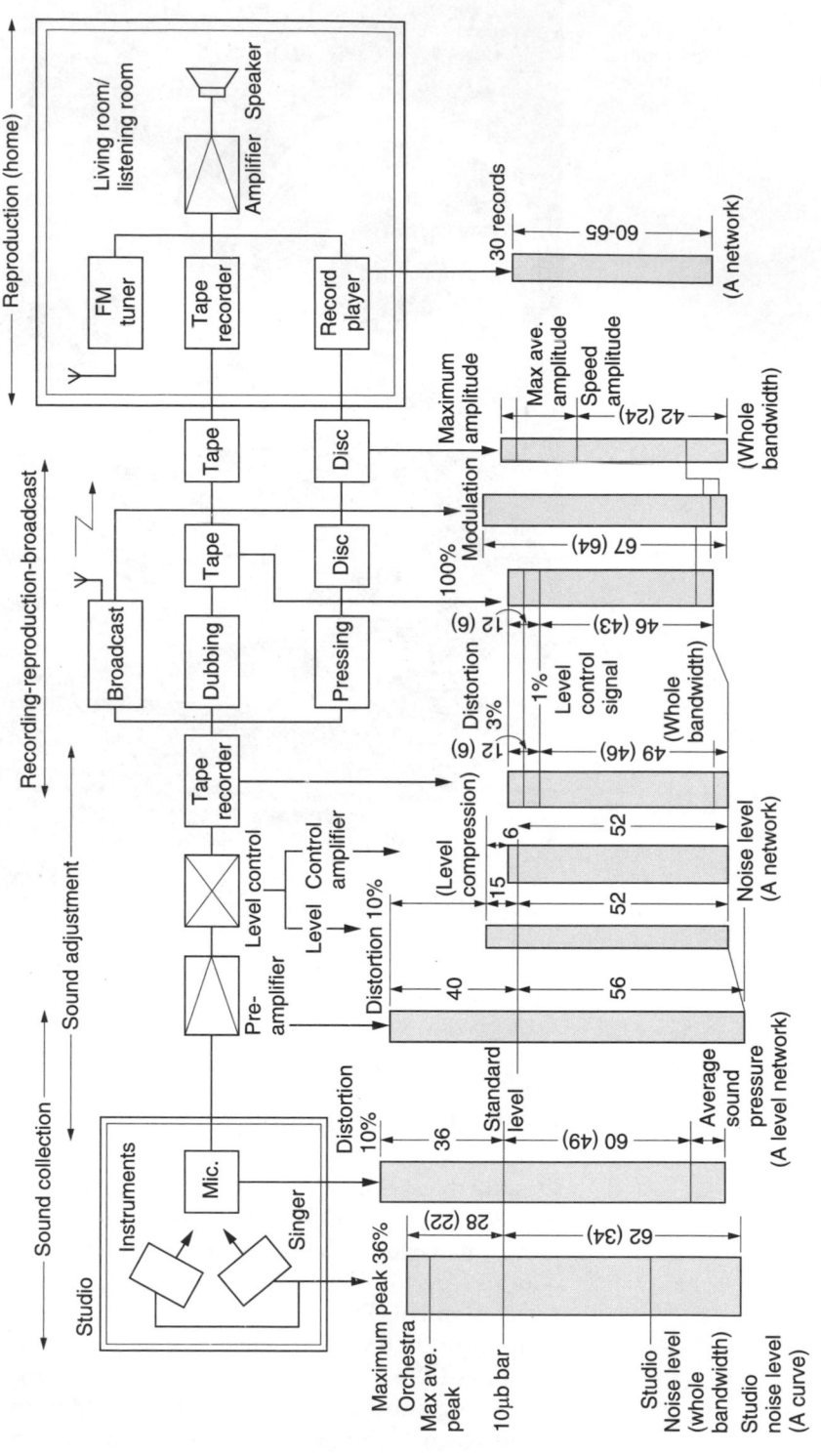

Figure V Typical analog audio systems, showing dynamic range.

Figure V represents a standard analog audio chain, from recording to reproduction, showing dynamic ranges in the three media: tape, record, broadcast.

The lower limit of dynamic range is determined by system noise and especially the lower frequency component of the noise. Distortion by system non-linearity generally sets the upper limit of dynamic range.

The strength and extent of a pick-up signal from a microphone is determined by the combination of the microphone sensitivity and the quality of the microphone preamplifier, but it is possible to maintain a dynamic range in excess of 90 dB by setting the levels carefully. However, the major problems in microphone sound pick-up are the various types of distortion inherent in the recording studio, which cause a narrowing of the dynamic range, i.e., there is a general minimum noise level in the studio, created by, say, artists or technical staff moving around, or the noise due to air currents and breath, and all types of electrically induced distortions.

Up to the pre-mixing and level amplifiers no big problems are encountered. However, depending on the equipment used for level adjustment, the low and high limits of dynamic range are affected by the use of equalization. The type and extent of equalization depend on the medium. Whatever, control amplification and level compression are necessary, and this affects the sound quality and the audio chain. Furthermore, if you consider the fact that for each of the three media (tape, disc, broadcast) master tape and mother tape are used, you can easily understand that the narrow dynamic range available from conventional tape recorders becomes a 'bottle neck' which affects the whole process.

To summarize, in spite of all the spectacular improvements in analog technology, it is clear that the original dynamic range is still seriously affected in the analog reproduction chain.

Similar limits to other factors affecting the system – frequency response, signal-to-noise ratio, distortion, etc. – exist simply due to the analog processes involved. These reasons prompted manufacturers to turn to digital techniques for audio reproduction.

First development of PCM recording systems

The first public demonstration of pulse code modulated (PCM) digital audio was in May 1967, by NHK (Japan Broadcasting Corporation) and the record medium used was a 1-inch, two-

7

head, helical scan VTR. The impression gained by most people who heard it was that the fidelity of the sound produced by the digital equipment could not be matched by any conventional tape recorder. This was mainly because the limits introduced by the conventional tape recorder simply no longer occurred.

As shown in Figure VIa, the main reason why conventional analog tape recorders cause such a deterioration of the original signal is firstly that the magnetic material on the tape actually contains distortion components before anything is actually recorded. Secondly, the medium itself is non-linear, i.e., it is not capable of recording and reproducing a signal with total accuracy. Distortion is, therefore, built in to the very heart of every analog tape recorder. In PCM recording (Figure VIb), however, the original bit value pattern corresponding to the audio signal, and thus the audio signal itself, can be fully recovered, even if the recorded signal is distorted by tape non-linearities and other causes.

After this demonstration at least, there were no grounds for doubting the high sound quality achievable by PCM techniques. The engineers and music lovers who were present at this first public PCM playback demonstration, however, had no idea when this equipment would be commercially available, and many of these people had only the vaguest concept of the effect which PCM recording systems would have on the audio industry. In

Figure VI Conventional analog (a) and PCM digital (b) tape recording.

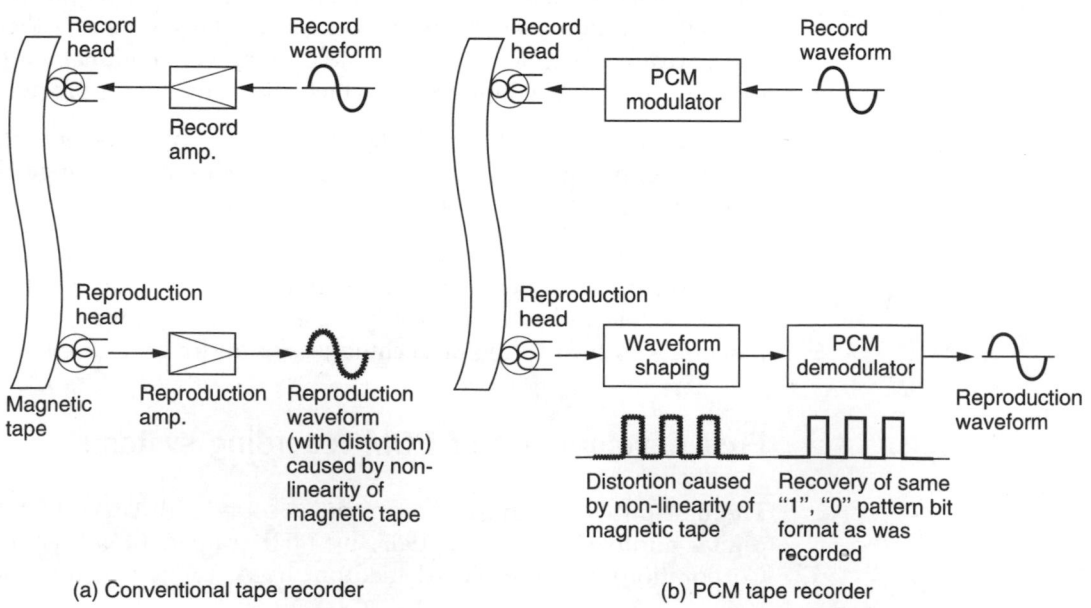

(a) Conventional tape recorder

(b) PCM tape recorder

fact, it would be no exaggeration to say that, owing to the difficulty of editing, the weight, size, price and difficulty in operation, not to mention the necessity of using the highest quality ancillary equipment (another source of high costs), it was at that time very difficult to imagine that any meaningful progress could be made.

Nevertheless, highest quality record production at that time was by the direct cutting method, in which the lacquer master is cut without using master and mother tapes in the production process: the live source signal is fed directly to the disc cutting head after being mixed. Limitations due to analog tape recorders were thus side-stepped. Although direct cutting sounds quite simple in principle, it is actually extremely difficult in practice. First of all, all the required musical and technical personnel, the performers, the mixing and cutting engineers, have to be assembled together in the same place at the same time. Then the whole piece to be recorded must be performed right through from beginning to end with no mistakes, because the live source is fed directly to the cutting head.

If PCM equipment could be perfected, high-quality records could be produced while solving the problems of time and value posed by direct cutting. PCM recording meant that the process after the making of the master tape could be completed at leisure.

In 1969, Nippon Columbia developed a prototype PCM recorder, loosely based on the PCM equipment originally created by NHK: a four-head VTR with 2-inch tape was used as recording medium, with a sampling rate of 47.25 kHz using a 13-bit linear analog-to-digital converter. This machine was the starting point for the PCM recording systems which are at present marketed by Sony, after much development and adaptation.

Development of commercial PCM processors

In a PCM recorder, there are three main parts: an encoder which converts the audio source signal into a digital PCM signal; a decoder to convert the PCM signal back into an audio signal; and, of course, there has to be a recording medium, using some kind of magnetic tape for record and reproduction of the PCM encoded signal.

The time period occupied by 1 bit in the stream of bits composing a PCM encoded signal is determined by the sampling frequency and the number of quantization bits. If, say, a sampling frequency of 50 kHz is chosen (sampling period 20 µs), and that a 16-bit quantization system is used, then the time period occupied by 1 bit when making a two-channel recording will be about 0.6 µs. In order to ensure the success of the recording, detection bits for the error-correction system will also have to be included. As a result, it is necessary to employ a record/reproduction system which has a bandwidth of between about 1 and 2 MHz.

Bearing in mind this bandwidth requirement, the most suitable practical recorder is a video tape recorder (VTR). The VTR was specifically designed for recording TV pictures, in the form of video signals. To successfully record a video signal, a bandwidth of several megahertz is necessary, and it is a happy coincidence that this makes the VTR eminently suitable for recording a PCM encoded audio signal.

The suitability of the VTR as an existing recording medium meant that the first PCM tape recorders were developed as two-unit systems comprising a VTR and a digital audio processor. The latter was connected directly to an analog hi-fi system for actual reproduction. Such a device, the PCM-1, was first marketed by Sony in 1977.

In the following year, the PCM-1600 digital audio processor for professional applications was marketed. In April 1978, the use of 44.056 kHz as a sampling frequency (the one used in the above-mentioned models) was accepted by the Audio Engineering Society (AES).

Figure VII PCM-1 digital audio processor.

At the 1978 Consumer Electronics Show (CES) held in the USA, an unusual display was mounted. The most famous names among the American speaker manufacturers demonstrated their latest products using a PCM-1 digital audio processor and a consumer VTR as the sound source. Compared with the situation only a few years ago, when the sound quality available from tape recorders was regarded as being of relatively low standard, the testing of speakers using a PCM tape recorder marked a total reversal of thought. The audio industry had made a major step towards true fidelity to the original sound source, through the total redevelopment of the recording medium which used to cause most degradation of the original signal.

At the same time, a committee for the standardization of matters relating to PCM audio processors using consumer VTRs was established, in Japan, by 12 major electronics companies. In May 1978, they reached agreement on the Electronics Industry Association of Japan (EIAJ) standard. This standard basically agreed on a 14-bit linear data format for consumer digital audio applications.

The first commercial processor for domestic use according to this EIAJ standard, which gained great popularity, was the now famous PCM-F1 launched in 1982. This unit could be switched from 14-bit into 16-bit linear coding/decoding format so, in spite of being basically a product designed for the demanding hi-fi enthusiast, its qualities were so outstanding that it was immediately used on a large scale in the professional audio recording business as well, thus quickening the acceptance of digital audio in the recording studios.

Figure VIII PCM-F1 digital audio processor.

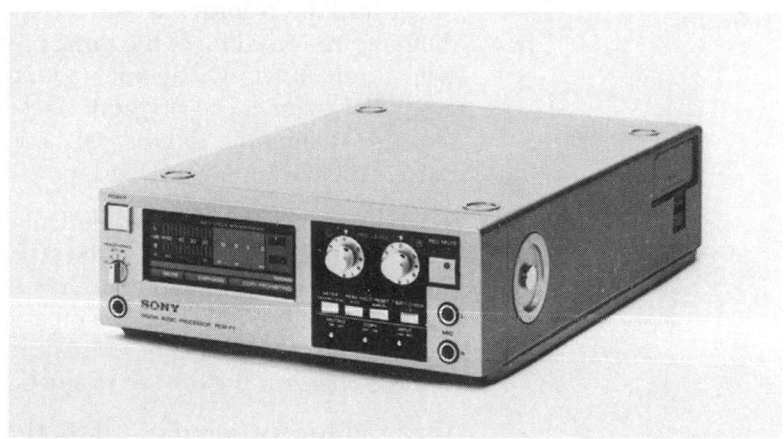

Figure IX PCM-1610 digital audio processor.

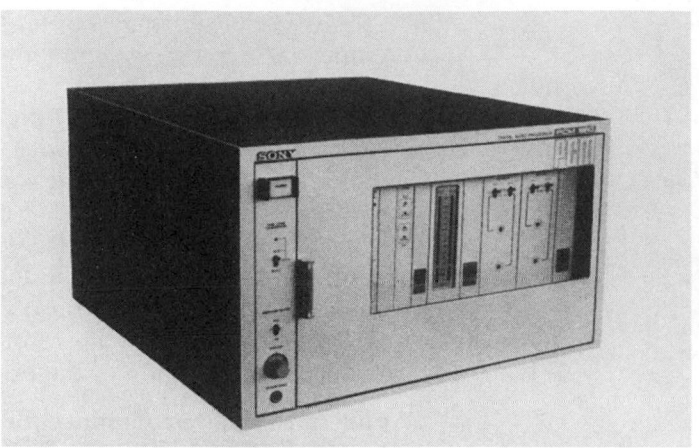

In the professional field the successor to the PCM-1600, the PCM-1610, used a more elaborate recording format than EIAJ and consequently necessitated professional VTRs based on the U-Matic standard. It quickly became a de facto standard for two-channel digital audio production and compact disc mastering.

Stationary-head digital tape recorders

The most important piece of equipment in the recording studio is the multi-channel tape recorder: different performers are recorded on different channels – often at different times – so that the studio engineer can create the required 'mix' of sound before editing and dubbing. The smallest number of channels used is generally four, the largest 32.

A digital tape recorder would be ideal for studio use because dubbing (re-recording of the same piece) can be carried out more or less indefinitely. On an analog tape recorder (Figure XI), however, distortion increases with each dub. Also, a digital tape recorder is immune to cross-talk between channels, which can cause problems on an analog tape recorder.

It would, however, be very difficult to satisfy studio standard requirements using a digital audio processor combined with a VTR. For a studio, a fixed head digital tape recorder would be the answer. Nevertheless, the construction of a stationary-head digital tape recorder poses a number of special problems. The most important of these concerns the type of magnetic tape and the heads used.

The head-to-tape speed of a helical scan VTR (Figure XI) used with a digital audio processor is very high, around 10 m s^{-1}.

Figure X PCM-3324 digital audio stationary-head (DASH) recorder.

However, on a stationary-head recorder, the maximum speed possible is around 50 cm s^{-1}, meaning that information has to be packed much more closely on the tape when using a stationary-head recorder; in other words, it has to be capable of much higher recording densities. As a result of this, a great deal of research was carried out in the 1970s into new types of modulation recording systems and special heads capable of handling high-density recording.

Another problem is generated when using a digital tape recorder to edit audio signals – it is virtually impossible to edit without introducing 'artificial' errors in the final result. Extremely powerful error-correcting codes were invented capable of eliminating these errors.

Figure XI Analog audio and video tape recording.

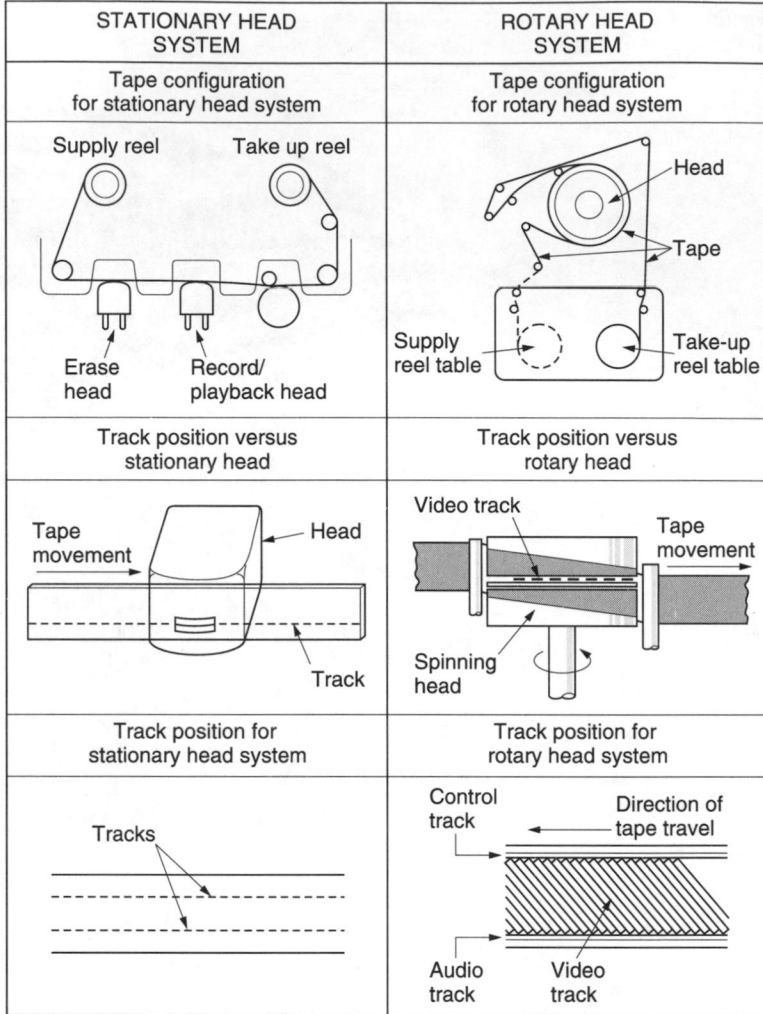

STATIONARY HEAD SYSTEM	ROTARY HEAD SYSTEM
Tape configuration for stationary head system	Tape configuration for rotary head system
Supply reel · Take up reel · Erase head · Record/playback head	Head · Tape · Supply reel table · Take-up reel table
Track position versus stationary head	Track position versus rotary head
Tape movement · Head · Track	Video track · Tape movement · Spinning head
Track position for stationary head system	Track position for rotary head system
Tracks	Control track · Direction of tape travel · Audio track · Video track

The digital multi-channel recorder had finally developed after all the problems outlined above had been resolved. A standard format for stationary-head recorders, called DASH (digital audio stationary head), was agreed upon by major manufacturers like Studer, Sony and Matsushita. An example of such a machine is the 24-channel Sony PCM-3324.

Development of the compact disc

In the 1970s, the age of the video disc began, with three different systems being pursued: the optical system, where the video signal is laid down as a series of fine grooves on a sort of record, and is read off by a laser beam; the capacitance system, which uses changes in electrostatic capacitance to plot the video signal; and the electrical system, which uses a transducer. Engineers then began to think that because the bandwidth needed to record a video signal on a video disc was more than the one needed to record a digitized sound signal, similar systems could be used for PCM/VTR recorded material. Thus, the digital audio disc (DAD) was developed, using the same technologies as the optical video discs: in September 1977, Mitsubishi, Sony and Hitachi demonstrated their DAD systems at the Audio Fair. Because everyone knew that the new disc systems would eventually become widely used by the consumer, it was absolutely vital to reach some kind of agreement on standardization.

Furthermore, Philips from the Netherlands, who had been involved in the development of video disc technology since the early 1970s, had by 1978 also developed a DAD, with a diameter of only 11.5 cm, whereas most Japanese manufacturers were thinking of a 30-cm DAD, in analogy with the analog LP. Such a large record, however, would hold up to 15 hours of music, so it would be rather impractical and expensive.

During a visit of Philips executives to Tokyo, Sony was confronted with the Philips idea, and they soon joined forces to develop what was to become the now famous compact disc, which was finally adopted as a worldwide standard. The eventual disc size was decided upon as 12 cm, in order to give it a capacity of 74 minutes: the approximate duration of Beethoven's Ninth Symphony.

The compact disc was finally launched on the consumer market in October 1982, and in a few years, it gained great popularity with the general public, becoming an important part of the audio business.

Peripheral equipment for digital audio production

It is possible to make a recording with a sound quality extremely close to the original source when using a PCM tape recorder, as digital tape recorders do not 'colour' the recording, a failing inherent in analog tape recorders. More important, a digital tape recorder offers scope for much greater freedom and flexibility during the editing process.

There follows a brief explanation of some peripheral equipment used in a studio or a broadcasting station as part of a digital system for the production of software.

- **Digital mixer**. A digital mixer which processed the signal fed to it digitally would prevent any deterioration in sound quality, and would allow the greatest freedom for the production of software. The design and construction of a digital mixer is an extremely demanding task. However, multi-channel mixers suitable for use in studios and broadcasting stations have been produced and are starting to replace analog mixing tables in demanding applications.
- **Digital editing console**. One of the major problems associated with using a VTR-based recording system is the difficulty in editing. The signal is recorded onto a VTR cassette, which means that normal cutting and splicing of the tape for editing purposes is impossible. Therefore, one of the most pressing problems after development of PCM recording systems is the design of an electronic editing console. The most popular editing console associated with the PCM-1610 recording system is Sony's DAE-1100.
- **Digital reverberator**. A digital reverberator is based on a totally different concept from conventional reverb units, which mostly use a spring or a steel plate to achieve the desired effect. Such mechanical reverbs are limited in the reverb effect and suffer significant signal degradation. The reverb effect available from a digital reverberation unit covers an extremely wide and precisely variable range, without signal degradation.
- **Sampling frequency conversion and quantization processing**. A sampling frequency unit is required to connect together two pieces of digital recording equipment which use

Figure XII Digital editing console.

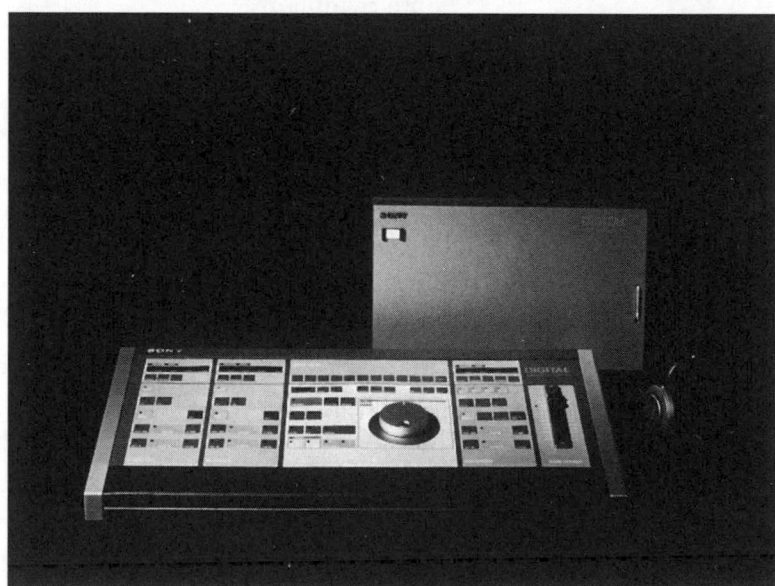

Figure XIII Digital
reverberator.

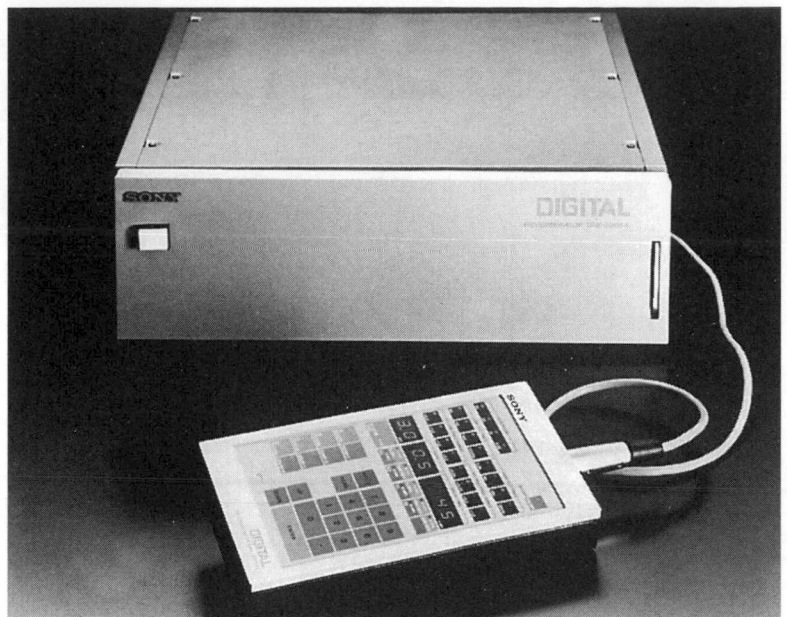

different sampling frequencies. Similarly, a quantization processor is used between two pieces of equipment using different quantization bit numbers. These two devices allow free transfer of information between digital audio equipment of different standards.

Outline of a digital audio production system

Several units from the conventional analog audio record production system can already be replaced by digital equipment in order to improve the quality of the end product, as shown in Figure XIV.

Audio signals from the microphone, after mixing, are recorded on a multi-channel digital audio recorder. The output from the digital recorder is then mixed into a stereo signal through the analog mixer, with or without the use of a digital reverberator.

The analog output signal from the mixer is then converted into a PCM signal by a digital audio processor and recorded on a VTR.

Editing of the recording is performed on a digital audio editor by means of digital audio processors and VTRs. The final result is stored on a VTR tape or cassette. The cutting machine used is a digital version.

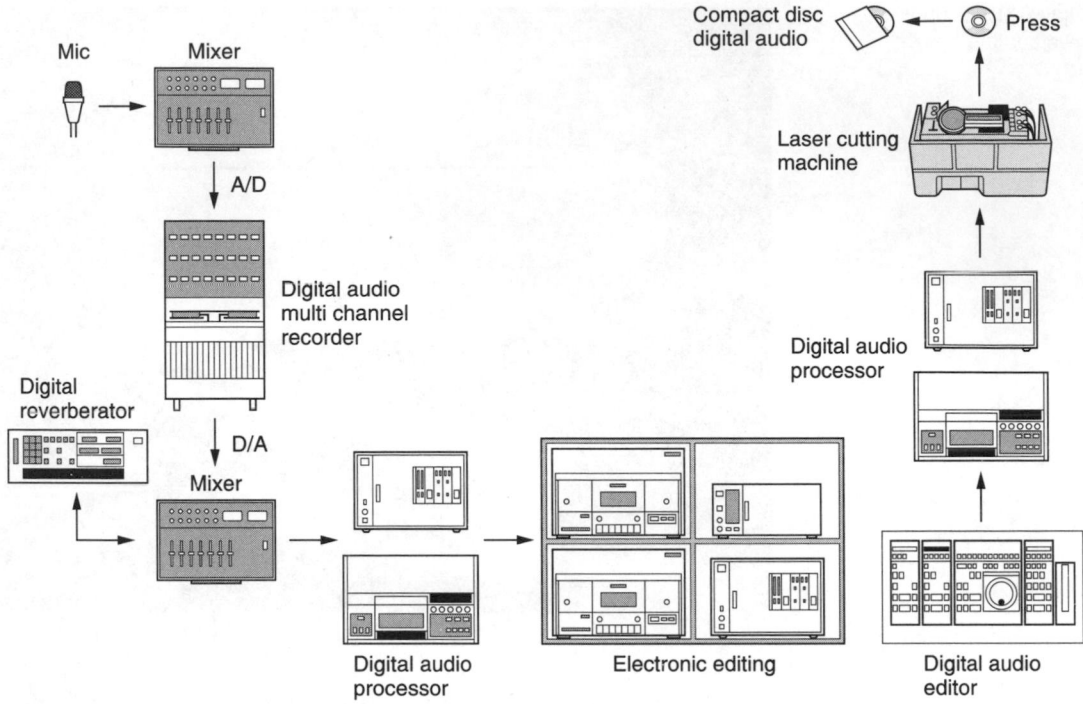

Figure XIV Digital audio editing and record production.

When the mixer is replaced by a digital version and – in the distant future – a digital microphone is used, the whole production system will be digitized.

Digital audio broadcasting

Since 1978, FM broadcasting stations have expressed a great deal of interest in digital tape recorders, realizing the benefits they could bring almost as soon as they had been developed. Figure XV shows an FM broadcast set-up using digital tape recorders to maintain high-quality broadcasts.

In 1987, the European Eureka-147 project was the starting point to develop Digital Audio Broadcasting (DAB). This full digital broadcasting system has meanwhile been fully developed, it is implemented in a number of countries around the world and gradually it is becoming a successor to the 'old' FM broadcasting system.

Basic specifications for DAB are a transmission bandwidth of 1.54 MHz; the audio is encoded according to ISO/MPEG layer II with possible sampling frequencies of 48 and 24 kHz. The

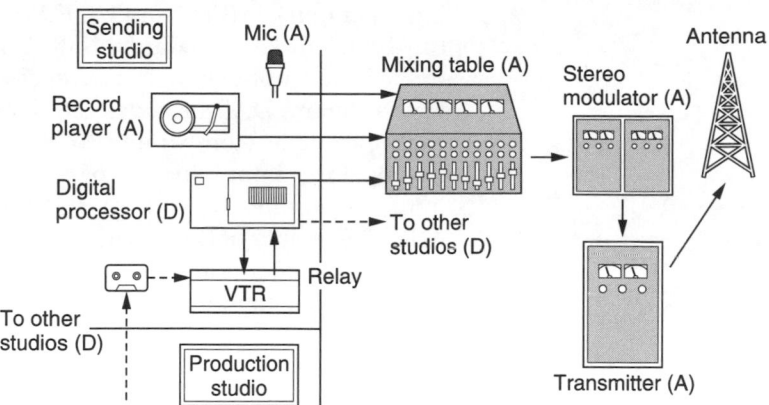

Figure XV Mixed analog and digital audio broadcasting system.

channel coding system is Coded Orthogonal Frequency Division Multiplex (COFDM) and the used frequencies are such that it can be used both on a terrestrial and satellite base.

A number of frequency blocks have been agreed: 12 blocks within the 87–108 MHz range, 38 blocks within VHF band III (174–240 MHz) and 23 blocks within the L-band (1.452–1.492 GHz). These are the terrestrial frequencies. For satellite use, a number of frequencies in the 1.5, 2.3 and 2.6 GHz band are possible, but this is still in the experimental stage.

Due to the use of the latest digital encoding, error protection and transmission techniques, this broadcasting system can drastically change our radio listening habits. First of all, due to the encod-

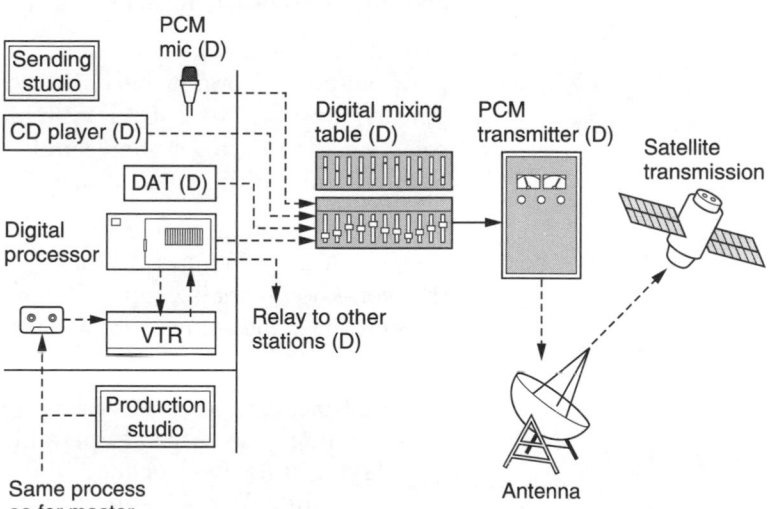

Figure XVI Digital audio broadcasting system.

ing techniques (MPEG and COFDM) used, the quality of sound can be up to CD standards, but also reception can be virtually free of interruptions or interferences. The contents and use are also different; DAB provides a number of so-called 'ensembles', each ensemble (comparable to an FM station frequency) containing a maximum number of 63 channels. In other words, tune in to one ensemble and there is a choice between 63 music channels. The quality of each channel is adaptable by the broadcasters to suit the contents and needs. A music programme, for example, can have the highest quality; a live transmission of a football match, on the other hand, can be lower quality, as full quality (needing full bandwidth) is not needed.

Another important aspect of the DAB system is the high data capacity. Apart from the contents (music), a lot of other information can be broadcast at the same time: control information, weather info, traffic info, multimedia data services, etc. are all perfectly possible and can still be developed further.

The experience gained from DAB will surely be useful for the next step: Digital Video Broadcasting (DVB).

Digital recording media: R-DAT and S-DAT

Further investigation to develop small, dedicated, digital audio recorders which would not necessitate a video recorder has led to the parallel development of both rotary-head and stationary-head approaches, resulting in the so-called R-DAT (rotary-head digital audio tape recorder) and S-DAT (stationary-head digital audio tape recorder) formats.

Like its professional counterpart, the S-DAT system relies on multiple-track thin-film heads to achieve a very high packing density, whereas the R-DAT system is actually a miniaturized rotary-head recording system, similar to video recording systems, optimized for audio applications.

The R-DAT system, launched first on the market, uses a small cassette of only 73 mm × 54 mm × 10.5 mm – about one-half of the well-known analog compact cassette. Tape width is 3.81 mm – about the same as the compact cassette. Other basic characteristics of R-DAT are:

- Multiple functions: a variety of quantization modes and sampling rates; several operating modes such as extra-long playing time, two- or four-channel recording, direct recording of digital broadcasts.

- Very high sound quality can be achieved using a 48 kHz sampling frequency and 16-bit linear quantization.
- High-speed search facilities.
- Very high recording density, hence reduced running cost (linear tape speed only 0.815 cm s^{-1}).
- Very small mechanism.
- Serial Copy Management System (SCMS), a copy system that allows the user to make a single copy of a copyrighted digital source.

A description of the R-DAT format and system is given in Chapter 15.

The basic specifications of the S-DAT system have been initially determined: in this case, the cassette will be reversible (as the analog compact cassette), using the same 3.81 mm tape.

Two or four audio channels will be recorded on 20 audio tracks. The width of the tracks will be only 65 µm (they are all recorded on a total width of only 1.8 mm!) and, logically, various technological difficulties hold up its practical realization.

New digital audio recording media

R-DAT and S-DAT have never established themselves in the mass consumer market. Expensive equipment and lack of pre-recorded tapes have led consumers to stay with the analog compact cassette, although broadcasters and professional users favoured the R-DAT format because of its high sound quality. This situation has led to the development of several new consumer digital audio formats, designed to be direct replacements for the analog compact cassette. The requirements for the new formats are high. The recording media must be small and portable with a sound quality approaching the sound quality of a CD. The formats must receive strong support from the software industry for pre-recorded material. Above all, the equipment must be cheap.

DCC

As an answer to customers' requests to have a cheap and easy to use recording system, Philips developed the Digital Compact Cassette (DCC) system. This system was launched on the market in the beginning of the 1990s.

As the name suggests, DCC has a lot in common with the analog compact cassette which was introduced almost 30 years ago by the same company. The DCC basically has the same dimensions as the analog compact cassette, which enables downward compatibility: DCC machines were indeed able to play back analog compact cassettes.

The DCC system uses a stationary thin-film head with 18 parallel tracks, comparable with the now obsolete S-DAT format. Tape width and tape speed are equal to those of the analog compact cassette.

The proposed advantage (backward compatibility with analog compact cassettes) turned out to be one of the main reasons why DCC never gained a strong foothold and was stopped in the middle of the 1990s. By that time, people were used to the advantages of disc-based media: easy track jumping, quick track search, etc. and as DCC was still tape based, it was not able to exploit these possibilities.

MiniDisc

With the advantages and success of CD in mind, Sony developed an optical disc-based system for digital recording: MiniDisc or MD. This was launched around the same time as DCC, these two formats being direct competitors.

One of the main advantages appreciated by CD users is the quick access. Contrary to tape-based systems, a number of editing possibilities can be exploited by MiniDisc, increasing the user friendliness: merging and dividing tracks, changing track numbers, inserting disc and track names, etc.

Also, the read-out and recording with a non-contact optical system makes the MD an extremely durable and long-lasting sound medium. The 4-Mbit semiconductor memory enables the listener to enjoy uninterrupted sound, even when the set vibrates heavily. This 'shockproof' memory underlines the portable nature of MiniDisc.

The MiniDisc system is explained in detail in Chapter 17 of this book.

CD-recordable

The original disc-based specifications as formulated jointly by Philips and Sony also described recordable CD formats (the so-called 'orange book' specification).

Originally, this techology was mostly used for data storage applications, but as of the middle of the 1990s, this technology also emerged for audio applications. Two formats have been developed: CD-recordable (CD-R) and CD-rewritable (CD-RW).

CD-R draws its technology from the original CD-WO (write-once) specification. This is a one-time recording format in this sense that, once a track is recorded, there is no way to erase or overwrite it. There are, however, possibilities to record one track after the other independently. The recording is similar to the first writing of a pre-recorded Audio CD. The same format is used.

The actual recording is done on a supplementary 'dye' layer. When this layer is heated beyond a critical temperature (around 250°C) a bump appears. In this way, the CD-like pit/bump structure is recreated. This process is irreversible; once a bump has been created by laser heat, it cannot be changed further.

The use of a dye layer is a significant difference from the original CD format. Another difference with audio CD is the use of a pre-groove (also called 'wobble') for tracking purposes. As this technique is also used in MiniDisc, it is explained in Chapter 17.

Since the recorded format and the dimensions of the CD-R are the same as conventional CD players, they can be played back on most players.

CD-rewritable

The main difference between CD-R and CD-RW is the unique recording layer, which is composed of a silver–indium–antimony–tellurium alloy, allowing recording and erasing over 1000 times. Recording is done by heating up the recording layer by the record laser to over 600°C, where the crystalline structure of the layer is changing to a less reflective surface. In this way, a 'pit' pattern is created. Erasing is done by heating up the surface to 200°C, returning it to a more reflective surface. This is the second main difference between CD-R and CD-RW; in the case of the latter, the process is reversible.

One major drawback of this system is that the reflectivity does not conform to the original CD (red book) standard. The CD standard states that at least 70% of the laser light which hits the reflective area of a disc must be reflected; the CD-RW only reflects about 20%. The CD-RW format can therefore not guarantee playback in any CD player, especially not the older CD players. Manufacturers of CD players cannot be held responsible if their players cannot handle the CD-RW format, but seen through users' eyes, this appeared as a contradictory situation. Obviously, this has caused some confusion on the market, but gradually most new CD players have a design which is adapted to this lower reflectivity.

High-quality sound reproduction: SACD and DVD-Audio

Even if the CD audio system provides a level of quality exceeding most people's expectations, the search for better sound never stops. Audiophiles sometimes complained that CD sound was still no match for the 'warmth' of an analog recording; technically speaking, this is related to the use of A/D and D/A converters, but also due to a rather sharp cut-off above 20 kHz. A human ear is not capable of hearing sounds above this frequency, but it is a known fact that these higher frequencies do influence the overall sound experience, as they will cause some harmonics within the audible area. Fairness, however, urges us to mention that part of what is considered the 'warmth' of an analog recording is in reality noise which is not related to the original sound; this noise is a result of the used technology and its limitations, as well as imperfections. The improved recording and filtering techniques eliminate most of this noise, so some people will always consider digital sound as being 'too clean'.

Two new formats for optical laser discs were developed in the second part of the 1990s. These two new systems are able to reproduce frequencies up to 100 kHz with a dynamic range of 120 dB in the audible range. Due to these higher frequency and dynamic ranges, the claims against CD audio can be addressed, but other advantages are also seen: the positioning of recorded instruments and artists in the sound field is far better; a listener might close his eyes and be able to imagine the artists at their correct position in front of him, not only from left to right but also from front to back.

As the disc media used can also contain much more data, both SACD and DVD-Audio can be used for multi-channel reproduction; as an example it is perfectly possible that if a rock band consists of five musicians, each musician is recorded on a separate track, which creates possibilities for the user to make his own channel mix.

Super Audio CD (SACD) was developed as a joint venture between Sony and Philips; this should not be a surprise since these two companies jointly developed all previous CD-based audio systems, and since SACD is a logical successor (also technically) of these highly successful systems.

SACD is a completely new 1-bit format, operating with a sampling frequency of 2.8224 MHz. Besides its high-quality reproduction capabilities, the hybrid disc is also compatible with the conventional CD player. Of course, the conventional multi-bit PCM CD can be played back on the SACD player.

The second format is based on the existing DVD optical read-out and CD technology. This new format produced on part 1 and part 2 of the DVD-ROM specifications is called DVD-Audio (DVD-A). To achieve a frequency band of nearly 100 kHz, a sampling frequency of 192 kHz with a resolution of 24 bits is adopted. Both SACD and DVD-A are explained within this book.

Digital audio compression

Already from the start of digital storage, the need to compress data became clear. In the early days of computers, memory was very expensive and storage capacity was limited. Also, from the production side, it was impossible to create the memory capacity as we know it now, at least not at an affordable price and on a usable scale. In particular, after the idea to convert the analog audio signals into a digital format originated, the impressive capacity required for storage became a problem.

The explosive growth of electronic communication and the Internet was also a major boost for distribution of data, but at the same time it was a major reason why data compression became a crucial matter.

To allow recording digital music signals to a limited storage medium or transferring it within an acceptable delay, compression becomes necessary. Consider the following example.

Suppose we want to record 1 minute of stereo CD quality music on our hard disk. The sampling frequency for CDs is 44.1 kHz, with 16 bits per sample. One minute of music represents an amount of data of 44 100 (samples/second) × 60 (samples/ minute) × 16 (bits/sample) × 2 (stereo) = 84 672 Mbits or about 10.6 Mbytes of memory on your hard disk.

Transferring this amount of data via the internet through an average 28.8 kb s^{-1} modem will take about 84 672 Mbits/28 800 (bits s^{-1}) × 60 (s/minute) or 49 minutes. When using conventional telephone lines, this is even an optimistic approach, as the modem speeds tend to vary (mostly downward) along with network loads, and the equation does not even take into account that data transfer through modems also uses error protection protocols, which also decrease the speed.

To achieve a reasonable delay when transferring or storing the data in a memory, the need of compressing becomes clear.

Over the years, different techniques have been developed. Most audio compression techniques are derived directly or indirectly from MPEG standards. MPEG stands for Motion Picture Expert Group; it is a worldwide organization where all interested parties like manufacturers, designers, producers, etc. can propose, discuss and agree on technologies and standards for motion pictures, but also the related audio and video technologies.

For consumer use, it started with the PASC system for DCC and ATRAC for MD – both MPEG1 layer 1-based compression techniques, they allow a four- to fivefold reduction in the amount of data. Thanks to the evolution of more accurate and faster working microcomputers, higher quality and higher compression has been achieved.

This same MPEG1, but now layer 3, is the basis for the widely used and sometimes contested MP3 format. More on this subject is given elsewhere within this book (see Chapter 20).

Summary of development of digital audio equipment at Sony

October 1974
First stationary digital audio recorder, the X-12DTC, 12-bit

September 1976
FM format 15-inch digital audio disc read by laser. Playing time: 30 minutes at 1800 rpm, 12-bit, two-channel

October 1976
First digital audio processor, 12-bit, two-channel, designed to be used in conjunction with a VTR

June 1977
Professional digital audio processor PAU-1602, 16-bit, two-channel, purchased by NHK (Japan Broadcasting Corporation)

September 1977
World's first consumer digital audio processor PCM-1, 13-bit, two-channel 15-inch digital audio disc read by laser. Playing time: 1 hour at 900 rpm

March 1978
Professional digital audio processor PCM-1600

April 1978
Stationary-head digital audio recorder X-22, 12-bit, two-channel,
using ¼-inch tape
World's first digital audio network

October 1978
Long-play digital audio disc read by laser. Playing time: 2½ hours
at 450 rpm
Professional, multi-channel, stationary-head audio recorder
PCM-3224, 16-bit, 24-channel, using 1-inch tape
Professional digital audio mixer DMX-800, eight-channel input,
two-channel output, 16-bit
Professional digital reverberator DRX-1000, 16-bit

May 1979
Professional digital audio processor PCM-100 and consumer dig-
ital audio processor PCM-10 designed to Electronics Industry
Association of Japan (EIAJ) standard

October 1979
Professional, stationary-head, multi-channel digital audio
recorder PCM-3324, 16-bit, 24-channel, using ½-inch tape
Professional stationary-head digital audio recorder PCM-3204,
16-bit, four-channel, using ¼-inch tape

May 1980
Willi Studer of Switzerland agrees to conform to Sony's digital
audio format on stationary-head recorder

June 1980
Compact disc digital audio system mutually developed by Sony
and Philips

October 1980
Compact disc digital audio demonstration with Philips, at Japan
Audio Fair in Tokyo

February 1981
Digital audio mastering system including digital audio processor
PCM-1610, digital audio editor DAE-1100 and digital reverbera-
tor DRE-2000

Spring 1982
PCM adapter PCM-F1, which makes digital recordings and play-
backs on a home VTR

October 1982
Compact disc player CDP-101 is launched onto the Japanese
market and as of March 1983 it is available in Europe

Figure XVII DTC-55ES
R-DAT player.

1983
Several new models of CD players with sophisticated features:
CDP-701, CDP-501, CDP-11S
PCM-701 encoder

November 1984
Portable CD player: the D-50
Car CD players: CD-X5 and CD-XR7

1985
Video 8 multi-track PCM recorder (EV-S700)

March 1987
First R-DAT player DTC-1000ES launched onto the Japanese market

July 1990
Second generation R-DAT player DTC-55ES available on the European market

March 1991
DAT Walkman: TCD-D3
Car DAT player: DTX-10

November 1992
Worldwide introduction of the MiniDisc (MD) format, first generation MD Walkman: MZ-1 (recording and playback) and MZ-2P (playback only)

1992–1999
Constant improvement of the ATRAC DSP for MiniDisc. Several versions were issued: ATRAC versions 1, 2, 3, 4, 4.5 and finally ATRAC-R. Evolution and expansion of the DAB network

May 1996
The International Steering Committee (ISC) made a list of recommendations for the next generation of high-quality sound formats

February 1999
First home DAB receiver

April 1999
ATRAC DSP Type-R in the fifth generation of MiniDisc players:
MDS-JA20ES

October 1999
Introduction of the first generation of SACD players: SCD-1 and
SCD-777ES

April 2000
Second generation SACD player: SCD-XB940 in the QS class

September 2000
Introduction of MDLP (Long Play function for MiniDisc) based
on ATRAC3

October 2000
Combined DVD/SACD player: DVP-S9000ES

February 2001
First multi-channel Hybrid SACD disc on the market

February 2001
Sony introduces its first multi-channel player: SCD-XB770

PART ONE
Principles of Digital Signal Processing

1 Introduction

For many years, two main advantages of the digital processing of analog signals have been known.

First, if the transmission system is properly specified and dimensioned, transmission quality is independent of the transmission channel or medium. This means that, in theory, factors which affect the transmission quality (noise, non-linearity, etc.) can be made arbitrarily low by proper dimensioning of the system.

Second, copies made from an original recording in the digital domain are identical to that original; in other words, a virtually unlimited number of copies, which all have the same basic quality as the original, can be made. This is a feature totally unavailable with analog recording.

A basic block diagram of a digital signal processing system is shown in Figure 1.1.

Digital audio processing does require an added circuit complexity and a larger bandwidth than analog audio processing systems, but these are minor disadvantages when the extra quality is considered.

Perhaps the most critical stages in digital audio processing are the conversions from analog to digital signals, and vice versa. Although the principles of A/D and D/A conversion may seem relatively simple, in fact they are technically speaking very

Figure 1.1 Basic digital signal processing system.

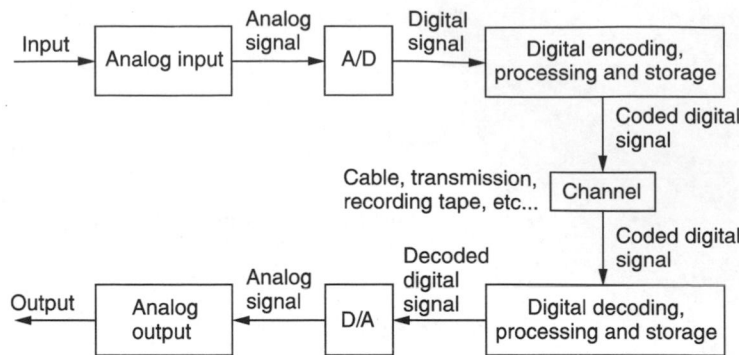

difficult and may cause severe degradation of the original signal. Consequently, these stages often generate a limiting factor that determines the overall system performance.

Conversion from analog to digital signals is done in several steps:

- filtering – this limits the analog signal bandwidth, for reasons outlined below
- sampling – converts a continuous-time signal into a discrete-time signal
- quantization – converts a continuous-value signal into a discrete-value signal
- coding – defines the code of the digital signal according to the application that follows.

Figure 1.2 shows a block diagram of an analog-to-digital conversion system.

Figure 1.2 An analog-to-digital conversion system.

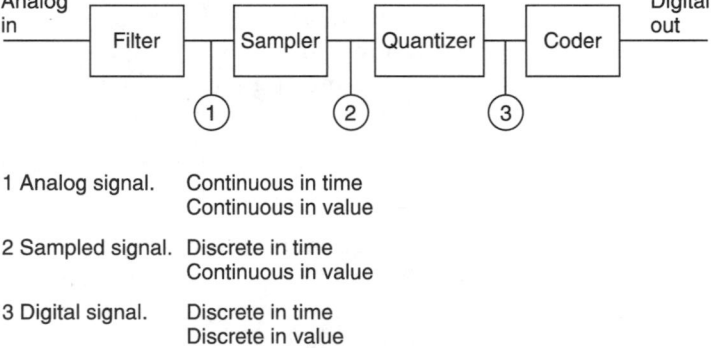

2 Principles of sampling

The Nyquist theorem

By definition, an analog signal varies continuously with time. To enable it to be converted into a digital signal, it is necessary that the signal is first sampled, i.e., at certain points in time a sample of the input value must be taken (Figure 2.1). The fixed time intervals between each sample are called sampling intervals (t_s).

Although the sampling operation may seem to introduce a rather drastic modification of the input signal (as it ignores all the signal changes that occur between the sampling times), it can be shown that the sampling process in principle removes no information whatsoever, as long as the sampling frequency is at least present in the input signal. This is the famous Nyquist theorem on sampling (also called the Shannon theorem).

The Nyquist theorem can be verified if we consider the frequency spectra of the input and output signals (Figure 2.2).

Figure 2.1

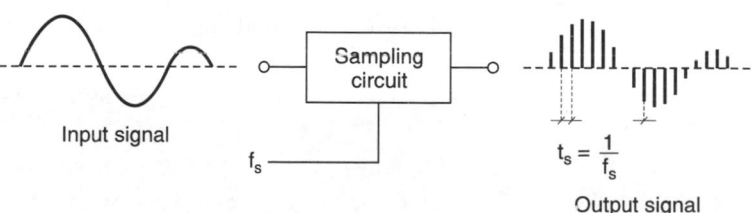

Input signal

f_s

Sampling circuit

$t_s = \dfrac{1}{f_s}$

Output signal

Figure 2.2 The sampling process in the time domain (a) and frequency domain (b)

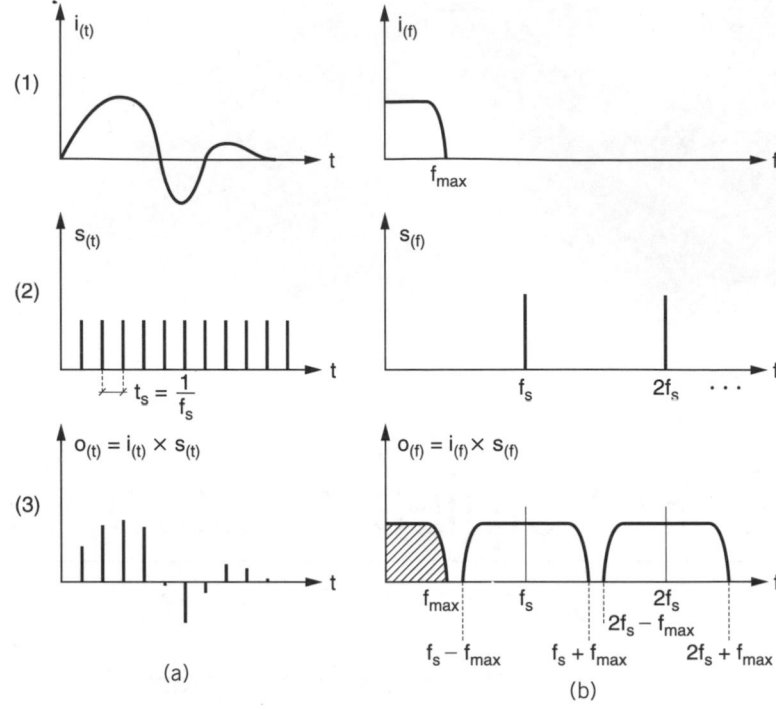

(a)

(b)

An analog signal $i_{(t)}$ which has a maximum frequency f_{max} will have a spectrum having any form between 0 Hz and f_{max} (Figure 2.2, 1a); the sampling signal $s_{(t)}$, having a fixed frequency f_s, can be represented by one single line at f_s (Figure 2.2, 2b). The sampling process is equivalent to a multiplication of $i_{(t)}$ and $s_{(t)}$, and the spectrum of the resultant signal (Figure 2.2, 3b) can be seen to contain the same spectrum as the analog signal, together with repetitions of the spectrum modulated around multiples of the sampling frequency. As a consequence, low-pass filtering can completely isolate and thus completely recover the analog signal.

The diagrams in Figure 2.3 show two sine waves (bottom traces) and their sampled equivalents (top traces).

Although sampling in Figure 2.3b seems much coarser than in Figure 2.3a, in both cases restitution of the original signal is perfectly possible.

Figure 2.2 (3b) also shows that f_s must be greater then $2f_{max}$, otherwise the original spectrum would overlap with the modulated part of the spectrum, and consequently be inseparable from it (Figure 2.4).

For example, a 20 kHz signal sampled at 35 kHz produces a 5 kHz difference frequency. This phenomenon is known as aliasing.

Figure 2.3 Two examples of sine waves together with sampled versions: (a) 1 kHz sine wave; (b) 10 kHz sine wave. Sampling frequency f_s is 44.056 kHz.

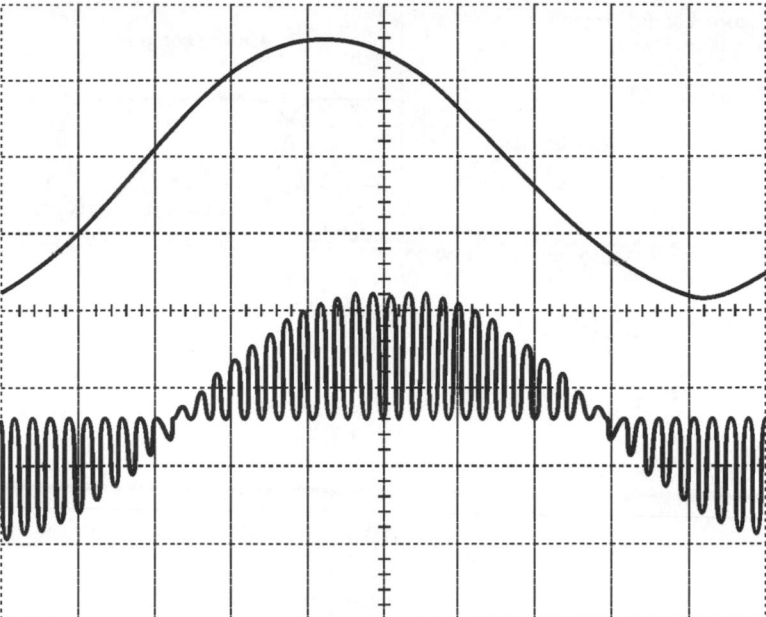

(a)

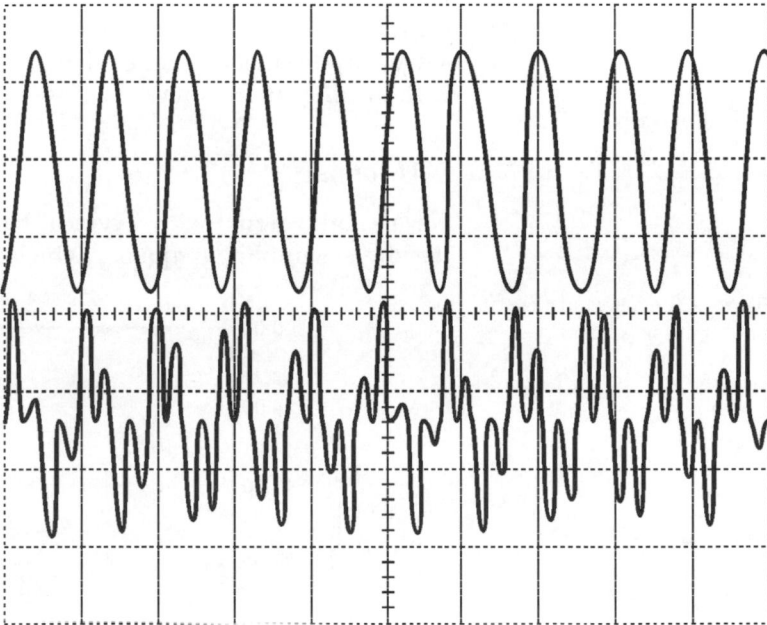

(b)

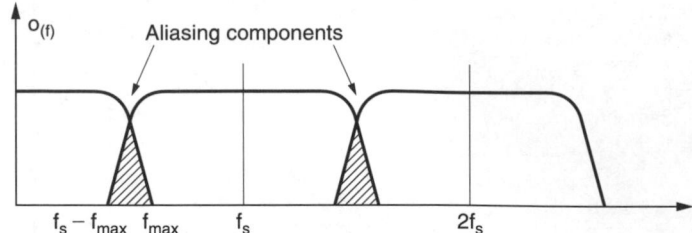

Figure 2.4 If sampling frequency is too low, aliasing occurs.

To avoid aliasing due to harmonics of the analog signal, a very sharp cut-off filter (known as an anti-aliasing filter) is used in the signal path to remove harmonics before sampling takes place. The characteristics of a typical anti-aliasing filter are shown in Figure 2.5.

Sampling frequency

From this, it is easy to understand that the selection of the sampling frequency is very important. On one hand, selecting too high a sampling rate would increase the hardware costs dramatically. On the other hand, since ideal low-pass filters do not exist, a certain safety margin must be incorporated in order to avoid any frequency higher than $\frac{1}{2}f_s$ passing through the filter with insufficient attenuation.

EIAJ format

For an audio signal with a typical bandwidth of 20 Hz–20 kHz, the lowest sampling frequency which corresponds to the Nyquist

Figure 2.5 Characteristics of an anti-aliasing filter.

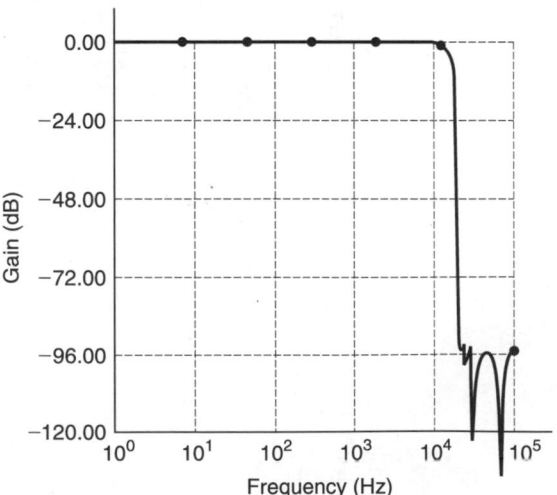

theorem is 40 kHz. At this frequency a very steep and, consequently, very expensive anti-aliasing filter is required. Therefore, a sampling frequency of approximately 44 kHz is typically used, allowing use of an economical 'anti-aliasing filter', flat until 20 kHz but with sufficient attenuation (60 dB) at 22 kHz to make possible aliasing components inaudible.

Furthermore, because the first commercially available digital audio recorders stored the digital signal using a standard helical scan video recorder, there had to be a fixed relationship between sampling frequency (f_s) and horizontal video frequency (f_h), so these frequencies could be derived from the same master clock by frequency division.

For the NTSC 525-line television system, a sampling frequency of 44 055944... Hz was selected, whereas for the PAL 625-line system, a frequency of 44 100 Hz was chosen. The difference between these two frequencies is only 0.1%, which is negligible for normal use (the difference translates as a pitch difference at playback, and 0.1% is entirely imperceptible).

Compact disc sampling rate

For compact disc, the same sampling rate as used in the PCM-F1 format, i.e., 44.1 kHz, was commonly agreed upon by the establishers of the standard.

Video 8 PCM

The Video 8 recording standard also has a provision for PCM recording. The PCM data are recorded in a time-compressed form, into a 30° tape section of each video channel track (Figure 2.6). However, sampling frequency must be reduced to allow the data to fit.

Nevertheless, as the sampling frequency is exactly twice the video horizontal frequency (f_h), the audio frequency range is still more than 15 kHz, which is still acceptable for hi-fi recording purposes. Table 2.1 lists the horizontal frequencies for PAL and NTSC television standards, giving resultant sampling frequencies.

Also, since only 300 of the tracks are used for the audio recording, a multi-channel (six channels) PCM recording is possible when the whole video recording area is used for PCM.

Sampling rate for professional application

A sampling frequency of 32 kHz has been chosen by the EBU for PCM communications for broadcast and telephone, since an

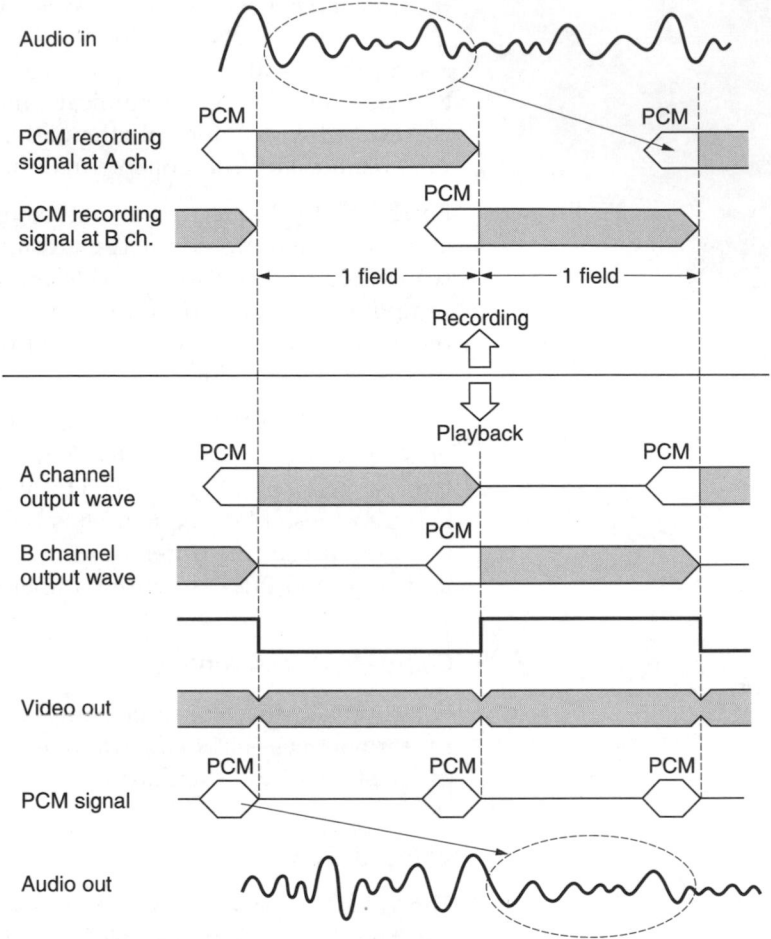

Figure 2.6 How PCM coded audio signals are recorded on a section of tape in a Video 8 track.

Table 2.1 Horizontal and sampling frequencies of Video 8 recorders, on PAL and NTSC television standards

	PAL	NTSC
f_h	15.625	15.734365
f_s	31.250	31.468530

audio frequency range of up to 15 kHz is considered to be adequate for television and FM radio broadcast transmissions.

As we saw in Chapter 1, stationary-head digital audio recorders are most suited to giving a multi-track recording capability in the studio. One of the aspects of such studio recorders is that the tape speed must be adjustable, in order to allow easy synchronization between several machines and correct tuning. Considering a speed tuning range of, say 10%, a sampling frequency of 44 kHz could increase to less than 40 kHz, which is too low to comply with the Nyquist theorem. Therefore, such machines should use a higher sampling rate, which at the lowest speed must still be above $2f_s$.

After a study of all aspects of this matter, 48 kHz has been selected as the recommended sampling frequency for studio recorders. This frequency is compatible with television and motion picture system frame frequencies (50 and 60 Hz), and has an integer relationship (3/2) with the 32 kHz sampling of the PCM network used by broadcast companies. The relationships between sampling frequency and frame frequencies (1/960, 1/800) enable the application of time coding, which is essential for editing and synchronization of the tape recorder. As there is no fixed relationship between 44.1 kHz (CD sampling frequency) and 48 kHz, Sony's studio recorders can use either frequency.

The reason why so much importance was attached to the integer relation of sampling frequencies is conversion: it is then possible to economically dub or convert the signals in the digital mode without any deterioration. Newly developed sampling rate converters, however, also allow conversion between non-related sampling rates, so that this issue has become less important than it used to be.

Sampling rates for R-DAT and S-DAT formats

In the DAT specification, multiple sampling rates are possible to enable different functions:

- 48 kHz, for highest-quality recording and playback.
- 32 kHz, for high-quality recording with longer recording time or for recording on four tracks and for direct (digital) recording of digital broadcasts.
- 44.1 kHz playback-only, for reproduction of commercial albums released on R-DAT or S-DAT cassettes.

Sample-hold circuits

In the practice of analog-to-digital conversion, the sampling operation is performed by sample-hold circuits, that store the sampled analog voltage for a time, during which the voltage can be converted by the A/D converter into a digital code. The principle of a sample-hold circuit is relatively simple and is shown in Figure 2.7.

Figure 2.7 Basic sample-
hold circuit.

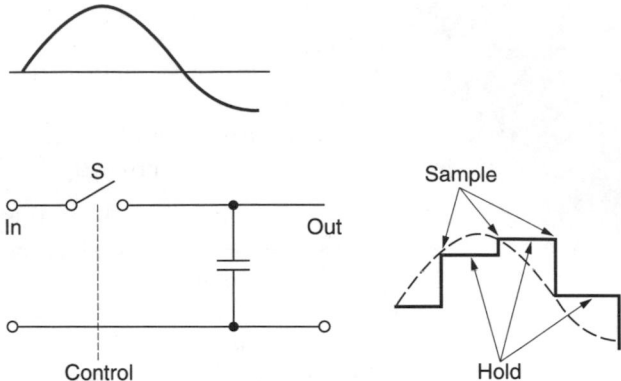

A basic sample-hold circuit is a 'voltage memory' device that
stores a given voltage in a high-quality capacitor. To sample the
input voltage, switch S closes momentarily: when S re-opens,
capacitor C holds the voltage until S closes again to pass the next
sample. Figure 2.8 is a diagram showing a sine wave at the input
(top) and the output (bottom) of a sample-hold circuit.

Practical circuits have buffer amplifiers at input, in order not to
load the source, and output, to be able to drive a load such as
an A/D converter. The output buffer amplifier must have a very
high input impedance, and very low bias current, so that the

Figure 2.8 Sine wave before
(top) and after (bottom)
sample-hold.

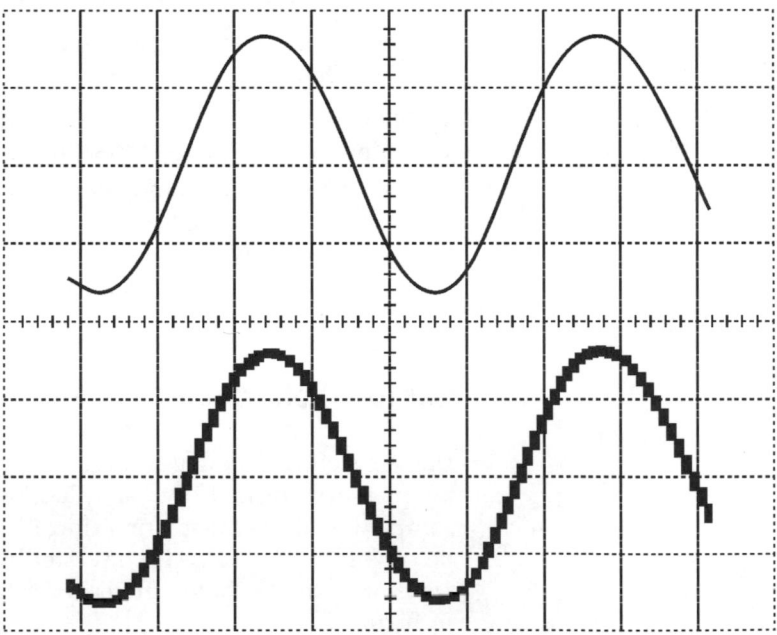

charge of the hold capacitor does not leak away. Also, the switch must be very fast and have low off-stage leakage. An actual sample-hold circuit may use an analog (JFET switch) and a high-quality capacitor, followed by a buffer amplifier (voltage follower), as shown in Figure 2.9.

Sample-hold circuits are not only used in A/D conversion but also in D/A conversion, to remove transients (glitches) from the output of the D/A converter. In this case, a sample-hold circuit is often called a deglitcher (Figure 2.10).

Aperture control

The output signal of a sampling process is in fact a pulse amplitude modulated (PAM) signal. It can be shown that, for sinusoidal input signals, the frequency characteristic of the sampled output is:

$$H(\omega_v) = \frac{t_o}{t_s} \cdot \frac{\sin \frac{t_o}{2} \omega_v}{\frac{t_o}{2} \omega_v}$$

in which ω_v = angular velocity of input signal ($= 2f_v$)

$\quad\quad t_o$ = pulse width of the sampling pulse

$\quad\quad t_s$ = sampling period.

Figure 2.9 FET input sample-hold circuit.

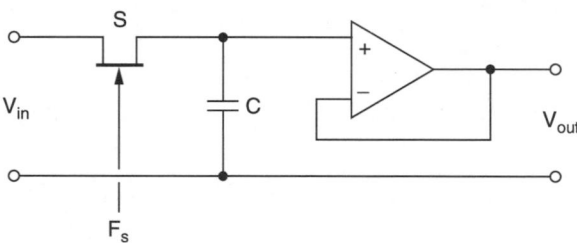

Figure 2.10 The effect of using a sample-hold circuit as a deglitcher.

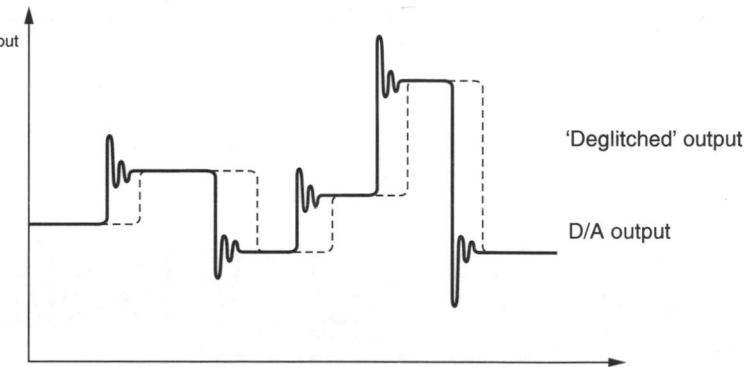

At the output of a sample-hold circuit or a D/A converter, however, $t_o = t_s$. Consequently:

$$H(\omega_v)_{t_o=t_s} = \frac{\sin\dfrac{t_s}{2}\omega_v}{\dfrac{t_s}{2}\omega_v}$$

This means that at maximum admissible input frequency (which is half the sampling frequency), $\omega_v = \pi/t_s$, and consequently:

$$H\left(\frac{\pi}{t_s}\right)_{t_o-t_s} = \frac{\sin\dfrac{\pi}{2}}{\dfrac{\pi}{2}} \approx 0.64$$

This decreased frequency response can be corrected by an aperture circuit, which decreases t_o and restores a normal PAM signal (Figure 2.11).

In most practical circuits $t_o = t_s/4$, which leads to:

$$H\left(\frac{\pi}{t_s}\right)_{t_s} = \frac{t_s}{4} = \frac{\sin\dfrac{\pi}{8}}{\dfrac{\pi}{8}} \approx 0.97$$

Figure 2.11 Basic circuit and waveforms of aperture control circuit.

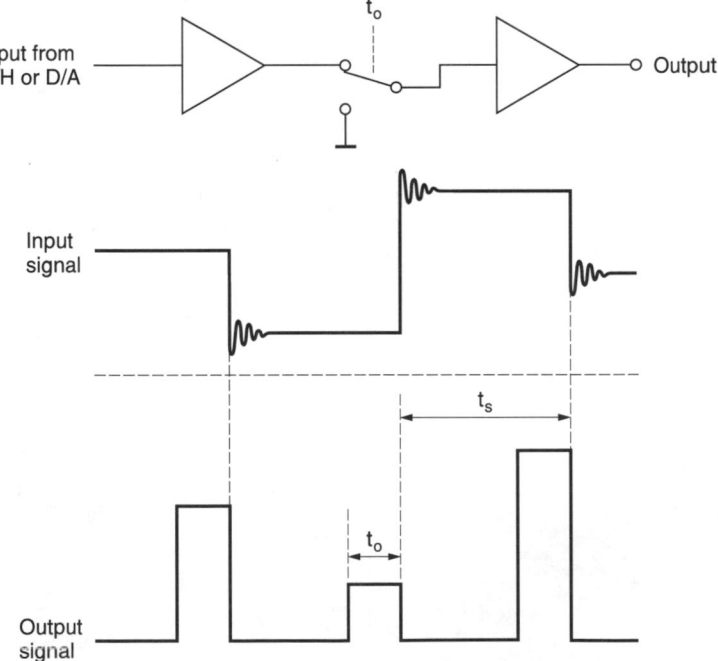

This is an acceptable value: reducing t_o further would also reduce the average output voltage too much and thus worsen the signal-to-noise ratio.

Figure 2.12 shows the frequency response for some values of t_o.

Characteristics and terminology of sample-hold circuits

In a sample-hold circuit, the accuracy of the output voltage depends on the quality of the buffer amplifiers, on the leakage current of the holding capacitor and of the sampling switch. Unavoidable leakage generally causes the output voltage to decrease slightly during the 'hold' period, in a process known as droop.

In fast applications, acquisition time and settling time are also important. Acquisition time is the time needed after the transition from hold to sample periods for the output voltage to match the input, within a certain error band. Settling time is the time needed after the transition from sample to hold periods to obtain a stable output voltage. Both times obviously define the maximum sampling rate of the unit.

Aperture time is the time interval between beginning and end of the transition from sample to hold periods; also, terms like aperture uncertainty and aperture jitter are used to indicate variations in the aperture time and consequently variations of the sample instant itself.

Figure 2.12 Output characteristics of an aperture control circuit, as functions of aperture time and frequency response.

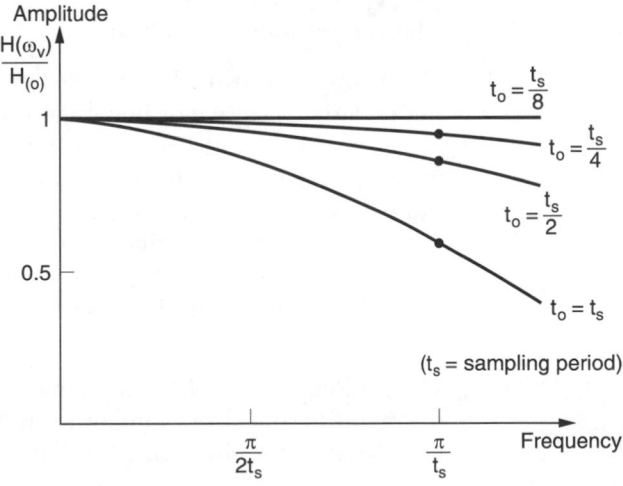

3 Principles of quantization

Even after sampling the signal is still in the analog domain: the amplitude of each sample can vary infinitely between analog voltage limits. The decisive step to the digital domain is now taken by quantization (see Figure 3.1), i.e., replacing the infinite number of voltages by a finite number of corresponding values.

In a practical system the analog signal range is divided into a number of regions (in our example, 16), and the samples of the signal are assigned a certain value (say, –8 to +7) according to the region in which they fall. The values are denoted by digital (binary) numbers. In Figure 3.1, the 16 values are denoted by a 4-bit binary number, as $2^4 = 16$.

The example shows a bipolar system in which the input voltage can be either positive or negative (the normal case for audio). In this case, the coding system is often the two's complement code, in which positive numbers are indicated by the natural binary code while negative numbers are indicated by complementing the positive codes (i.e., changing the state of all bits) and adding one. In such a system, the most significant bit (MSB) is used as a sign bit, and is 'zero' for positive values but 'one' for negative values.

The regions into which the signal range is diverted are called quantization intervals, sometimes represented by the letter Q. A series of n bits representing the voltage corresponding to a quan-

Figure 3.1 Principle of quantization.

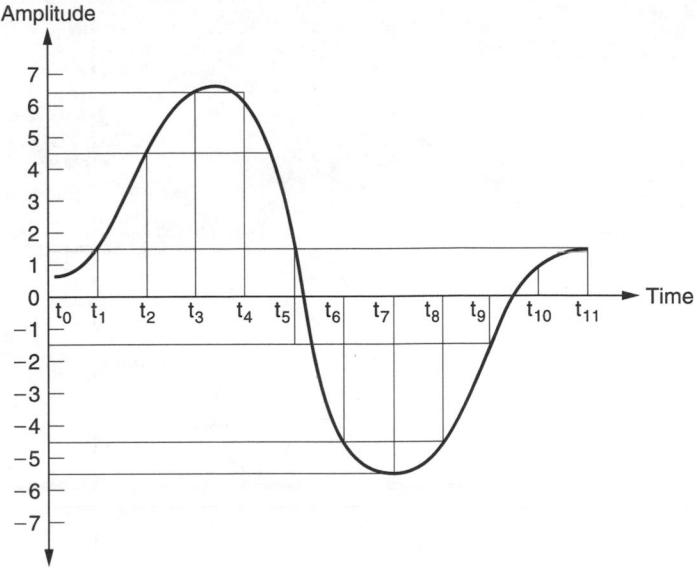

Sample	t_0	t_1	t_2	t_3	t_4	t_5	t_6	t_7	t_8	t_9	t_{10}	t_{11}
Value	1	2	5	7	6	0	5	6	5	2	1	1
4-bit code (2's complement)	0001	0010	0101	0111	0110	0000	1011	1010	1011	1110	0001	0001

tization interval is called a word. In our simple example a word consists of 4 bits. Figure 3.2 shows a typical quantization stage characteristic.

In fact, quantization can be regarded as a mechanism in which some information is thrown away, keeping only as much as necessary to retain a required accuracy (or fidelity) in an application.

Quantization error

By definition, because all voltages in a certain quantization interval are represented by the voltage at the centre of this interval, the process of quantization is a non-linear process and creates an error, called quantization error (or, sometimes, round-off error). The maximum quantization error is obviously equal to half the quantization interval Q, except in the case that the input voltage widely exceeds the maximum quantization levels (+ or $-V_{max}$), when the signal will be rounded to these values. Generally,

Figure 3.2 Quantization stage characteristics.

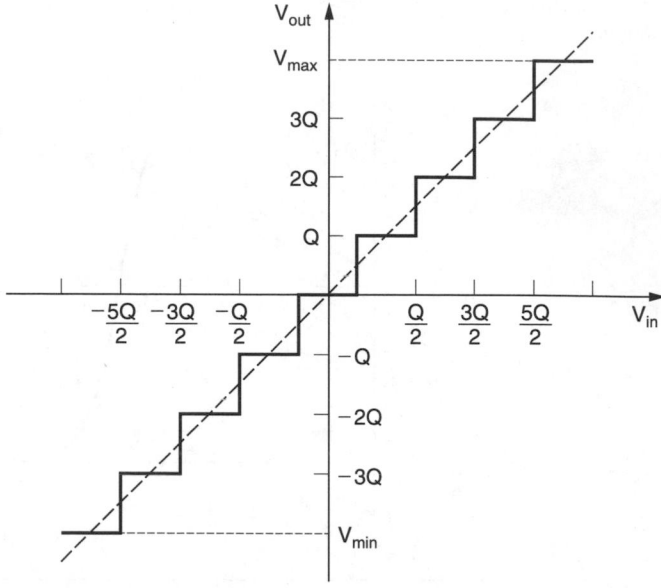

however, such overflows or underflows are avoided by careful scaling of the input signal.

So, in the general case, we can say that:

$$-Q/2 < e_{(n)} < Q/2$$

where $e_{(n)}$ is the quantization error for a given sample n.

It can be shown that, with most types of input signals, the quantization errors for the several samples will be randomly distributed between these two limits, or in other words, its probability density function is flat (Figure 3.3).

There is a very good analogy between quantization error in digital systems and noise in analog systems: one can indeed

Figure 3.3 Probability density function of quantization error.

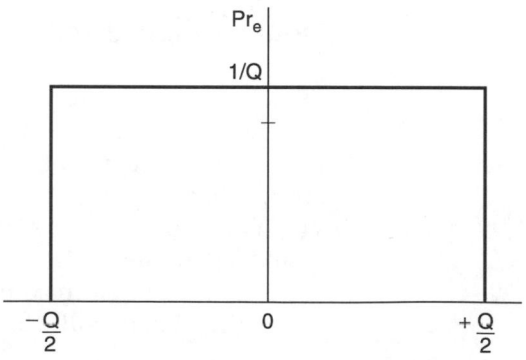

consider the quantized signal as a perfect signal plus quantization error (just like an analog signal can be considered to be the sum of the signal without noise plus a noise signal; see Figure 3.4). In this manner, the quantization error is often called quantization noise, and a 'signal-to-quantization noise' ratio can be calculated.

Calculation of theoretical signal-to-noise ratio

In an n-bit system, the number of quantization intervals N can be expressed as:

$$N = 2^n \tag{3.1}$$

If the maximum amplitude of the signal is V, the quantization interval Q can be expressed as:

$$Q = \frac{V}{N-1} \tag{3.2}$$

As the quantization noise is equally distributed within $\pm Q/2$, the quantization noise power N_a is:

$$N_a = \frac{2}{Q}\int_0^{Q/2} x^2 \mathrm{d}x = \frac{2}{Q}\left(\frac{(Q/2)^3}{3}\right) = \frac{1}{12}Q^2 \tag{3.3}$$

If we consider a sinusoidal input signal with peak-to-peak amplitude V, the signal power is:

Figure 3.4 Analogy between quantization error and noise.

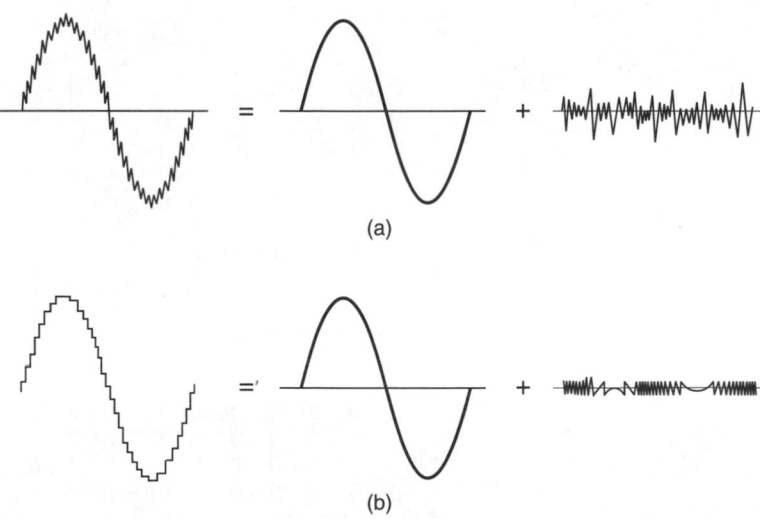

(a)

(b)

$$S = \frac{1}{2} \int_0^2 \left(\frac{2}{Q} \sin x \right)^2 dx = \frac{1}{8} V^2$$

(3.4)

Consequently, the power ratio of signal-to-quantization noise is:

$$\frac{S}{N_a} = \frac{V^2/8}{Q^2/12} = \frac{V^2/8}{V^2/(N-1)^2 \cdot 12} \approx \frac{3}{2} N^2 \ \ (\text{for} \ N \gg 1)$$

(3.5)

Or, by substituting Equation 3.1:

$$\frac{S}{N_a} = \frac{3}{2} \left(2^{2n} \right) = 3 \left(2^{2n-1} \right)$$

Expressed in decibels, this gives:

S/N (dB) = 10 log (S/N_a) = 10 log $3(2^{2n-1})$

Working this out gives:

S/N (dB) = 6.02 × n + 1.76

A 16-bit system, therefore, gives a theoretical signal-to-noise ratio of 98 dB; a 14-bit system gives 86 dB.

In a 16-bit system, the digital signal can take 2^{16} (i.e., 65 535) different values, according to the truth table shown in Table 3.1.

Table 3.1 Truth table for 16-bit two's complement binary system

0	1	1	1	1	1	1	1	1	1	1	1	1	1	1	1	=	+32767
0	1	1	1	1	1	1	1	1	1	1	1	1	1	1	0	=	+32766
0	1	1	1	1	1	1	1	1	1	1	1	1	1	0	1	=	+32765
0	1	1	1	1	1	1	1	1	1	1	1	1	1	0	0	=	+32764
							:										
							:										
							:										
0	0	0	0	0	0	0	0	0	0	0	0	0	0	1	1	=	+3
0	0	0	0	0	0	0	0	0	0	0	0	0	0	1	0	=	+2
0	0	0	0	0	0	0	0	0	0	0	0	0	0	0	1	=	+1
0	0	0	0	0	0	0	0	0	0	0	0	0	0	0	0	=	0
1	1	1	1	1	1	1	1	1	1	1	1	1	1	1	1	=	−1
1	1	1	1	1	1	1	1	1	1	1	1	1	1	1	0	=	−2
1	1	1	1	1	1	1	1	1	1	1	1	1	1	0	1	=	−3
1	1	1	1	1	1	1	1	1	1	1	1	1	1	0	0	=	−4
							:										
							:										
							:										
1	0	0	0	0	0	0	0	0	0	0	0	0	0	1	1	=	−32765
1	0	0	0	0	0	0	0	0	0	0	0	0	0	1	0	=	−32766
1	0	0	0	0	0	0	0	0	0	0	0	0	0	0	1	=	−32767
1	0	0	0	0	0	0	0	0	0	0	0	0	0	0	0	=	−32768

Masking of quantization noise

Although, generally speaking, the quantization error is randomly distributed between + and $-Q/2$ (see Figure 3.1) and is consequently similar to analog white noise, there are some cases in which it may become much more noticeable than the theoretical signal-to-noise figures would indicate.

The reason is mainly that, under certain conditions, quantization can create harmonics in the audio passband which are not directly related to the input signal, and audibility of such distortion components is much higher than in the 'classical' analog cases of distortion. Such distortion is known as granulation noise, and in bad cases it may become audible as beat tones.

Auditory tests have shown that to make granulation noise just as perceptible as 'analog' white noise, the measured signal-to-noise ratio should be up to 10–12 dB higher. To reduce this audibility, there are two possibilities:

(a) To increase the number of bits sufficiently (which is very expensive).
(b) To 'mask' the digital noise by a small amount of analog white noise, known as 'dither noise'.

Although such an addition of 'dither noise' actually worsens the overall signal-to-noise ratio by several decibels, the highly audible granulation effect can be very effectively masked by it. The technique of adding 'dither' is well known in the digital signal processing field, particularly in video applications, where it is used to reduce the visibility of the noise in digitized video signals.

Conversion codes

In principle, any digital coding system can be adopted to indicate the different analog levels in A/D or D/A conversion, as long as they are properly defined. Some, however, are better for certain applications than others. Two main groups exist: unipolar codes and bipolar codes. Bipolar codes give information on both the magnitude and the sign of the signal, which makes them preferable for audio applications.

Unipolar codes

Depending upon application, the following codes are popular.

Natural binary code

The MSB has a weight of 0.5 (i.e., 2^{-1}), the second bit has a weight of 0.25 (2^{-2}), and so on, until the least significant bit (LSB), which has a weight of 2^{-n}. Consequently, the maximum value that can be expressed (when all bits are one) is $1 - 2^{-n}$, i.e., full scale minus one LSB.

BCD code

The well-known 4-bit code in which the maximum value is 1001 (decimal 9), after which the code is reset to 0000. A number of such 4-bit codes is combined in case we want, for instance, a direct read-out on a numerical scale such as in digital voltmeters. Because of this maximum of 10 levels, this code is not used for audio purposes.

Gray code

Used when the advantage of changing only 1 bit per transition is important, for instance in position encoders where inaccuracies might otherwise give completely erroneous codes. It is easily convertible to binary. Not used for audio.

Bipolar codes

These codes are similar to the unipolar natural binary code, but one additional bit, the sign bit, is added. Structures of the most popular codes are compared in Table 3.2.

Sign magnitude

The magnitude of the voltage is expressed by its normal (unipolar) binary code, and a sign bit is simply added to express polarity.

An advantage is that the transition around zero is simple; however, it is more difficult to process and there are two codes for zero.

Offset binary

This is a natural binary code, but with zero at minus full scale; this makes it relatively easy to implement and to process.

Table 3.2 Common bipolar codes

Decimal fraction						
Number	Positive reference	Negative reference	Sign + magnitude	Two's complement	Offset binary	One's complement
+7	+⅞	–⅞	0 1 1 1	0 1 1 1	1 1 1 1	0 1 1 1
+6	+6/8	–6/8	0 1 1 0	0 1 1 0	1 1 1 0	0 1 1 0
+5	+⅝	–⅝	0 1 0 1	0 1 0 1	1 1 0 1	0 1 0 1
+4	+4/8	–4/8	0 1 0 0	0 1 0 0	1 1 0 0	0 1 0 0
+3	+⅜	–⅜	0 0 1 1	0 0 1 1	1 0 1 1	0 0 1 1
+2	+2/8	–2/8	0 0 1 0	0 0 1 0	1 0 1 0	0 0 1 0
+1	+⅛	–⅛	0 0 0 1	0 0 0 1	1 0 0 1	0 0 0 1
0	0+	0–	0 0 0 0	0 0 0 0	1 0 0 0	0 0 0 0
0	0–	0+	1 0 0 0	(0 0 0 0)	(1 0 0 0)	1 1 1 1
–1	–⅛	+⅛	1 0 0 1	1 1 1 1	0 1 1 1	1 1 1 0
–2	–2/8	+2/8	1 0 1 0	1 1 1 0	0 1 1 0	1 1 0 1
–3	–⅜	+⅜	1 0 1 1	1 1 0 1	0 1 0 1	1 1 0 0
–4	–4/8	+4/8	1 1 0 0	1 1 0 0	0 1 0 0	1 0 1 1
–5	–⅝	+⅝	1 1 0 1	1 0 1 1	0 0 1 1	1 0 1 0
–6	–6/8	+6/8	1 1 1 0	1 0 1 0	0 0 1 0	1 0 0 1
–7	–⅞	+⅞	1 1 1 1	1 0 0 1	0 0 0 1	1 0 0 0
–8	–8/8	+8/8		(1 0 0 0)	(0 0 0 0)	

Two's complement

Very similar to offset binary, but with the sign bit inverted. Arithmetically, a two's complement code word is formed by complementing the positive value and adding 1 LSB. For example:

+2 = 0010
–2 = 1101 + 1 = 1110

It is a very easy code to process; for instance, positive and negative numbers added together always give zero (disregarding the extra carry). For example:

$$\begin{array}{r} 0010 \\ +\underline{1110} \\ 0000 \end{array}$$

This is the code used almost universally for digital audio; there is, however (as with offset binary), a rather big transition at zero – all bits change from 1 to 0.

One's complement

Here negative values are full complements of positive values. This code is not commonly used.

4 Overview of A/D conversion systems

Linear (or uniform) quantization

In all the examples seen so far, the quantization intervals Q were identical. Such quantization systems are commonly termed linear or uniform. Regarding simplicity and quality, linear systems are certainly best. However, linear systems are rather costly in terms of required bandwidth and conversion accuracy. Indeed, a 16-bit audio channel with a sampling frequency of 44.056 kHz gives a bit stream of at least $16 \times 44.056 = 705 \times 10^3$ bit s^{-1}, which requires a bandwidth of 350 kHz – 17.5 times the bandwidth of the original signal. In practice, a wider bandwidth than this is required because more bits are needed for synchronization, error correction and other purposes.

Since the beginning of PCM telephony, ways to reduce the bandwidths that digitized audio signals require have been developed. Most of these techniques can also be used for digital audio.

Companding systems

If, in a quantizer, the quantization intervals Q are not identical, we talk about non-linear quantization. It is, for instance, perfectly possible to change the quantization intervals according to the level of the input signal. In general, in such systems, small-level signals will be quantized with more closely spaced intervals, while larger signals can be quantized with bigger quantization intervals. This is possible because the larger signals more or less mask the unavoidably higher noise levels of the coarser quantization.

Such a non-linear quantization system can be thought to consist of a linear system, to which a compander has been added. In such a system, the input signal is first compressed, following some non-linear law $F(x)$, then linearly quantized, processed and then, after reconversion, expanded by the reverse non-linearity $F^{-1}(y)$ (see Figure 4.1). The overall effect is analogous to companders used in the analog field (e.g., Dolby, dBx and others).

The non-linear laws which compressors follow can be shown in graphical forms as curves. One compressor curve used extensively in digital telephony in North America for the digitization of speech is the 'μ-law' curve. This curve is characterized by the formula:

$$F(x) = V \frac{V_{\log}(1 + \mu x / V)}{\log(1 + \mu)}$$

Curves for this equation are shown in Figure 4.2 for several values of μ. In Europe, the 'A-law' curve is more generally used (Figure 4.3).

Figure 4.1 Non-linear quantization.

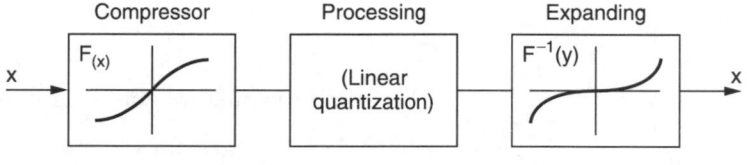

Figure 4.2 Characteristics of μ-law compressors.

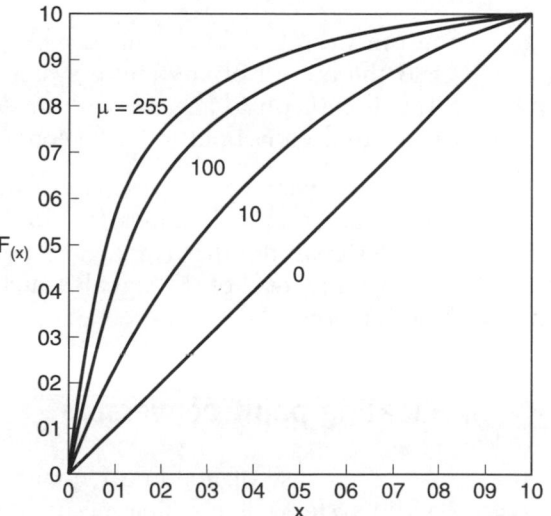

Figure 4.3 *A*-law
characteristic curve.

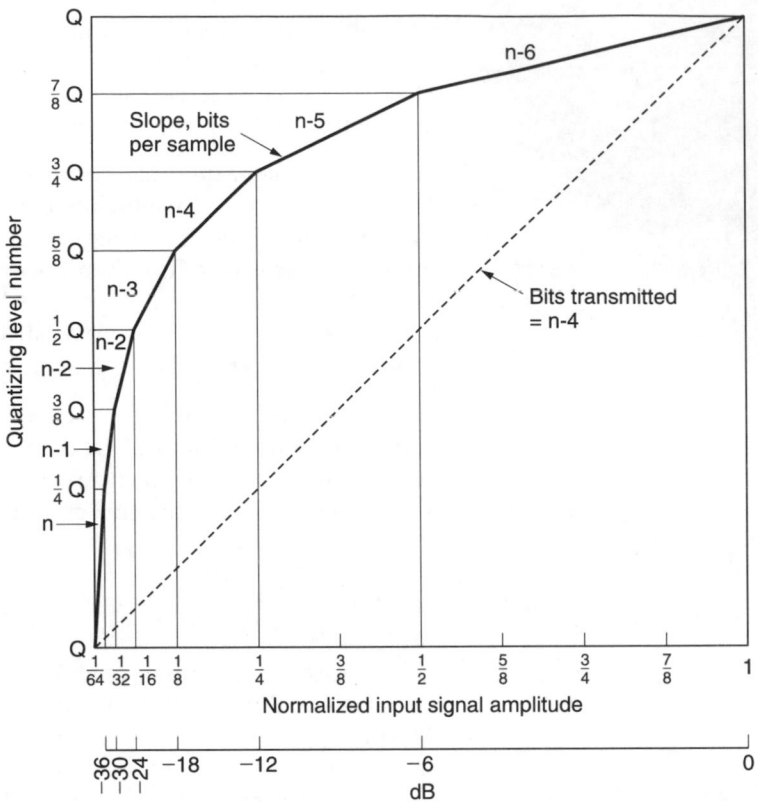

The (dual) formula for the 'A law' is:

$F(x) = Ax/1 + \log A$ for $0 < x < V/a$
$F(x) = V + V \log (Ax/V)/1 + \log A$ for $V/a < x < V$

In practice, it is important that the non-linearities at the input
and the output of any audio system are very closely matched.
This is difficult to achieve with analog techniques, so compressors are usually built in to the conversion process.

The big advantage of these companded systems is that the signal-to-noise ratio becomes less dependent on the level of the
input signal; the disadvantage, however, is that the noise level
follows the level of the signal, which may lead to audible noise
modulation.

Floating-point conversion

A special case of non-linear quantization, used in professional
audio systems, is the 'floating-point converter' (Figure 4.4).

Figure 4.4 Floating-point
converter principle.

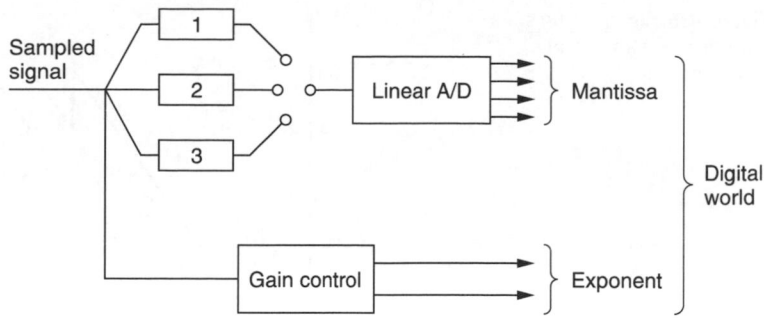

Sampled signal is sent through several selectable paths, each with a different gain, depending on the input level of the signal. Path, and hence gain, is selected by a logic monitor circuit in order to make maximum use of the linear A/D converter without overloading it. The output from the A/D converter, called the 'mantissa' in an analogy with logarithmic annotation, is meaningless without a way to indicate the gain that was originally selected. This information is provided by a logic output from the monitor circuit, called the 'exponent'. The exponent and mantissa, taken together, give an unambiguous digital word that can be reconverted to the original signal by selecting the corresponding (inverse) gains in the decoding stage. In this way, 2 bits of exponent can indicate four different gains. If we select these gains as 0, 6, 12 and 18 dB for instance, the two additional bits provide an increase of 18 dB over the dynamic range of the basic system.

Because the signal level determines basic system gain, noise modulation is unavoidable. This may become audible, for instance, with a high-level, low-frequency signal: in this case, noise modulation will not be masked by the signal.

Due to the effects of noise modulation, a distinction must be made between the dynamic range and the signal-to-noise ratio. The dynamic range can be defined as:

$$\frac{\text{maximum signal level (RMS)}}{\text{RMS level of quantization noise } without \text{ signal}}$$

whereas the signal-to-noise ratio is:

$$\frac{\text{signal level (RMS)}}{\text{RMS level of quantization noise } with \text{ signal}}$$

A curve for the signal-to-noise ratio of a typical floating-point converter with a 10-bit mantissa, a 3-bit exponent and 6-dB gain steps is shown in Figure 4.5. Although, theoretically, this system

Figure 4.5 Signal-to-noise ratio of a floating-point converter.

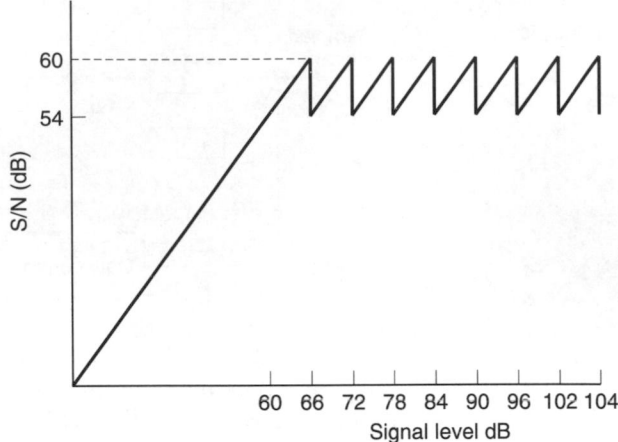

provides the same dynamic range as a 17-bit linear system (i.e., over 100 dB), the signal-to-noise ratio is unacceptable for high-quality purposes.

In spite of this, high-quality floating-point converters having, say, a 13-bit mantissa and 3-bit exponent are still considered for digital audio purposes, as they are considerably cheaper than linear systems.

Block floating-point conversion

When a low bandwidth is of utmost importance, block conversion can be used. This technique is also known as near-instantaneous companding (in contrast to basic floating-point or other companding systems). The term 'near-instantaneous' is used to describe the fact that not every sample is scaled by an exponent, but a number of successive samples (usually 32). Each block of samples is then followed by a scale factor word, so that, at the receiving end, each block can be correctly scaled up again (Figure 4.6).

Figure 4.6 A block floating-point converter.

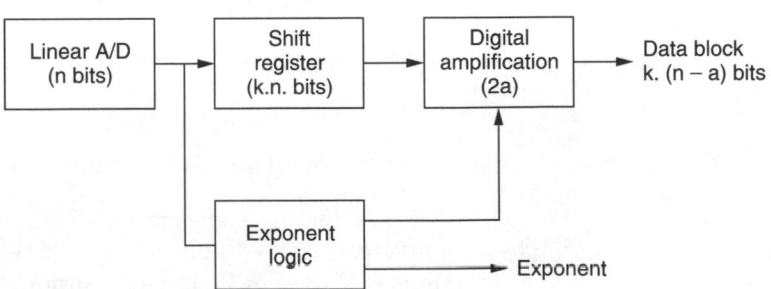

This system is rather expensive as far as hardware is concerned, but permits significant reductions in bit rates. Consequently, a typical application is digital transmission of audio signals in radio networks.

Subjective listening tests have shown that an original 14-bit system compressed to 10 bits is almost indistinguishable from a 13-bit linear system, although the signal-to-noise ratio limitations of a floating-point converter remain valid.

An example of such a system is the BBC's NICAM-3 (near-instantaneous companding audio multiplex), which permits transmission of six audio channels over one (standard) telephony 2048 kbit s^{-1} circuit.

Differential PCM and delta modulation

Instead of transmitting the exact binary value of each sample, it is possible to transmit only the difference between the current sample and the previous one. As this difference is generally small, a smaller number of bits can be used with no apparent degradation in performance. Operation is fairly straightforward: one sample is stored for the complete sample period, then added to the received difference signal to obtain the next sample. This sample is then stored until the next received difference signal.

Differential PCM, in fact, is a special type of predictive encoding. In such encoding schemes, a prediction is generated for the current sample, based upon past data; the correcting signal is simply the difference between the prediction and the actual signal.

As sampling rate increases, the differences between previous and present samples become smaller, so that, in the extreme for very high sampling rates, only 1 bit is needed for the error signal to indicate the sign of the error; in this case we talk about delta modulation.

Figure 4.7 shows a basic single-bit A/D converter. The input signal is compared with the output of a 1-bit D/A converter; the resulting voltage is then compared with a reference and the output used to increment or decrement the DAC value. For any input signal the system needs to perform a certain number of iterations to obtain the required resolution. Each iteration results in a high or low signal at the output of the A/D converter. Looking at the output we see a pulse train whose mean value equals the level of the input signal. The analog input has been converted to a

Figure 4.7 One-bit A/D converter.

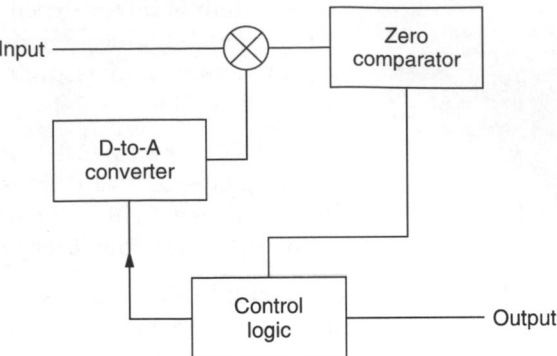

binary bit stream. The typical sampling frequency in 1-bit converters is several megahertz.

Because the serial bit stream is of little practical use, it is mostly converted to a multi-bit format (e.g., 16-bit) with a much lower sampling rate. This is done in a digital filter, a so-called decimation filter, which includes noise shaping (Figure 4.8).

In a further step, the transmitted data can be used to indicate not only the sign of the error, but also the step size. For example, a continuous series of ones means that the signal is quickly increasing, so the step size can be increased; if ones and zeros are alternating, step size can be reduced. Such strategies are called adaptive differential PCM (the quantization interval is changed) or adaptive delta modulation (the step size is changed). Although these techniques have some interesting theoretical and practical properties, it is presently difficult to use them for high-quality applications.

Super bit mapping (SBM)

SBM is an enhanced A/D conversion technique used for disc mastering and also on some recent DAT recorders (as from 1994). When the CD technology started, the amount of 16 bits per

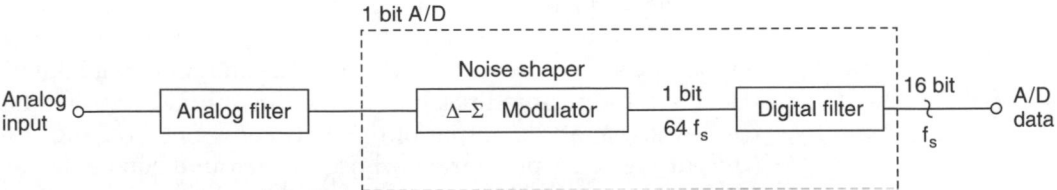

Figure 4.8 Block diagram of 1-bit A/D converter.

sample was specified. At that time 16 bits seemed ambitious, as the available technology only allowed about 14–15 bits.

Now, however, it is no problem to convert substantially more than 16 bits per sample, and A/D conversion in professional sound studios is performed at 18 or 20 bits, if not higher. Of course, if a conversion is made at 18 or 20 bits and the disc specification allows 'only' 16 bits, there is a need for reduction. SBM uses the extra bits to increase the accuracy of the 16-bit signal.

A very important point to note is that the use of such techniques poses no compatibility problems at all. A disc to which the SBM principle has been applied can be played back without any problem in any CD player. If a digital recording was made using a 20-bit A/D converter, the last 4 bits (bits 16–19) cannot be implemented in the 16-bit data stream on the disc, but these bits can be used to increase the accuracy of the least significant bits of the 16-bit samples and thereby maintain compatibility, decrease the noise level and increase the sound quality. Figure 4.9 illustrates the SBM effect.

SBM operation is highly complex and involves higher order mathematics, which is beyond the scope of this publication. It should always be remembered, however, that these calculations are applied to digital audio data not to analog signals. Figure 4.10 presents an SBM block diagram.

The input is audio data (more than 16 bits per sample), each channel is calculated separately, but it is obvious that in order to maintain sound integrity, the calculations performed on one channel must have at some points input from the other channel.

Audio data will be calculated by blocks of 512 samples (512 samples at a sampling rate of 44.1 kHz equals a sound block of 11.6 ms). Any DC component will be subtracted from the input.

The next block is the simultaneous masking curve calculation. The input blocks (512 samples left and right channel) are processed by Fast Fourier Transforms in order to analyse the input audio. This calculation determines the signal power in each frequency band, the total audio frequency spectrum being split into a number of well-defined sub-spectra. Note that this is a known psychoacoustic fact: the human ear works according to similar principles, analysing signal power at predetermined subdivisions of the total audio spectrum.

Figure 4.9 SBM effect.

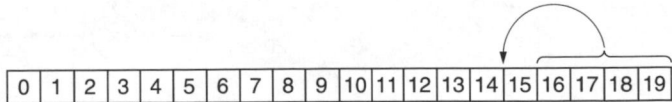

Figure 4.10 SBM block diagram.

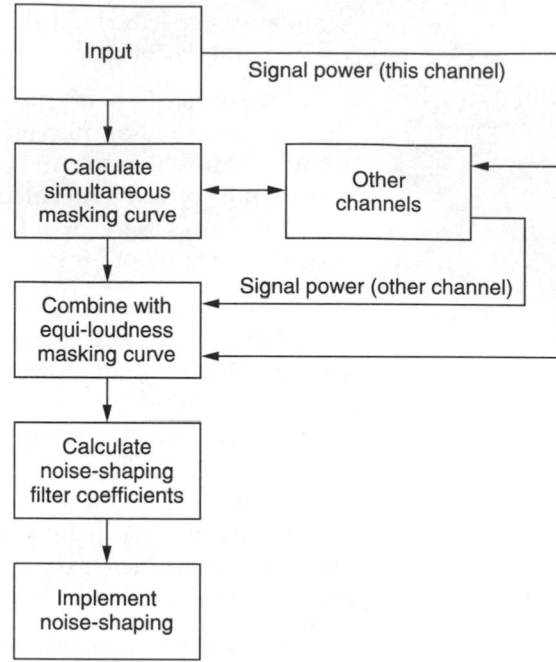

At this point, and also in the next block, the equi-loudness masking block, the SBM system determines whether one part of the incoming signal can mask another part and where and at what level quantization noise is found.

The knowledge gained in previous blocks will be used to calculate filter coefficients which will be used to perform noise shaping. Noise shaping (see also Figure 5.11) is a technology able to eliminate quantization noise contents from the audible spectrum.

In the next stage, noise shaping is performed based upon all the previously performed calculations. Figure 4.11 illustrates the SBM noise shaping.

Figure 4.11 SBM noise shaping.

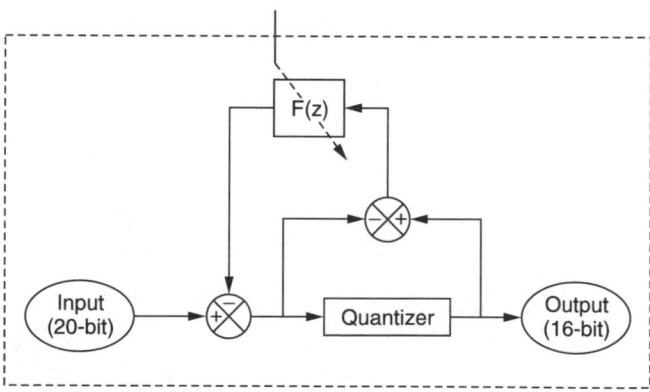

In comparison with noise shaping filters used at the D/A side of CD players, it should be noted that in this case the filter coefficients are adapted continuously, so the operation is far more complex and needs a very high calculation speed.

The result of SBM is a 16-bit signal with a lower quantization noise, a higher linearity and therefore a better sound definition.

Direct stream digital (DSD)

The latest approach to digital recording for the next generation of optical disc readers is DSD. Compared to the conventional CD, which makes use of linear PCM, DSD offers the possibility to encode and reproduce extreme high-quality sounds. The basic concept of the DSD format is a 1-bit A/D–D/A conversion system operating at a very high frequency of 2.8224 MHz. The basic principles of 1-bit A/D–D/A conversion have already been explained in this chapter; the state of technology and design of circuitry have now allowed these principles to be put into practice up to a rather extreme level. Together with the more powerful and faster operating microprocessors, which allow more complex mathematics, the DSD format is capable of reproducing sounds beyond 100 kHz.

In spite of the progress of multi-bit PCM and even if bit rates used in recording studios have been raised to 24 bits, one major disadvantage will always be present: filtering. As seen in Chapter 2, the maximum frequency that can be recorded is half the sampling frequency. In the case of the CD format with a sampling frequency of 44.1 kHz (f_s), steep 'Brick Wall' filters have to be used to filter out 22.05 kHz but leaving 20 kHz. This format is not only limited in frequency, but also the influence that these filters have on the lower part of the audible range was seen as a system constraint.

Professional analog tape recorders used in studios are able to record signals up to 50 kHz, including the higher harmonic frequencies produced by musical instruments. These harmonic frequencies have their influence on the original music. To archive these recordings with the important high-frequency information, a new higher performing A/D conversion system had to be developed. The logical answer to these constraints of conventional CD processing, and the need to incorporate the advantages of analog recording, was indeed DSD.

DSD is in theory a 'simple' 1-bit conversion system operating at a sample frequency of 2.8224 MHz ($64 \times f_s$). A simple block

Figure 4.12 DSD block diagram.

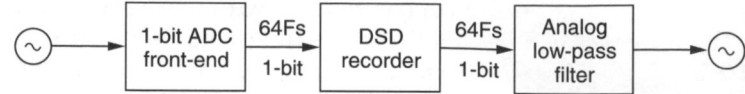

diagram of the DSD process is shown in Figure 4.12. When CD systems were originally designed, the theoretical basis for 1-bit high-frequency sampling was of course already available; at that time, however, it was not yet technologically possible to produce such high-speed A/D converters. The exponential speed of technological progress has now made it possible to produce the technology needed at an affordable price.

After 1-bit sampling at 64 f_s (2.8224 MHz), the signal is encoded and formatted similar to a CD; at this point, there is also some extended encoding and encryption which will be explained in subsequent chapters, but the basic ideas as used in CD have been retained. The most interesting part of DSD is that, after read-out of a disc, and after decoding/decrypting, the signal stream is such that it can be presented virtually directly to an amplifier and speakers. The main reason is again the use of a 1-bit signal at a very high frequency, which might even be compared to the signal coming from the needle of a pick-up, the only difference being the low-pass filter.

One of the reasons mentioned to quit the conventional multi-bit PCM system was the need for complex decimation and interpolation.

As can be seen in the most simple presentation in Figure 4.13, multi-bit PCM requires more steps to convert the analog input signal into the digital format. Also, to convert the digital signal read out by the laser to the original analog signal, complex cir-

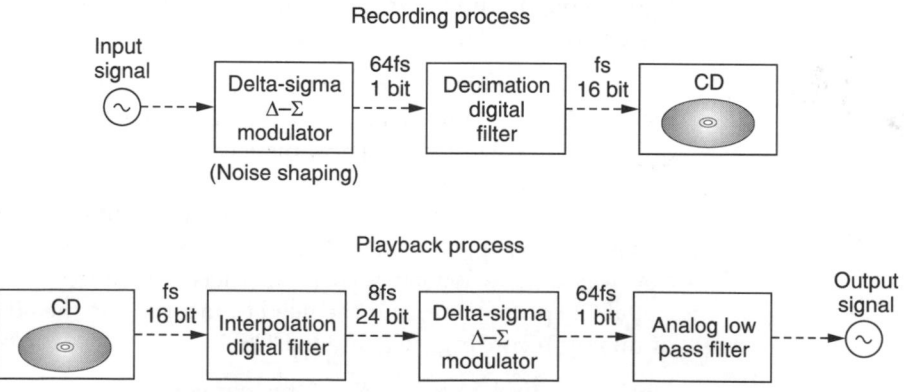

Figure 4.13 Multi-bit PCM.

cuitry is required. More circuitry needed for conversion results in a decreased quality of the sound.

The main differences between conventional CD processing and DSD processing are:

- 16-bit sampling versus 1-bit sampling;
- 44.1 kHz sampling rate versus 2.8224 MHz sampling rate;
- bandwidth limited to just above audible range versus bandwidth well beyond audible range;
- the need for noise shaping, digital filters, interpolation versus much reduced need for such circuitry;
- the need for high-precision D/A converter circuitry versus the possibility to even replace D/A circuitry by simple low-pass filtering.

One possible negative factor for DSD is the amount of data space needed for this higher bit rate: where a conventional CD requires around 650 Mbytes for a 74-minute disc, the DSD equivalent needs roughly five to six times more (around 3.5 Gbytes). The use of new and improved laser and disc techniques (see also Chapter 18 on SACD) has made this possible.

The main part of the DSD encoder is the Delta–Sigma (Δ–Σ) modulator. Figure 4.14 shows the simple block diagram of the Δ–Σ modulator.

The theory of this 1-bit Δ–Σ modulator has already been described previously; the basic design is unchanged.

The output of D switches between +1 and −1 when the input is 1 or 0, respectively. Suppose we have a situation with 0 V at the analog input and the output of Q was zero before. At the time of sampling, the output of D becomes −1 V. Connected to the inverting input of Δ, the output of Δ will become 1 V. Passing this 1 V through Σ will change the output of Q to 1. At the next sample, the 1 is returned to Δ, resulting in −1 V at its output. This −1 V is added to the previous value of Σ, giving a 0 at its output, making the output of Q return to 0. In this stage the converter is back to its start position.

Figure 4.14 One-bit Δ–Σ modulator.

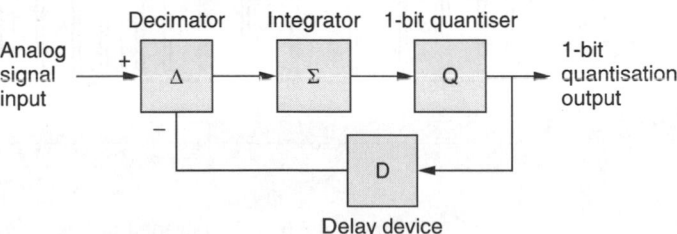

65

Consequently, the output of the DSD converter will alter between 0 and 1 when 0 V is applied at the input.

When the input level is at maximum, suppose this is 1 V; at this time the output of Δ is 2 V (suppose there was 0 V at the previous sample on the output of Q so there was –1 V at the output of D). This 2 V is added to the previous output of Σ. If it was 0 V in the previous sample, the output will become 2 V. Q will then output a '1'. This '1' is returned to the inverting input of Δ, resulting in a '0' on its output. This '0' will not change the output of Σ, so at this moment the modulator is in a stable situation with '1' always at the output.

A similar explanation can be made when applying the minimum level or changing signals at the input.

Figure 4.15 shows the output signal from the Δ–Σ modulator in the case where a sine wave is on the input. In the case of the positive-going waveform the density of the ones is much higher than in case of the negative-going waveform. In fact, if the input signal is at the maximum level the digital output stream will be all '1' and if the input signal is at minimum level it will be all '0'.

As a logical conclusion to the name Direct Stream Digital, the analog signal is directly converted to a pulse train of zeros and ones. To convert this continuous bit stream back to an analog signal, in order to feed it into an amplifier and/or speakers, there is no real need for a dedicated D/A converter, although the decision to use one or not can still be taken on grounds of specific needs or designs. A low-pass filter can be used instead of the D/A converter.

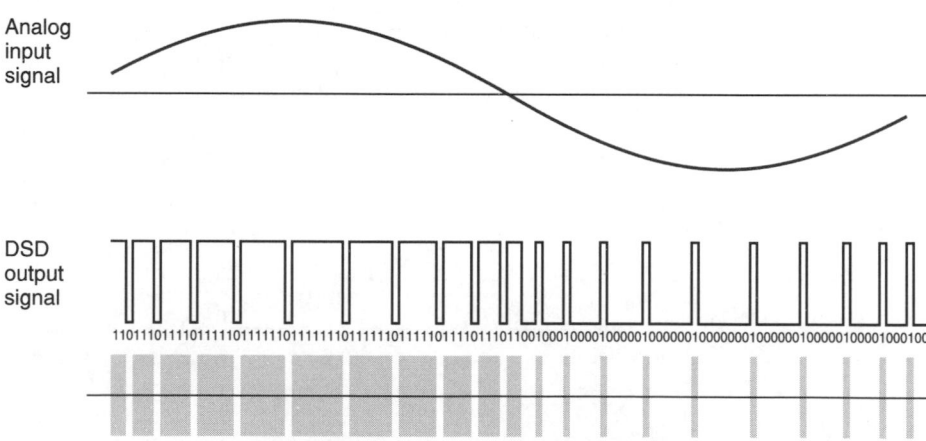

Figure 4.15 DSD converted signal.

The use of a low-pass filter for this purpose can be explained in a simplified – but valid – way: a low-pass filter is an integrator function, and an integrator function is a way to calculate the surface. If we now refer back to the previous illustration, it can be seen that the sine wave is the integrated function of the underlying rectangular patterns as caused by the '1' and '0' bits, keeping in mind that the '0' bits should be considered as '−1' to create the negative part of the sine wave. The greater the concentration of logic ones, the higher the sine wave rises; the greater the concentration of logic zeros, the lower it falls. In other words, by integrating the output bit pattern we are effectively recreating the corresponding audio spectrum.

Additionally, because of the high sampling frequency of 2.8224 MHz, DSD has the noise level shifted to higher frequencies (remember that one of the original purposes of the 1-bit oversampling technique was noise shaping). This is contrary to the linear PCM without noise shaping, where the noise has a constant level depending on the bit resolution. These frequencies are in the inaudible part of the spectrum and can be filtered out. Such circuitry does need some attention to the analog parts; these need to be of good quality, but these days it should not be a problem.

5 Operation of A/D–D/A converters

Some of the most important components in digital audio systems are the converters. Previous chapters have shown the need for high resolution to obtain a satisfactory signal-to-noise ratio.

In video applications, an 8-bit conversion is more than sufficient. A 14-bit conversion (or an equivalent) seems a minimum for good audio performance and, for professional use, 16-bit conversion is required to leave a margin for further processing (e.g., filtering, mixing).

In the PCM-F1 and compact disc system, 16-bit converters are used, while the PCM Video 8 system uses a 10-bit converter.

A/D converters

Fundamentally, A/D converters operate in one of two general ways. They either convert the analog input signal to a frequency or a set of pulses whose time is measured to provide a representative digital output, or compare the input signal with a variable reference, using an internal D/A converter to obtain the digital output.

Basic types of A/D converters

Voltage-to-frequency, ramp and integrating-ramp methods are the three leading conversion processes that use the time-measurement method. Successive approximation and parallel/modified parallel circuits rely on comparison methods.

Dual-slope integrating A/D converters

The dual-slope integrating A/D converter contains an integrator, some control logic, a clock, a comparator and an output counter, as shown in Figure 5. 1. A graph of integrator output voltage against time is shown in Figure 5.2.

The input analog signal is initially switched to the integrator, and the output of the integrator ramps up for a time t_1. The slope of the ramp, and hence the integrator output voltage at the end of this time, depends on the amplitude of the analog input signal and the time constant τ of the integrator:

$$\frac{V_{IN}}{\tau}$$

So the integrator output voltage V_0 at the end of time t_1 is:

$$\frac{V_{IN}}{\tau}t_1$$

Figure 5.1 Block diagram of a dual-slope integrating A/D converter.

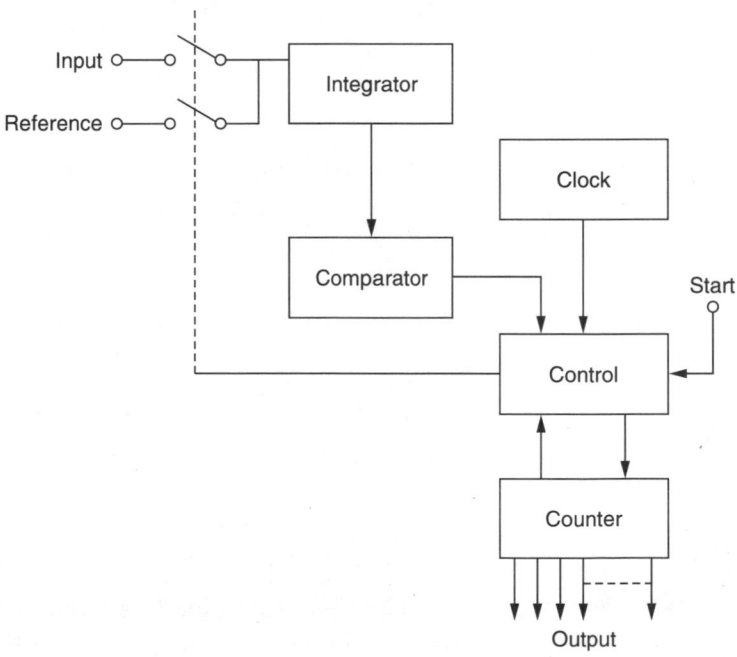

Figure 5.2 Integrator output voltage as a function of time, during conversions.

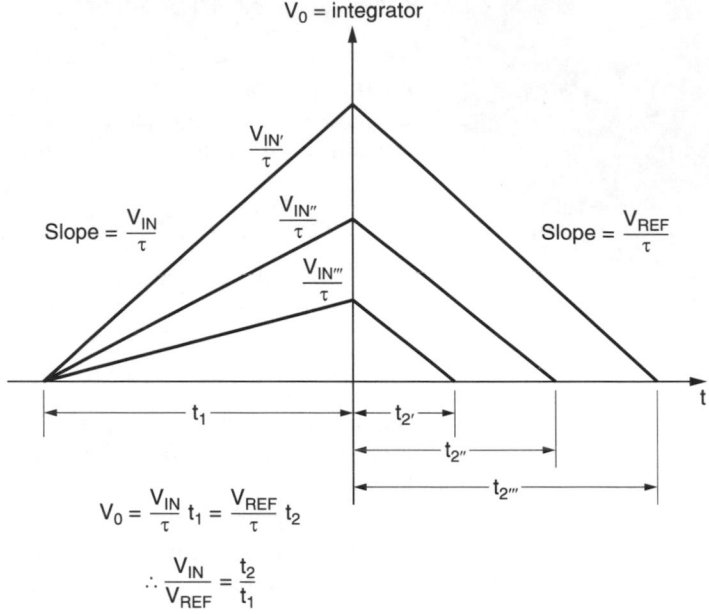

V_0 = integrator

$\dfrac{V_{IN'}}{\tau}$

$\dfrac{V_{IN''}}{\tau}$

$\dfrac{V_{IN'''}}{\tau}$

Slope = $\dfrac{V_{IN}}{\tau}$

Slope = $\dfrac{V_{REF}}{\tau}$

t_1 $t_{2'}$ $t_{2''}$ $t_{2'''}$ t

$$V_0 = \frac{V_{IN}}{\tau} t_1 = \frac{V_{REF}}{\tau} t_2$$

$$\therefore \frac{V_{IN}}{V_{REF}} = \frac{t_2}{t_1}$$

The reference signal is then switched to the integrator input, and the integrator output voltage ramps down until it returns to the starting voltage. The slope of the ramp during time t_2 similarly depends on the integrator time constant and the integrator input voltage, this time the reference signal amplitude:

$$\frac{V_{REF}}{\tau}$$

So the integrator output voltage at the initial time t_2 is:

$$\frac{V_{REF}}{\tau} t_2$$

However, as these voltages are the same:

$$\frac{V_{IN}}{\tau} t_1 = \frac{V_{REF}}{\tau} t_2$$

Therefore:

$$\frac{V_{IN}}{V_{REF}} = \frac{t_2}{t_1}$$

which shows that time t_2 is totally dependent on the input signal amplitude, and independent of the integrator time constant.

By counting clock pulses during time t_2, a digital measure of the analog input signal's amplitude is made.

Average conversion time, i.e., the time the converter takes to perform the conversion of an applied input signal, is two clock periods times the number of quantization levels. Thus, for a 12-bit converter with a 1 MHz clock, the average conversion time is: $2 \times 1\ \mu s \times 4096$, or 8.192 ms. The precise conversion time, however, depends on the applied input signal amplitude.

Due to this long conversion time, integrating converters are not useful for digitizing high-speed, rapidly varying signals, although they are useful to 14-bit accuracy, offering high noise rejection and excellent stability with both time and temperature. They can be modified to increase conversion speeds and are used mostly in 8- to 12-bit converters for digital voltmeters (DVMs), digital panel meters (DPMs) and digital multimeters (DMMs). However, basic dual-slope integrating A/D converters are too slow for general computer applications.

Successive-approximation A/D converters

The main reasons that the successive-approximation technique is used almost universally in A/D conversion systems are the reliability of the conversion technique, simplicity and inherent high-speed data conversion. Conversion time is equal to the clock period times the number of bits being converted. Thus, for a 1 MHz clock, a 12-bit converter would take 12 μs to convert an applied analog signal.

A successive approximation converter consists of a comparator, a register, control logic and a D/A converter. The output of the D/A converter is compared with the input analog voltage (Figure 5.3). Each bit line in the D/A converter corresponds to a bit position in the register. Initially, the converter is clear.

Figure 5.3 Successive-approximation A/D converter.

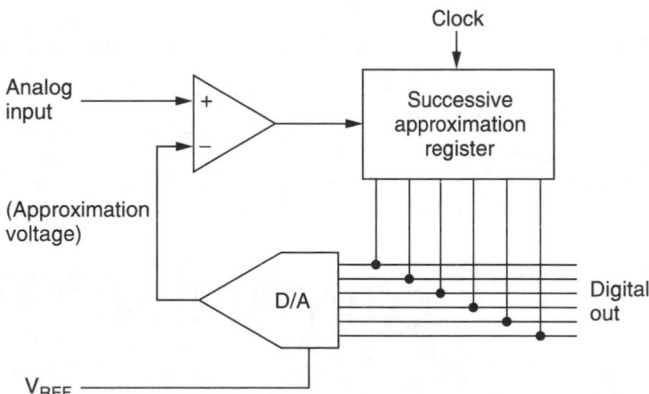

When an input signal is applied the control logic instructs the register to change its MSB to 1. This is changed by the D/A converter to an analog voltage equivalent to one-half the converter's full-scale range. If the input voltage is greater than this, the next most significant bit of the register becomes 1. If, however, the input is less, the next most significant bit remains 0. Then the circuit 'tries' the following bits through to the LSB, at which stage the conversion is complete. Thus, the number of approximations occurring in any conversion equals the number of bits in the digital output. Figure 5.4 shows the operation of the successive-approximation A/D converter graphically.

The main advantage of the successive-approximation converter is speed and this is limited by the settling time of the DAC. Accuracy is limited by the accuracy of the DAC, and a high susceptibility to noise is its major drawback. As only one comparator is used and ancillary hardware is limited to logic, register and D/A converter, the successive-approximation technique provides an inexpensive A/D converter.

Other types of A/D converters

Voltage-to-frequency converters

Figure 5.5 shows a typical voltage-to-frequency converter. Here, the input analog signal is integrated and fed to a comparator. When the comparator changes its state, the integrator is reset and the process repeats itself. The counter counts the number of integration cycles for a given time to provide a digital output.

The principal advantage of this type of conversion is its excellent noise rejection due to the fact that the digital output represents the average value of the input signal. Voltage-to-frequency conversion, however, is too slow for use in data-acquisition system applications because it operates bit-serially (with a maximum

Figure 5.4 Illustration of successive-approximation conversion. The digitally generated voltage gets closer to the analog input voltage in a series of approximations; each approximation is half the preceding one.

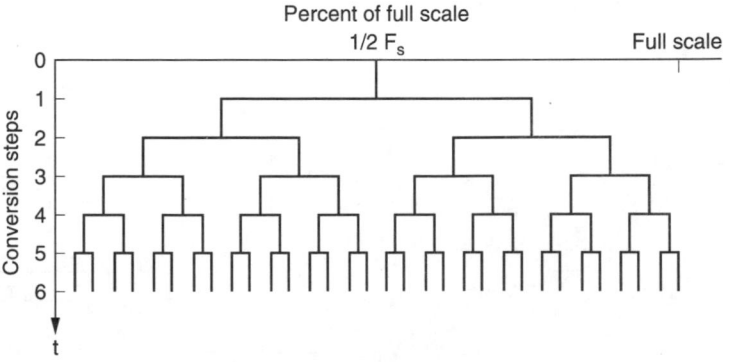

Figure 5.5 Voltage-to-frequency A/D converter.

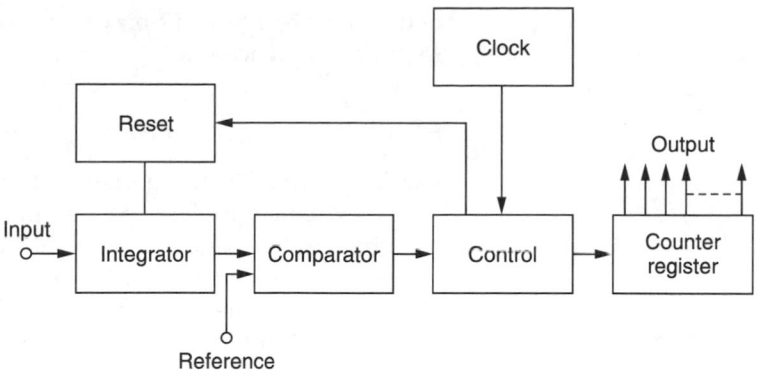

of approximately 1000 conversions/s). Its applications are mostly in digital voltmeters (DVMs) using converters with resolutions of 10 bits or less.

Ramp converters

Ramp conversion works by continuously comparing a linear reference ramp signal with the input signal using a comparator (Figure 5.6). The comparator initiates a counter when changing state and the counter counts clock pulses during the time the comparator is logically HIGH; the count is therefore proportional to the magnitude of the input signal. The counter output is the digital representation of the analog input.

This method is slightly faster than the previous one, but it requires a highly linear ramp source in order to be effective. It

Figure 5.6 Block diagram of a ramp converter.

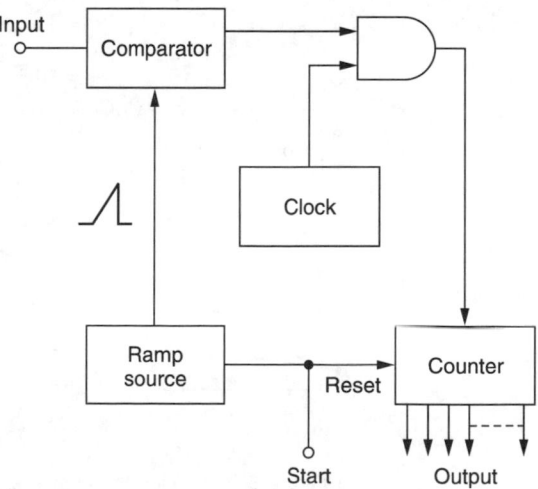

does offer good 8- to 12-bit differential linearity for applications requiring high accuracy.

Parallel A/D converters

Parallel-series and straight parallel converters are used primarily where extremely high speed is required, taking advantage of the fact that the propagation time through a chain of amplifiers is equal to the square root of the number of stages times the individual setting time, as opposed to adding up the times of each stage. By adding a comparator for every binary-weighted network, as shown in Figure 5.7, it is possible to take advantage of this higher speed. Parallel A/D converters are often called flash converters because of their high operating speeds.

The parallel A/D converter of Figure 5.7 uses one comparator for each input quantization level (i.e., a 6-bit converter would have six comparators). Conversion is straightforward; all that is required besides the comparators is logic for decoding the comparator outputs.

Because only comparators and logic gates stand between the analog inputs and digital outputs, extremely high speeds of up to 50 000 000 samplings/s can be obtained at low resolutions of 6 bits or less. The fact that the number of comparators and logic elements increases with resolution obviously makes this converter increasingly impractical for resolutions greater than 6 bits.

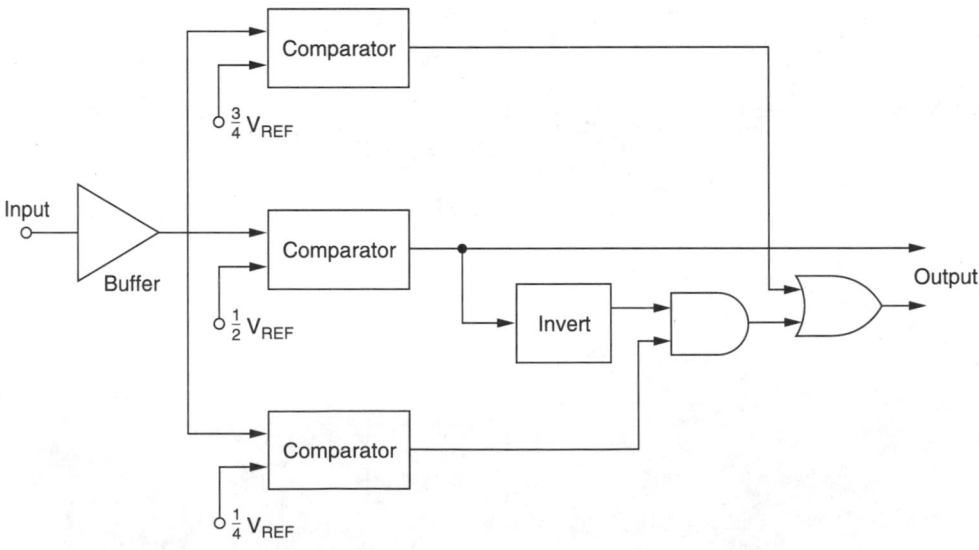

Figure 5.7 Parallel A/D converter.

Modified parallel designs can provide a good tradeoff between hardware complexity and the resolution/speed combination at a slight addition in hardware and a sacrifice in speed. They can provide up to 100 000 conversions/s for up to 14-bit resolutions. Sequential conversion (Figure 5.8), for example, is often used for such applications. However, because of the increase in the number of comparators and the need to use an amplifier for every weighting network, cost is considerably more than that of a successive approximation.

The first 4-bit converter in the circuit in Figure 5.8 provides the four most significant bits in parallel. These outputs are converted back to an analog voltage which is subtracted from the input. The difference is applied to the next converter and the process is continued until the required 10 bits are obtained. This approach gives a reasonable tradeoff among speed, cost and accuracy.

Delta–sigma modulator

A delta–sigma modulator is the key device in a 1-bit A/D converter. Figure 5.9 shows a first-order delta–sigma modulator. Operation is performed at each clock cycle, which corresponds to the oversampling frequency. At the beginning of each clock cycle, the differential amplifier outputs the difference between the input voltage V and the output voltage of the single-bit D/A converter. The integrator adds the voltage a to its own output from the preceding clock cycle. This voltage b is provided to the zero comparator. The output of the comparator will be logically HIGH or LOW, depending on voltage b being higher or lower than 0 V. The output then becomes a piece of single-bit A/D data, which is also used to determine the output of the 1-bit DAC for

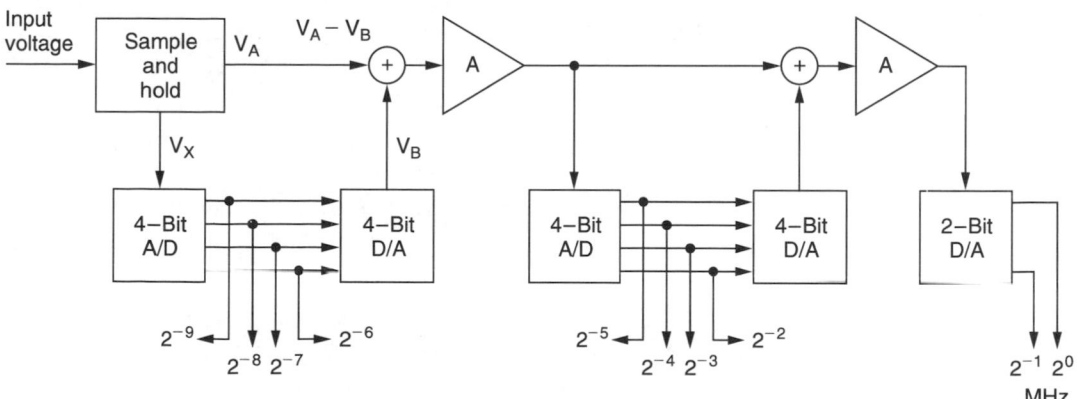

Figure 5.8 Sequential parallel A/D conversion.

the next clock cycle. The 1-bit DAC outputs a positive full-scale voltage if its input is HIGH and a negative full-scale voltage if its input is LOW. Table 5.1 shows an example of actual operation in which the input is 0.6 V, with the full-scale voltage being ±1 V and all initial values 0.

The 1-bit A/D converter outputs only HIGH or LOW, which has no meaning in itself; this only becomes meaningful when a string of 1-bit data is averaged.

Because of the high sampling frequency (64 times oversampling) a very gentle low-pass filter can be used, resulting in low phase distortion. Compared to successive approximation A/D convert-

Figure 5.9 First-order Δ–Σ modulator.

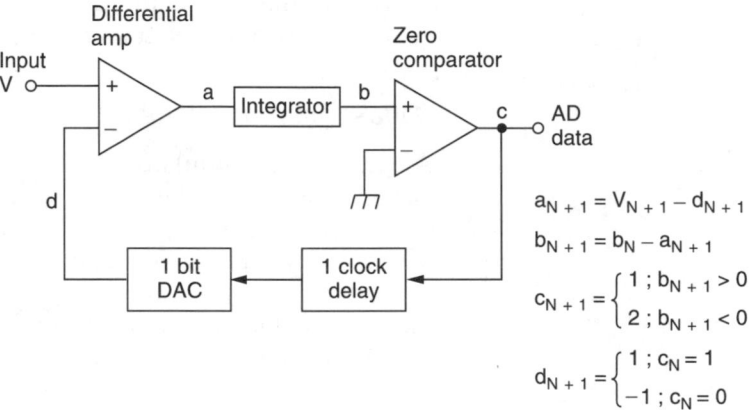

$$a_{N+1} = V_{N+1} - d_{N+1}$$

$$b_{N+1} = b_N - a_{N+1}$$

$$c_{N+1} = \begin{cases} 1 \; ; b_{N+1} > 0 \\ 2 \; ; b_{N+1} < 0 \end{cases}$$

$$d_{N+1} = \begin{cases} 1 \; ; c_N = 1 \\ -1 \; ; c_N = 0 \end{cases}$$

Table 5.1 Operation example of Δ–Σ modulator

Clock	d	a	b	c
0	0	0.6	0.6	1
1	1	−0.4	0.2	1
2	1	−0.4	−0.2	0
3	−1	1.6	1.4	1
4	1	−0.4	1.0	1
5	1	−0.4	0.6	1
6	1	−0.4	0.2	1
7	1	−0.4	−0.2	0
8	−1	1.6	1.4	1
9	1	−0.4	1.0	1
10	1	−0.4	1.0	1
11	1	−0.4	0.2	1
12	1	−0.4	−0.2	0
13	−1	1.6	1.4	1
14	1	−0.4	1.0	1
15	1	−0.4	0.6	1

0.6V fixed input, ±1V full-scale voltage.

ers, single-bit A/D converters provide better performance while circuit complexity and cost remain equal.

Noise shaping

A delta–sigma modulator is sometimes also called a noise shaper because it passes signals and noise according to different transfer functions (Figure 5.10). The signal transfer function for the modulator simplifies to:

$$\frac{Y(s)}{X(s)} = \frac{1}{s+1}$$

This is the s-domain representation of a first-order low-pass filter. Deriving the noise transfer function for the same modulator produces:

$$\frac{Y(s)}{N(s)} = \frac{s}{s+1}$$

This is the s-domain representation of a simple high-pass filter. Plotting the transfer functions gives the result shown in Figure 5.11. The signal is attenuated at higher frequencies, while the noise is shaped so that very little of its content is in the low-frequency region. By using higher order delta–sigma modulators, the in-band noise can even be reduced further; however, out-of-band noise will increase. In practice, a third- or fourth-order delta–sigma modulator is used to avoid stability problems while still using most of the noise shaping capabilities.

High-density linear A/D converter

The A/D converter currently used in Sony R-DAT recorders and some MiniDisc recorders is the high-density linear converter. This converter uses two fourth-order delta–sigma modulators for simultaneous sampling of two audio channels. Its output is a serial data signal with 16-bit resolution coded in two's complement at a sampling frequency of 32, 44.1 or 48 kHz. The delta–sigma modulators operate at 64 times oversampling; for a

Figure 5.10 Δ–Σ modulator or noise shaper.

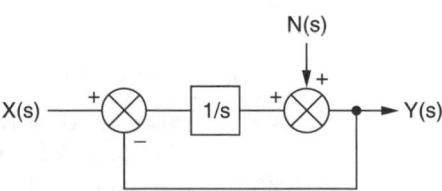

Figure 5.11 Transfer functions of noise shaper.

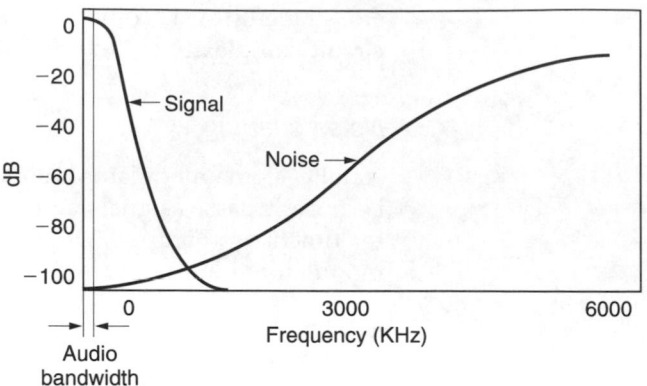

system sampling frequency of 48 kHz, the analog audio signals are actually sampled at:

$$f_s = 48 \text{ kHz} \times 64 - 3.072 \text{ MHz}$$

This high sampling frequency eliminates the need for a sample-hold circuit and a complex analog low-pass filter. A first-order RC network can be used as anti-aliasing filter because the audio signals do not normally contain frequencies above $\frac{1}{2} f_s$ (approx. 1.5 MHz). As a consequence, circuit complexity is greatly reduced.

The 1-bit stream, output from the delta–sigma modulators is hardly usable for further application. A built-in digital filter recalculates the single-bit input data at $64 f_s$ to 16-bit words at f_s. The so-called requantization is performed by the decimation filters.

Figure 5.12 shows a block diagram of the high-density linear converter. This converter is clearly separated into an analog and a digital section. The analog part contains the voltage reference source and the delta–sigma modulators, while the digital part includes the digital filters, a controller and the output interface. Each section has its own power supply to prevent digital noise from entering the analog signals.

During initialization, the converter performs an automatic calibration to compensate for possible offset errors in the converter itself. The analog inputs are grounded while the resulting output is measured and its value stored in SRAM as an offset. After calibration, the actual data are corrected by this offset value before being output.

With a harmonic distortion (THD) of less than 0.002% and a signal-to-noise ratio of more than 94 dB (EAIJ), the high-density linear converter is ideally suited for high-quality digital audio

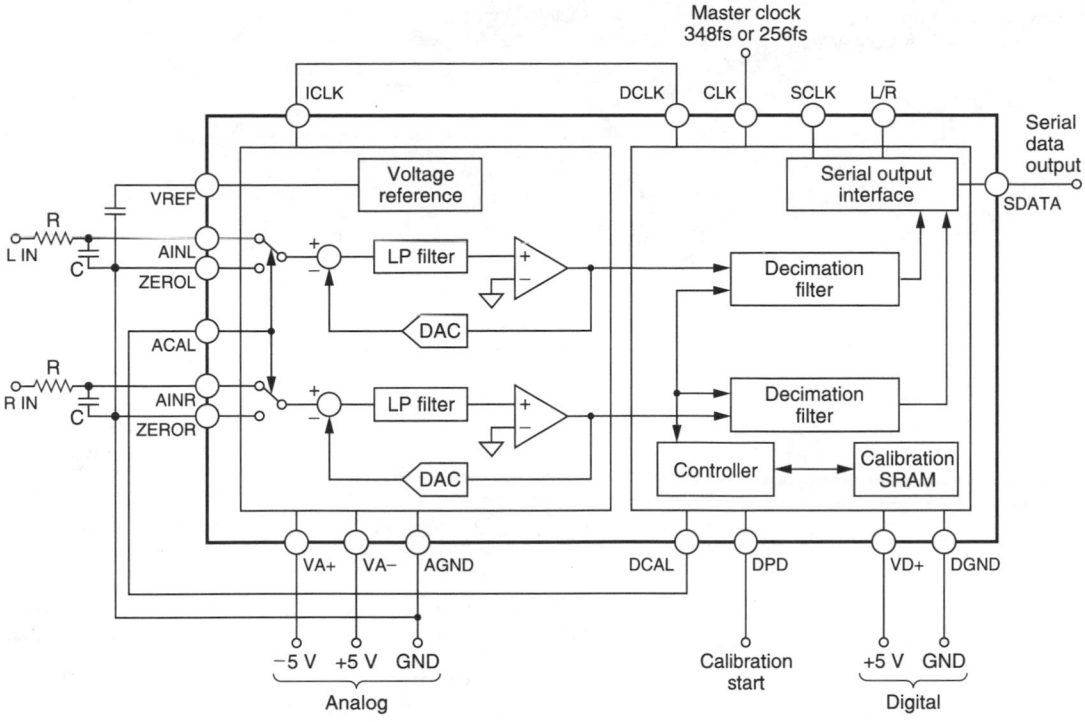

Figure 5.12 High-density linear converter as used by Sony in digital audio equipment.

signal processing. This type of converter is used in all Sony's R-DAT recorders, such as DTC-77ES and DTC-59ES, as well as in the MDS-101 MiniDisc recorder.

D/A conversion in digital audio equipment

Weighted current D/A converter

The most common D/A converter in digital audio is the weighted current type. It consists of a series of electronic switches, each of which is connected to a current source. In an n-bit D/A converter there are n weighted current sources with a current value of $2^{(n-1)}$ times the original current value 1. The current sources corresponding to the weight of the bit in the digital data signal are added to obtain a current that represents the value of the digital data. In fact, the digital data input directly controls the electronic switches that turn the current sources on or off. A current-to-voltage amplifier converts the obtained current into a voltage before being output. Figure 5.13 shows a typical block diagram

Figure 5.13 Current summing type D/A converter.

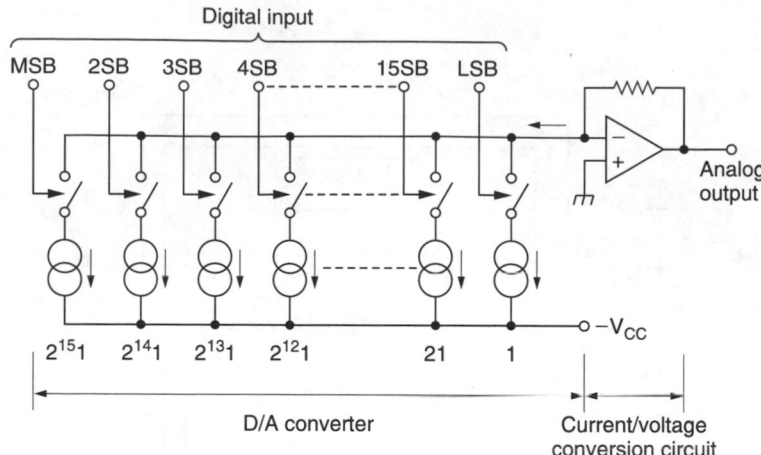

of the weighted current type or current summing type D/A converter.

Although this is basically a very simple converter, it has some serious drawbacks. The constant-current sources must be very accurately matched to prevent non-linear distortion. Suppose that the current 1 equals 0.1 mA, the constant-current source for the MSB should then deliver $2^{(n-1)} \times 1 = 3276.8$ mA or more than 3 A! Moreover, to maintain 16-bit resolution the accuracy of the MSB constant-current source must be better than the current delivered by the LSB constant-current source.

Another type of D/A converter, more or less based on the same principles, is the ladder network type D/A converter as shown in Figure 5.14. The so-called 'ladder network' is composed of resistors with values R and $2R$ to form a voltage divider. The electronic switches, controlled by the digital input data, change the output voltage of the voltage divider. The output voltage therefore represents the digital input signal. Accuracy is also a problem because all the resistors must be perfectly matched to ensure linearity. Temperature variations and ageing inevitably have a bad influence on the D/A converter linearity. The ON-resistance of the electronic switches must be sufficiently low compared to the resistor value R to prevent it from having too much influence on the output signal.

Actual D/A converters are in fact a combination of a ladder network type converter with constant-current sources added for the upper and lower bits.

Figure 5.14 Ladder network type D/A converter.

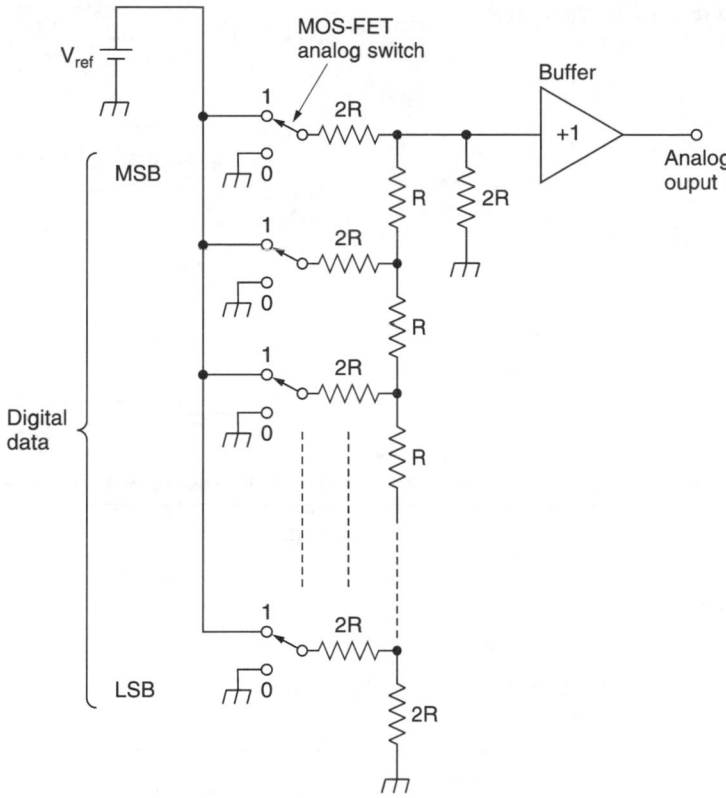

Single-bit D/A converter

The disadvantages of the weighted current D/A converter can be overcome by using advanced laser trimming techniques to accurately match resistors and current sources. However, this has a negative influence on manufacturing costs. The single-bit converter makes high-precision D/A conversion possible without the need for expensive matched components.

Figure 5.15 shows a block diagram of the single-bit converter. The first step in the conversion process is the requantization of the multi-bit digital audio signal at f_s1 to 1-bit signal with a much higher sampling frequency f_s2. Operation of the requantizer is explained in Figure 5.16. The requantizer will determine whether the input data D_i is higher or lower than the reference data (D_r), the reference data being the requantized result of the previous input data. Whenever the reference data is lower than the input data, the 1-bit output of the requantizer will be 1 and the reference data will be incremented. This process is repeated for every block cycle of the sampling frequency f_s1. At a certain time the reference data will be higher than the input data. The

Figure 5.15 Single-bit D/A converter.

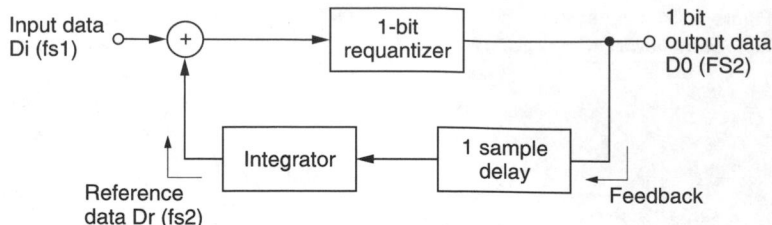

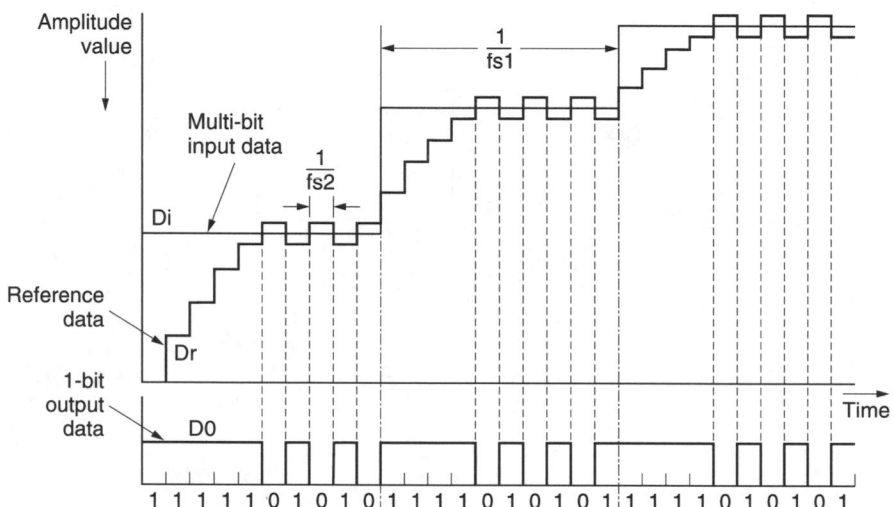

Figure 5.16 Operation of the 1-bit requantizer.

resulting 1-bit data now becomes 0 and the reference data decremented. The output sequence 0–1–0 continues until new input data are applied to the converter.

The 1-bit output is converted to a pulse width modulation (PWM) or pulse density modulation (PDM) signal which, after low-pass filtering, represents the analog output voltage (Figure 5.17). Because of feedback of the reference data, the single-bit D/A converter acts as a noise shaper, similar to the single-bit A/D converter (Figure 5.11). It shifts the requantization noise to the higher frequency regions, thereby lowering the quantization noise in the audible frequency range. Hence, the 1-bit D/A converter easily achieves a resolution of more than 16 bits in the audio frequency range. Non-linear distortion and DC offset are non-existing problems. The only major disadvantage is high-frequency noise radiation caused by the high sampling frequency f_s2, which is usually several megahertz. A carefully designed

Figure 5.17 Output signal of a conventional D/A converter (a), a 1-bit PWM converter (b) and a 1-bit PDM converter (c).

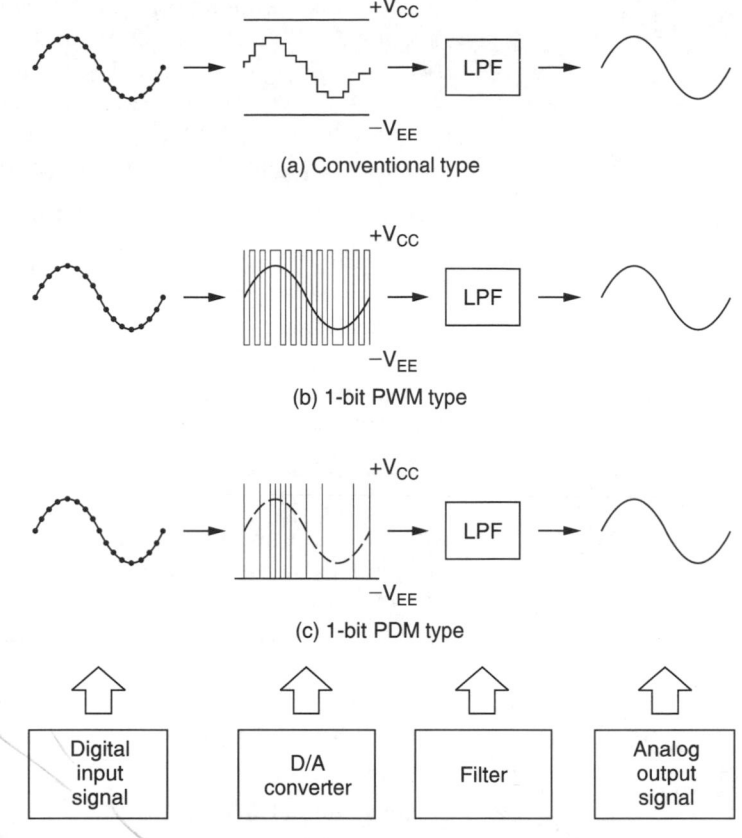

(a) Conventional type

(b) 1-bit PWM type

(c) 1-bit PDM type

| Digital input signal | D/A converter | Filter | Analog output signal |

single-bit D/A converter is therefore an inexpensive alternative for high-precision D/A conversion.

Oversampling

The output of a digital-to-analog converter cannot be used directly; filtering is necessary. The converter output produces the frequency spectrum shown in Figure 5.18, where the baseband audio signal ($0–f_m$) is reproduced symmetrical around the sampling frequency (f_s) and its harmonics. The low-pass reconstruction filter must reject everything except the baseband signal.

Figure 5.18 Spectrum of a sample baseband audio signal. A filter must reject all frequencies above a cut-off frequency f_m.

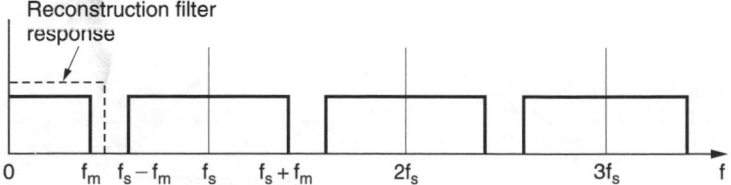

Reconstruction filter response

0   f_m $f_s - f_m$ f_s $f_s + f_m$ $2f_s$ $3f_s$ f

A sampling frequency (f_s) of 44 100 Hz and a maximum audio frequency (f_m) of 20 000 Hz mean that a low-pass filter with a flat response to 20 kHz and a high attenuation at $f_s - f_m$ (44 100 – 20 000 = 24 100 Hz) is needed. An analog filter can be made to have such a sharp roll-off, but the phase response will introduce an audible phase distortion and group delay.

One approach to getting round this problem is **oversampling**.

Oversampling is the use of a sampling rate greatly in excess of that stipulated by the Nyquist theorem. Practical implementations use a ×2 oversampling (f_s = 88.2 kHz) or a ×4 oversampling frequency (f_s = 176.5 kHz). The output spectrum of the D/A converter in a ×2 oversampling system is shown in Figure 5.19, where the large separation between baseband and sidebands allows a low-pass filter with a gentle roll-off to be used. This improves the phase response of the filter.

Digital words are input at the standard sampling rate of 44.1 kHz (i.e., no extra samples need be taken at the A/D conversion stage), and extra samples are generated at a rate of 88.2 kHz (Figure 5.20). The missing samples are computed by digital simulation of the analog reconstruction process. A digital transversal filter (also known as a finite impulse response filter) is well suited for this purpose.

Analog versus digital filters

The discrete-time signal produced by sampling an analog input signal (Figure 5.21) is defined as an infinite series of numbers,

Figure 5.19 In a ×2 oversampling system the effective sampling frequency becomes twice that of the actual sampling frequency. A simple low-pass filter can be used to reject all unwanted signal frequencies.

Figure 5.20 Timing diagram of an oversampling system. Words at a sampling frequency of 44.1 kHz have interpolated samples added, such that the effective sampling rate is 88.2 kHz.

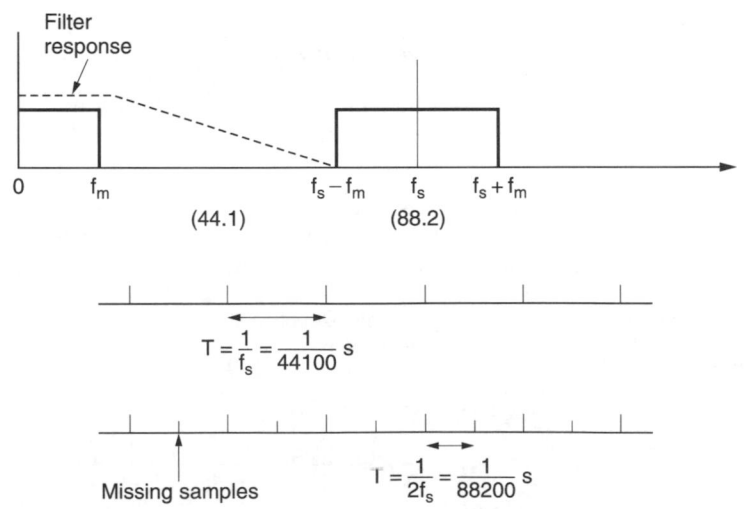

Figure 5.21 How a continuous value and continuous time analog signal is first converted to a discrete time but continuous value set of signals.

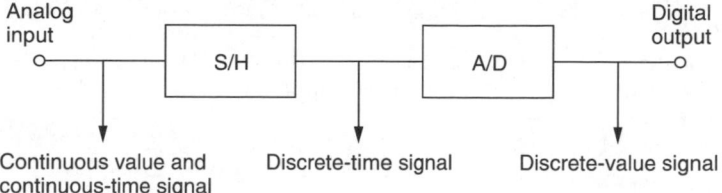

each corresponding to a sampling point at time $t = T_n$ for $-\infty < n < +\infty$. Such a series is always referred to by its value at $t = T_n$ which is $x(n)$.

The series $x(n)$ is defined as:

$$x(n) = ..., x(-2), x(-1), x(0), x(1), x(2), ...$$

with element $x(n)$ occurring at time $t = T_n$.

Analog filters

The first-order low-pass analog filter shown in Figure 5.22 is often described as a function of s, the independent variable in the complex frequency domain. The transfer function of such a filter is given by:

$$f(s) = \frac{V_o}{V_i} = \frac{1}{1+s} = \frac{1}{1+j\dfrac{\omega}{\omega_0}}$$

where $\omega =$ angular frequency $= 2\pi f$ and ω_0 is the angular frequency at the filter's cut-off frequency $f_c = 1/RC$.

Knowing this, the cut-off frequency of the filter can be calculated as follows:

$$2\pi f_c = \frac{1}{RC}$$

so:

$$f_c = \frac{1}{2\pi RC}$$

Figure 5.22 Simple first-order low-pass filter.

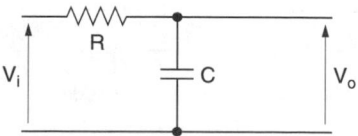

Digital filters

A digital filter is a processing system which generates the output sequence, $y(n)$, from an input sequence, $x(n)$, where:

$$y(n) = \sum_{i=0}^{M} a_i x(n-i) - \sum_{j=0}^{N} b_j y(n-j)$$

$$\underbrace{\phantom{\sum_{i=0}^{M} a_i x(n-i)}}_{\substack{\text{present and past} \\ \text{output samples}}} \underbrace{\phantom{\sum_{j=0}^{N} b_j y(n-j)}}_{\substack{\text{past input} \\ \text{samples}}}$$

The coefficients $a_0, a_1, \ldots, a_M$ and $b_0, b_1, \ldots, b_N$ are constants which describe the filter response.

When $N > 0$, indicating that past output samples are used in the calculation of the present output sample, the filter is said to be **recursive** or **cyclic**. An example is shown in Figure 5.23. When only present and past input samples are used in the calculation of the present output sample, the filter is said to be **non-recursive** or **non-cyclic**: because no past output samples are involved in the calculation, the second term then becomes zero (as $N = 0$). An example is shown in Figure 5.24.

Generally, digital audio systems use non-recursive filters and an example, used in the CDP-102 compact disc player, is shown in Figure 5.25 as a block diagram. IC309 is a CX23034, a 96th-order filter which contains 96 multipliers. The constant coefficients are contained in an ROM look-up table. Also note that the CX23034 operates on 16-bit wide data words, which means that all adders and multipliers are 16-bit devices.

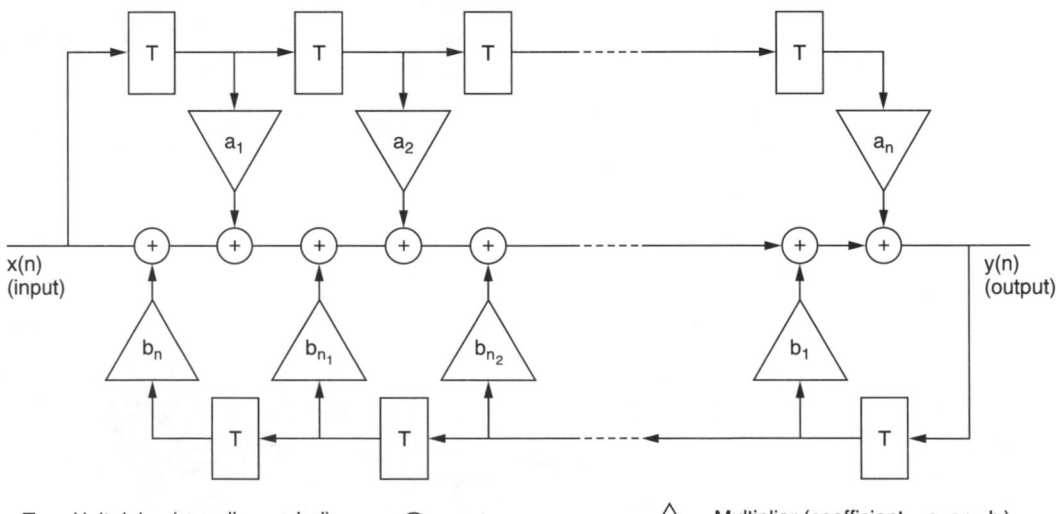

T = Unit delay (sampling period)   ⊕ = Adder   △ = Multiplier (coefficient = a_i or $-b_j$)

Figure 5.23 Recursive digital filter.

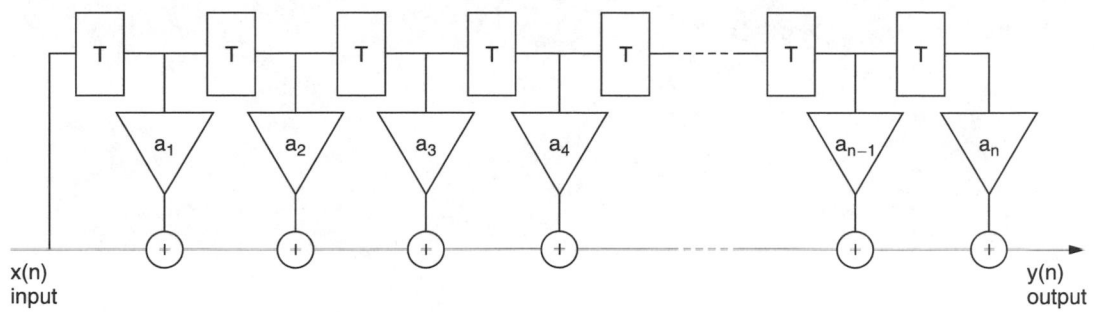

Figure 5.24 Non-recursive digital filter.

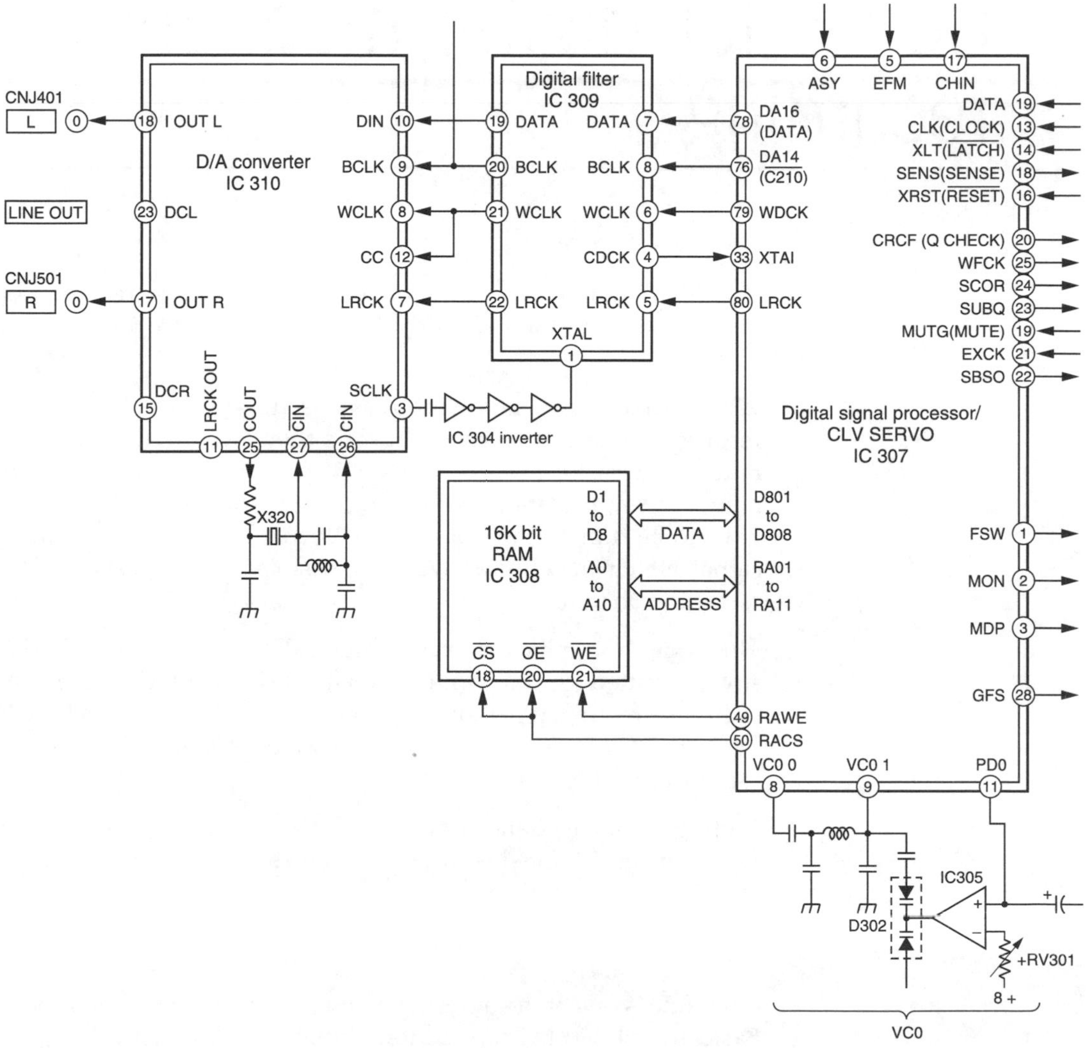

Figure 5.25 CDP-102 digital filter.

6 Codes for digital magnetic recording

The binary data representing an audio signal can be recorded on tape (or disc) in two ways: either directly, or after frequency modulation.

When frequency modulation is used, say, in helical-scan recorders, data can be modulated as they are, usually in a non-return-to-zero format (see below). If they are recorded directly, however, say, in stationary-head recorders and compact disc, they have to be transformed to some new code to obtain a recording signal which matches as well as possible the properties of the recording channel.

This code should have a format which allows the highest bit density permitted by the limiting characteristics of the recording channel (frequency response, dropout rate, etc.) to be obtained. Also, its DC content should be eliminated, as magnetic recorders cannot reproduce DC.

Coding of binary data in order to comply with the demands of the recording channel is often referred to as channel coding.

Non-return to zero (NRZ)

This code is one of the oldest and best known of all channel codes. Basically, a logic 1 is represented by positive magnetization, and a logic 0 by negative magnetization.

A succession of the same logic levels, though, presents no change in the signal, so that there may be a significant low-frequency content, which is undesirable for stationary-head recording.

In helical-scan recording techniques, on the other hand, the data are FM-converted before being recorded, so this property is less important. NRZ is commonly used in such formats as PCM-1600 and the EIAJ format PCM-100 and PCM-10 recorders.

Several variations of NRZ also exist for various applications.

Bi-phase

Similar to NRZ, but extra transitions are added at the beginning of every data bit interval. As a result, DC content is eliminated and synchronization becomes easier, but the density of signal transitions increases.

This code (and its variants) is also known as Manchester code, and is used in the Video 8 PCM recording format, where bits are modulated as a 2.9 MHz signal for a logic 0 and as 5.8 MHz for a logic 1.

Modified frequency modulation (MFM)

Also called Miller code or delay modulation. Ones are coded with transitions in the middle of the bit cell, isolated zeros are ignored, and between pairs of zeros a transition is inserted. It requires almost the same low bandwidth as NRZ, but has a reduced DC content. The logic needed for decoding is more complicated.

A variation is the so-called modified modified frequency modulation (M^2FM).

Three-position modulation (3PM)

This is a code which permits very high packing densities, but which requires rather complicated hardware. In principle, 3PM code is obtained by dividing the original NRZ data into blocks of three; each block is then converted to a 6-bit 3PM code, which is designed to optimize the maximum and minimum run lengths. In this way, the minimum possible time between transitions is twice the original (NRZ) clock period, whereas the maximum is six times the original.

For detection, on the other hand, a clock frequency twice that of the original signal is needed, consequently reducing the jitter margin of the system. This clock is normally recovered from the data itself, which have a high harmonic content around the clock frequency.

High-density modulation-1 (HDM-1)

This is a variation upon the 3PM system. The density ratio is the same as 3PM, but clock recovery is easier and the required hardware simpler. It is proposed by Sony for stationary-head recording.

Eight-to-fourteen modulation (EFM)

This code is used for the compact disc digital audio system. The principle is again similar to 3PM, but each block of eight data bits is converted into 14 channel bits, to which three extra bits are added for merging (synchronization) and low-frequency suppression. In this way, a good compromise is obtained between clock accuracy (and possible detection errors), minimum DC current (in disc systems, low frequencies in the signal give noise in the servo systems), and hardware complexity. Also, this modulation system is very suitable for combination with the error-correction system used in the same format.

EFM+

Based upon the knowledge gained on EFM used in CD players, a new and better performing modulating system EFM+ has been developed for SACD and DVD-Audio. The advantage of EFM+ is the more efficient use of the memory. Instead of eight-to-fourteen modulation, where three merging bits are additionally required for synchronization and low-frequency suppression, making it an eight-to-seventeen modulation in reality, EFM+ is modulating the 8-bit to a 16-bit word including low-frequency suppression.

Keeping in mind that to avoid reading problems by the optical system, a minimum three and maximum 11 subsequent zeros are allowed, 351 words of 16 bits meet the specifications. Only 256 words are required to convert the 8-bit word, leaving us about 95 additional or spare words. Depending on the DC value of the digital signal, different codes from these 'spare' words are used to minimize this DC component. The gain of memory is due to the fact that instead of the conversion from eight-to-seventeen (14 + 3 merging bits), only 16 bits are used for EFM+.

7 Principles of error correction

Types of code errors

When digital signals are sent through a transmission channel, or recorded and subsequently played back, many types of code errors can occur. Now, in the digital field, even small errors can cause disastrous audible effects: even one incorrectly detected bit of a data word will create an audible error if it is, say, the MSB of the word. Such results are entirely unacceptable in the high quality expected from digital audio, and much effort must be made to detect, and subsequently correct, as many errors as possible without making the circuit over-complicated or the recorded bandwidth too high.

There are a number of causes of code errors:

Dropouts

Dropouts are caused by dust or scratches on the magnetic tape or CD surface, or microscopic bubbles in the disc coating.

Tape dropout causes relatively long-time errors, called bursts, in which long sequences of related data are lost together. On discs dropouts may cause either burst or single random errors.

Jitter

Tape jitter causes random errors in the timing of detected bits and, to some extent, is unavoidable, due to properties of the tape transportation mechanism. Jitter margin is the maximum amount of jitter which permits correct detection of data. If the minimum run length of the signal is τ, then the jitter margin will be $\tau/2$. Figure 7.1 shows this with an NRZ signal.

Intersymbol interference

In stationary-head recording techniques, a pulse is recorded as a positive current followed by a negative current (see Figure 7.2). This causes the actual period of the signal that is read on the tape (T_1) to be longer than the bit period itself (T_0). Consequently, if the bit rate is very high, the detected pulse will be wider than the original pulse. Interference, known as intersymbol interference or time crosstalk, may occur between adjacent bits. Intersymbol interference causes random errors, depending upon the bit situation.

Figure 7.1 Jitter margin.

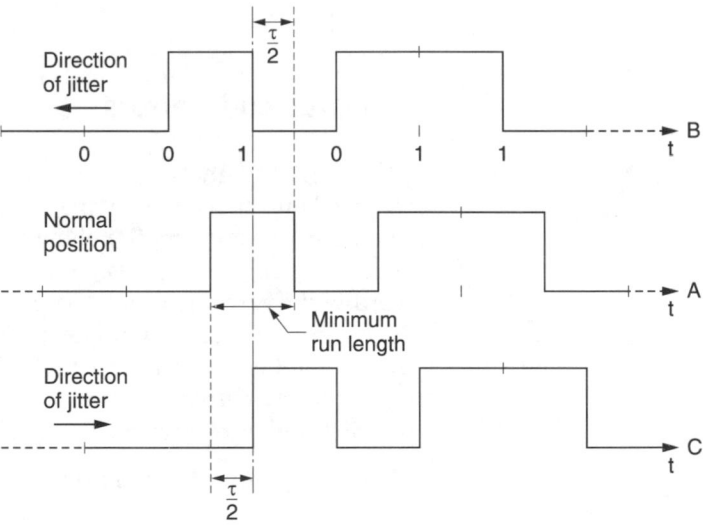

Figure 7.2 The cause of intersymbol interference.

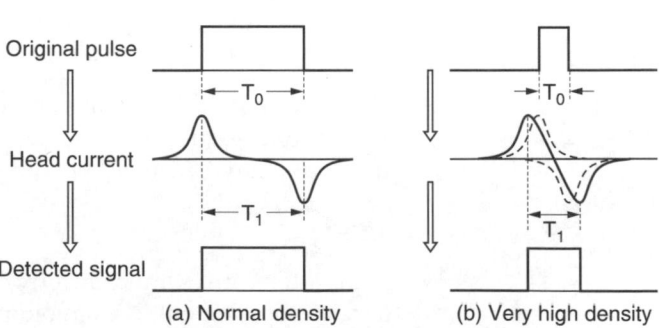

Noise

Noise may have similar effects to dropouts (differentiation between both is often difficult), but random errors may also occur in the case of pulse noise.

Editing

Tape editing always destroys some information on the tape, which consequently must be corrected. Electronic editing can keep errors to a minimum, but tape-cut editing will always cause very long and serious errors.

Error compensation

Errors must be detected by some error-detection mechanism: if misdetection occurs, the result is audible disturbance. After detection, an error-correction circuit attempts to recover the correct data. If this fails, an error-concealment mechanism will cover up the faulty information so, at least, there is no audible disturbance.

These three basic functions – detection, correction and concealment – are illustrated in Figure 7.3.

Figure 7.3 Three basic functions of error compensation: detection, correction and concealment.

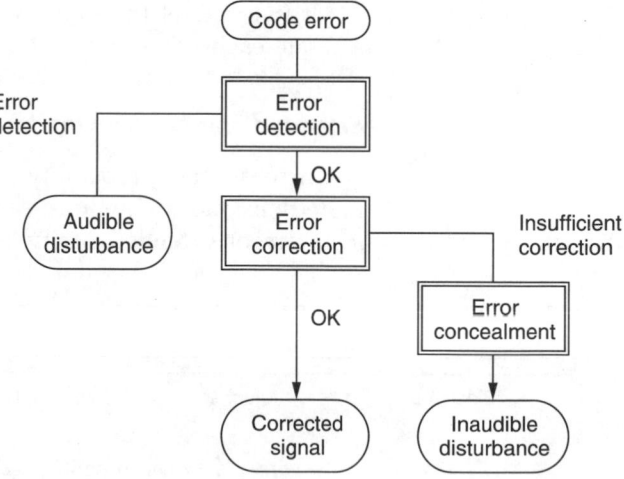

Error detection

Simple parity checking

To detect whether a detected data word contains an error, a very simple way is to add one extra bit to it before transmission. The extra bit is given the value 0 or 1, depending upon the number of ones in the word itself. Two systems are possible: odd parity and even parity checking.

- Odd parity: where the total number of ones is odd. Example:

  |1110|　　　|0|
  data　　　　parity

  |1001|　　　|1|
  data　　　　parity

- Even parity: where the total number of ones is even. Example:

  |1110|　　　|1|
  data　　　　parity

  |1001|　　　|0|
  data　　　　parity

The detected word must also have the required number of ones. If this is not the case, there has been a transmission error.

This rather elementary system has two main disadvantages:

1　Even if an error is detected, there is no way of knowing which bit was faulty.
2　If two bits of the same word are faulty, the errors compensate each other and no errors are detected.

Extended parity checking

To increase the probability of detecting errors, we can add more than one parity bit to each block of data. Figure 7.4 shows a system of extended parity, in which *M*-1 blocks of data, each

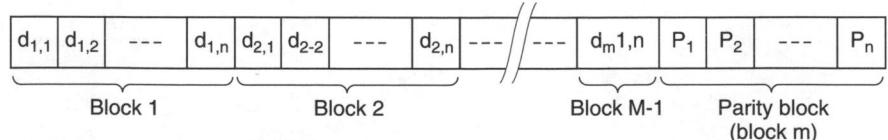

Figure 7.4 Extended parity checking.

of n bits, are followed by block m – the parity block. Each bit in the parity block corresponds to the relevant bits in each data block, and follows the odd or even parity rules outlined previously.

If the number of parity bits is n, it can be shown that (for reasonably high values of n) the probability of detecting errors is $\frac{1}{2}^n$.

Cyclic redundancy check code (CRCC)

The most efficient error-detection coding system used in digital audio is cyclic redundancy check code (CRCC), which relies on the fact that a bit stream of n bits can be considered an algebraic polynomial in a variable x with n terms. For example, the word 10011011 may be written as follows:

$$M(x) = 1x^7 + 0x^6 + 0x^5 + 1x^4 + 1x^3 + 0x^2 + 1x^1 + 1x^0$$
$$= x^7 + x^4 + x^3 + x + 1$$

Now to compute the cyclic check on $M(x)$, another polynomial $G(x)$ is chosen. Then, in the CRCC encoder, $M(x)$ and $G(x)$ are divided:

$$M(x)/G(x) = Q(x) + R(x)$$

where $Q(x)$ is the quotient of the division and $R(x)$ is the remainder.

Then (Figure 7.5), a new message $U(x)$ is generated as follows:

$$U(x) = M(x) + R(x)$$

so that $U(x)$ can always be divided by $G(x)$ to produce a quotient with no remainder.

It is this message $U(x)$ that is recorded or transmitted. If, on playback or at the receiving end, an error $E(x)$ occurs, the message $V(x)$ is detected instead of $U(x)$, where:

Figure 7.5 CRCC checking principle.

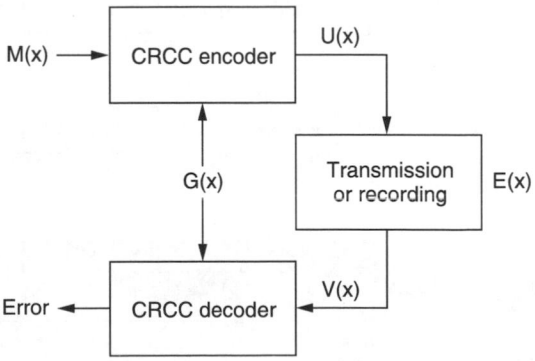

$$V(x) = U(x) + E(x)$$

In the CRCC decoder, $V(x)$ is divided by $G(x)$, and the resultant remainder $E(x)$ shows there has been an error.

Example (illustrated in Figure 7.6):

the message is $\qquad M(x) = x^9 + x^5 + x^2 + 1$
the check polynomial is $\qquad G(x) = x^5 + x^4 + x^2 + 1$

Now, before dividing by $G(x)$, we multiply $M(x)$ by x^5; or, in other words, we shift $M(x)$ five places to the left, in preparation of the five check bits that will be added to the message:

$$x^5 M(x) = x^{14} + x^{10} + x^7 + x^5$$

Then the division is made:

$$x^5 M(x) / G(x) = \underbrace{\frac{\left(x^9 + x^8 + x^7 + x^3 + x^2 + x + 1\right)}{\text{quotient}}} + \underbrace{\frac{(x+1)}{\text{remainder}}}$$

So that:

$$U(x) = x^5 M(x) + (x + 1)$$
$$= x^{14} + x^{10} + x^7 + x^5 + x + 1$$

which can be divided by $G(x)$ to leave no remainder.

Figure 7.6 shows that, in fact, the original data are unmodified (only shifted), and that the check bits follow at the end.

CRCC checking is very effective in detection of transmission error. If the number of CRCC bits is n, detection probability is $1 - 2^{-n}$. If, say, n is 16, as in the case of the Sony PCM-1600, detection probability is $1 - 2^{-16} = 0.999985$ or 99.9985%. This means that the CRCC features almost perfect detection capability. Only if $E(x)$, by coincidence, is exactly divisible by $G(x)$ will no error

Figure 7.6 Generation of a transmission polynomial.

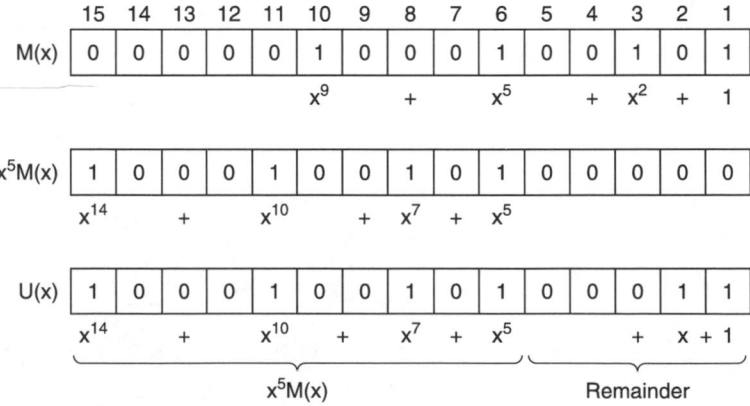

96

be detected. This obviously occurs only rarely and, knowing the characteristics of the transmission (or storage) medium, the polynomial $G(x)$ can be chosen such that the possibility is further minimized.

Although CRCC error checking seems rather complex, the divisions can be done relatively simply using modulo-2 arithmetic. In practical systems LSIs are used which perform the CRCC operations reliably and fast.

Error correction

In order to ensure later correction of binary data, the input signal must be encoded by some encoding scheme. The data sequence is divided into message blocks, then each message block is transformed into a longer one, by adding additional information, in a process called redundancy.

The ratio (data + redundant data)/data is known as the code rate.

There is a general theory, called the coding theorem, which states that the probability of decoding an error can be made as small as possible by increasing the code length and keeping the code rate less than the channel capacity.

When errors are to be corrected, we must not only know that an error has occurred, but also exactly which bit or bits are wrong. As there are only two possible bit states (0 or 1), correction is then just a matter of reversing the state of the erroneous bits.

Basically, correction (and detection) of code errors can be achieved by adding to the data bits an additional number of redundant check bits. This redundant information has a certain connection with the actual data, so that, when errors occur, the data can be reconstructed again. The redundant information is known as the error-correction code. As a simple example, all data could be transmitted twice, which would give a redundancy of 100%. By comparing both versions, or by CRCC, errors could easily be detected, and if some word were erroneous, its counterpart could be used to give the correct data. It is even possible to record everything three times; this would be even more secure. These are, however, rather wasteful systems, and much more efficient error-correction systems can be constructed.

The development of strong and efficient error-correction codes has been one of the key research points of digital audio technology. A wealth of experience has been used from computer technology, where the correction of code errors is equally important, and where much research has been carried out in order to optimize correction capabilities of error-correction codes. Design of strong codes is a very complex matter, however, which requires thorough study and use of higher mathematics: algebraic structure has been the basis of the most important codes. Some codes are very strong against 'burst' errors, i.e., when entire clusters of bits are erroneous together (such as during tape dropouts), whereas others are better against 'random' errors, i.e., when single bits are faulty.

Error-correction codes are of two forms, in which:

1 Data bits and error-correction bits are arranged in separate blocks; these are called block codes. Redundancy that follows a data block is only generated by the data in that particular block.

2 Data and error correction are mixed in one continuous data stream; these are called convolutional codes. Redundancy within a certain time unit depends not only upon the data in that same time unit, but also upon data occurring a certain time before. They are more complicated than, and often superior in performance to, block codes.

Figure 7.7 illustrates the main differences between block and convolutional error-correcting codes.

Combinational (horizontal/vertical) parity checking

If, for example, we consider a binary word or message consisting of 12 bits, these bits could be arranged in a 3×4 matrix, as shown in Figure 7.8.

Then, to each row and column one more bit can be added to make parity for that row or column even (or odd). Then, in the lower right-hand corner, a final bit can be added that will give

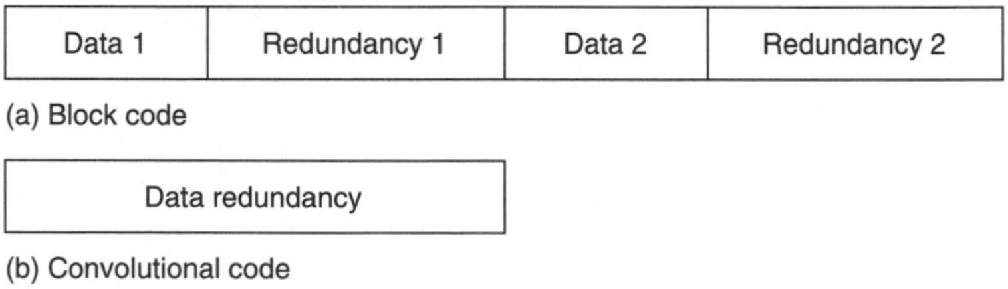

Figure 7.7 Main differences between block and convolutional error-correcting codes.

Data 1	Redundancy 1	Data 2	Redundancy 2

(a) Block code

Data redundancy

(b) Convolutional code

Figure 7.8 Combinational parity checking.

```
1  0  1  0 | 0
1  0  1  1 | 1
0  0  1  0 | 1
───────────────
0  0  1  1 | 0
```

the last column an even parity as well; because of this the last row will also have even parity.

If this entire array is transmitted in sequence (row by row, or column by column), and if during transmission an error occurs to one bit, a parity check on one row and on one column will fail; the error will be found at the intersection and, consequently, it can be corrected. The entire array of 20 bits, of which 12 are data bits, forms a code word, which is referred to as a (20, 12) code. There are 20 – 12 = 8 redundant digits.

All error-correcting codes are more or less based on this idea, although better codes, i.e., codes using fewer redundant bits than the one given in our example, can be constructed.

Crossword code

Correction of errors to single bits is just one aspect of error-correcting codes. In digital recording, very often errors come in bursts, with complete clusters of faulty bits. It will be obvious that, in view of the many possible combinations, the correction of such bursts or errors is very complicated and demands powerful error-correcting codes.

One such code, developed by Sony for use in its PCM-1600 series, is the crossword code. This uses a matrix scheme similar to the previous example, but it carries out its operations with whole words instead of individual bits, with the resultant advantage that large block code constructions can be easily realized so that burst error correction is very good. Random error correction, too, is good. Basically, the code allows detection and correction of all types of errors with a high probability of occurrence, and that only errors with a low probability of occurrence may pass undetected.

A simple illustration of a crossword code is given in Figure 7.9, where the decimal numbers represent binary values or words.

Figure 7.10 shows another simple example of the crossword code, in binary form, in which four words M_1–M_4 are complemented in the coder by four parity or information words R_5–R_8, so that:

$$R_5 = M_1 \oplus M_2$$
$$R_6 = M_3 \oplus M_4$$
$$R_7 = M_1 \oplus M_3$$
$$R_8 = M_2 \oplus M_4$$

where the symbol $\oplus$ denotes modulo-2 addition.

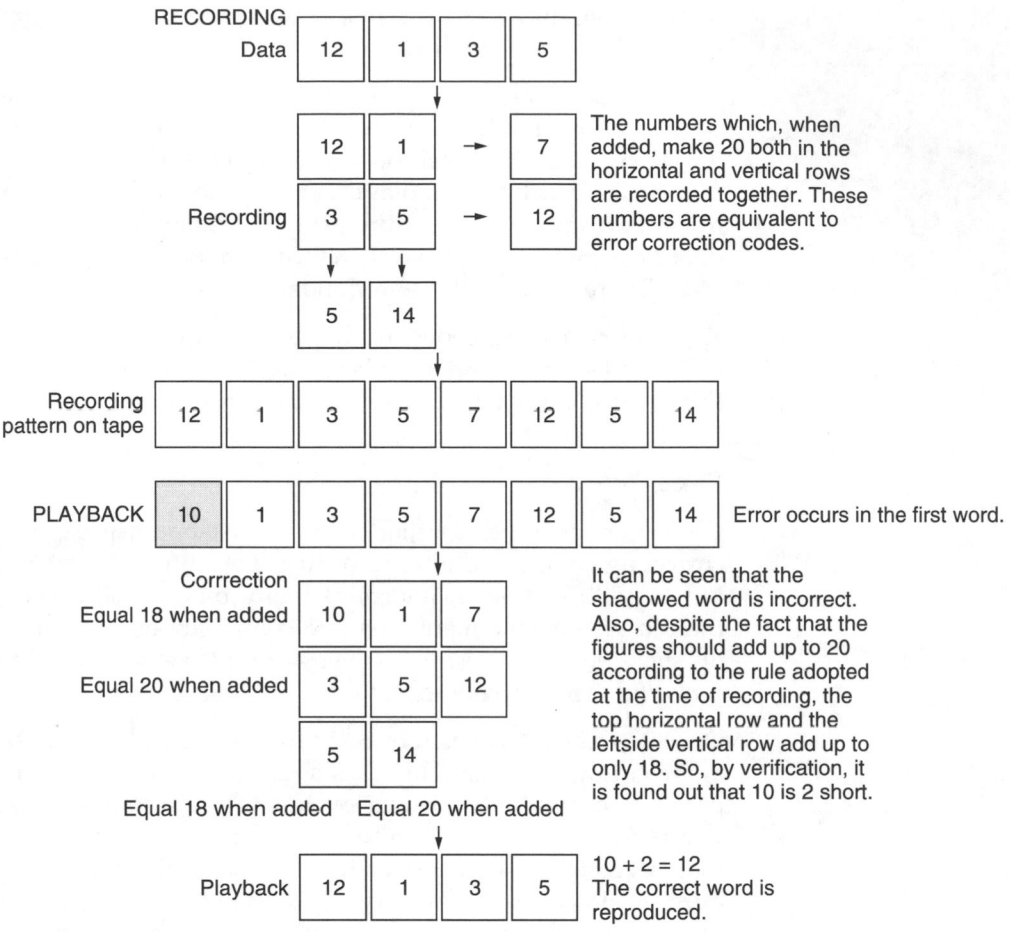

RECORDING
Data | 12 | 1 | 3 | 5 |

Recording
12 | 1 | → | 7
3 | 5 | → | 12
5 | 14

The numbers which, when added, make 20 both in the horizontal and vertical rows are recorded together. These numbers are equivalent to error correction codes.

Recording
pattern on tape | 12 | 1 | 3 | 5 | 7 | 12 | 5 | 14 |

PLAYBACK | 10 | 1 | 3 | 5 | 7 | 12 | 5 | 14 | Error occurs in the first word.

Corrrection
Equal 18 when added | 10 | 1 | 7
Equal 20 when added | 3 | 5 | 12
5 | 14

Equal 18 when added   Equal 20 when added

It can be seen that the shadowed word is incorrect. Also, despite the fact that the figures should add up to 20 according to the rule adopted at the time of recording, the top horizontal row and the leftside vertical row add up to only 18. So, by verification, it is found out that 10 is 2 short.

Playback | 12 | 1 | 3 | 5 |

10 + 2 = 12
The correct word is reproduced.

NOTE: All words and the correction codes are expressed by ordinary decimal figures instead of binary codes to facilitate understanding.

Figure 7.9 Illustration of a crossword code.

All eight words are then recorded, and at playback received as U_1–U_8.

In the decoder, additional words are constructed, called syndrome words, as follows:

$$S_1 = U_1 \oplus U_2 \oplus U_5$$
$$S_2 = U_3 \oplus U_4 \oplus U_6$$
$$S_3 = U_1 \oplus U_3 \oplus U_7$$
$$S_4 = U_2 \oplus U_4 \oplus U_8$$

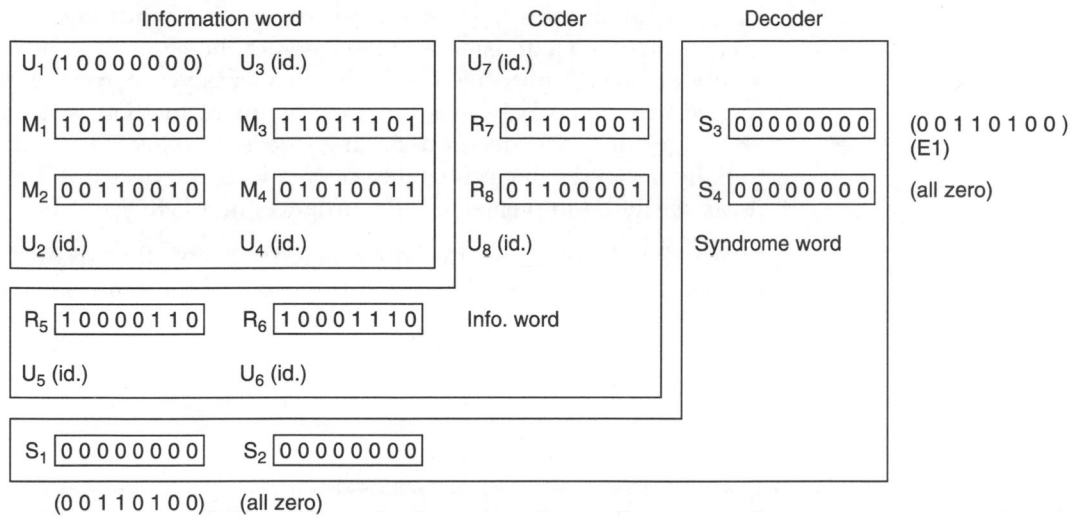

Figure 7.10 Crossword code using binary data.

By virtue of this procedure, if all received words U_1–U_8 are correct, all syndromes must be zero. On the other hand, if an error E occurs in one or more words, we can say that:

$$U_i = M_i \oplus E_i \quad \text{for } i = 1\text{–}4$$
$$U_i = R_i \oplus E_i \quad \text{for } i = 5\text{–}8$$

Now:

$$S_1 = U_1 \oplus U_2 \oplus U_5$$
$$= M_1 \oplus E_1 \oplus M_2 \oplus E_2 \oplus R_5 \oplus E_5$$
$$= E_1 \oplus E_2 \oplus E_5$$

as we know that $M_1 \oplus M_2 \oplus R_5 = 0$. Similarly:

$$S_2 = E_3 \oplus E_4 \oplus E_6$$
$$S_3 = E_1 \oplus E_3 \oplus E_7$$
$$S_4 = E_2 \oplus E_4 \oplus E_8$$

Correction can then be made, because:

$$U_1 \oplus S_3 = M_1 \oplus E_1 \oplus E_1 = M_1$$

Of course, there is still a possibility that simultaneous errors in all words compensate each other to give the same syndrome patterns as in our example. The probability of this occurring, however, is extremely low and can be disregarded.

In practical decoders, when errors occur, the syndromes are investigated and a decision is made whether there is a good probability of successful correction. If the answer is yes, correction is carried out; if the answer is no, a concealment method is used. The algorithm the decision-making process follows must be initially decided using probability calculations, but once it is fixed it can easily be implemented, for instance, in a P-ROM.

Figure 7.11 shows the decoding algorithm for this example crossword code. Depending upon the value of the syndrome(s),

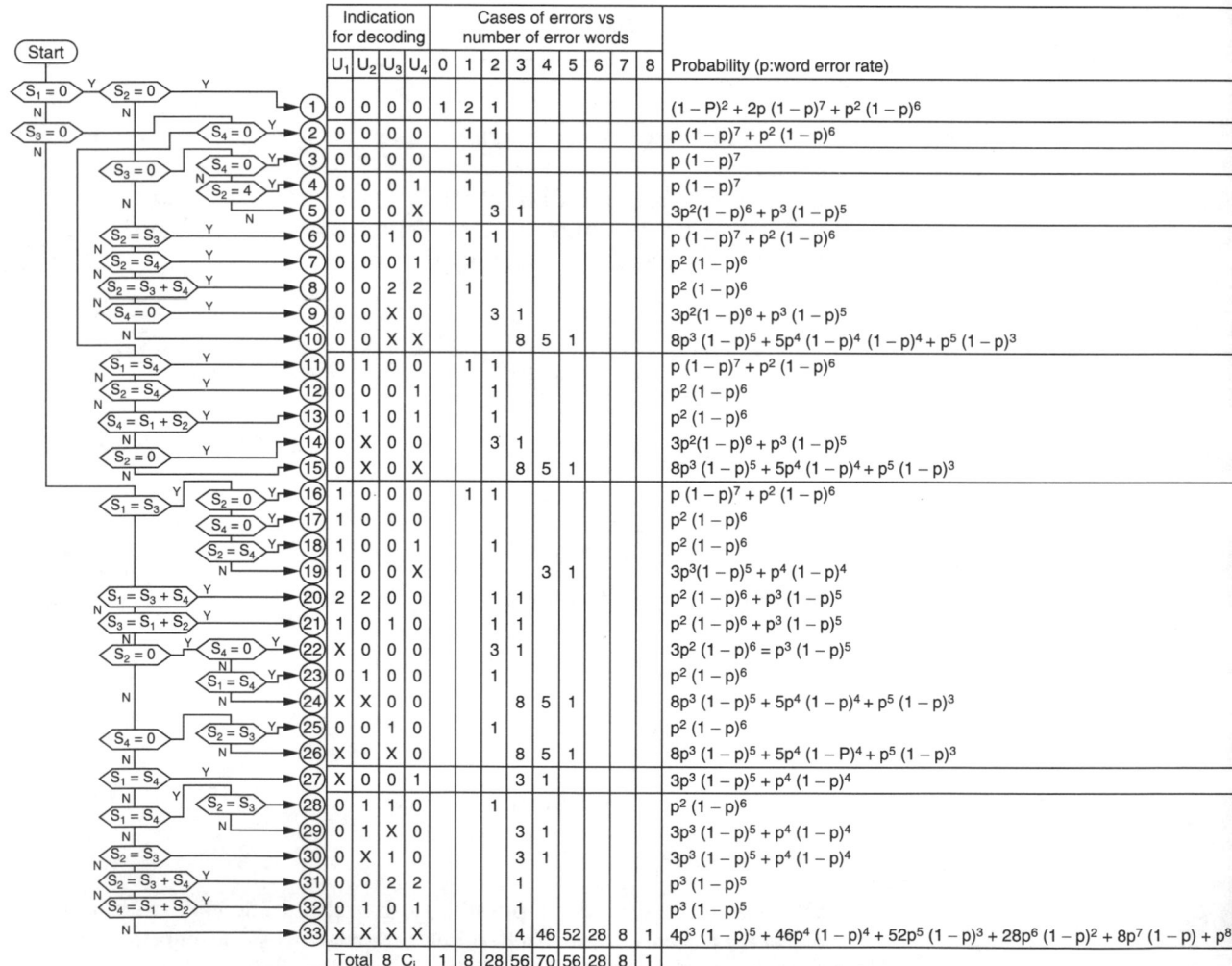

Figure 7.11 Decoding algorithm for crossword code.

decisions are made for correction according to the probability of miscorrection; the right column shows the probability of each situation occurring.

b-Adjacent code

A code which is very useful for correcting both random and burst errors has been described by D. C. Bossen of IBM, and called b-adjacent code. The b-adjacent error-correction system is used in the EIAJ format for home-use helical-scan digital audio recorders.

In this format two parity words, called P and Q, are constructed as follows:

$$P_n = L_n \oplus R_n \oplus L_{n+1} \oplus R_{n+1} \oplus R_{n+2}$$
$$Q_n = T^6 L_n \oplus T^5 R_n \oplus T^4 L_{n+1} \oplus T^3 R_{n+1} \oplus T^2 L_{n+2} \oplus T R_{n+2}$$

where T is a specific matrix of 14 words of 14 bits; L_n, R_n, etc. are data words from, respectively, the left and the right channel.

CIRC code and other codes

Many other error-correcting codes exist, as most manufacturers of professional audio equipment design their own preferred error-correction system.

Most, however, are variations on the best-known codes, with names such as Reed–Solomon code and BCH (Bose–Chaudhuri–Hocquenghem) code, after the researchers who invented them.

Sony (together with Philips) developed the cross-interleave Reed–Solomon code (CIRC) for the compact disc system. R-DAT tape format uses a double Reed–Solomon code, for extremely high detection and correction capability. The DASH format for professional stationary-head recorders uses a powerful combination of correction strategies, to allow for quick in/out and editing (both electric and tape splice).

Error concealment

Next comes a technique which prevents the uncorrected code errors from affecting the quality of the reproduced sound. This is known as concealment, and there are four typical methods.

1   **Muting**: the erroneous word is muted, i.e., set to zero (Figure 7.12a). This is a rather rough concealment method, and consequently used rarely.

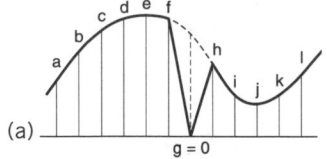

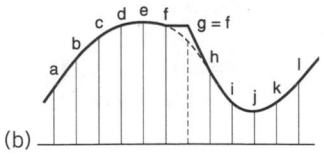

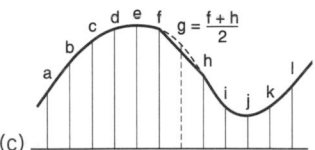

Figure 7.12 Three types of error concealment: (a) muting; (b) previous word holding; (c) linear interpolation.

2 **Previous word holding**: the value of the word before the erroneous word is used instead (Figure 7.12b), so that there is usually no audible difference. However, especially at high frequencies, where sampling frequency is only two or three times the signal frequency, this method may give unsatisfactory results.

3 **Linear interpolation** (also called **averaging**): the erroneous word is replaced by the average value of the preceding and succeeding words (Figure 7.12c). This method gives much better results than the two previous methods.

4 **Higher-order polynomial interpolation**: gives an estimation of the correct value of the erroneous word by taking into account a number of preceding and following words. Although much more complicated than the previous methods, it may be worthwhile using it in very critical applications.

Interleaving

In view of the high recording density used to record digital audio on magnetic tape, dropouts could destroy many words simultaneously, as a burst error. Error correction that could cope with such a situation would be prohibitively complicated and, if it failed, concealment would not be possible as methods like interpolation demand that only one sample of a series is wrong.

For this reason, adjacent words from the A/D converter are not written next to each other on the tape, but at locations a sufficient distance apart to make sure that they cannot suffer from the same dropout. In effect, this method converts possible long burst errors into a series of random errors. Figure 7.13 illustrates this in a simplified example. Words are arranged in interleave blocks, which must be at least as long as the maximum burst error. Practical interleaving methods are much more complicated than this example, however.

Figure 7.13 How interleaving of data effectively changes burst errors into random errors.

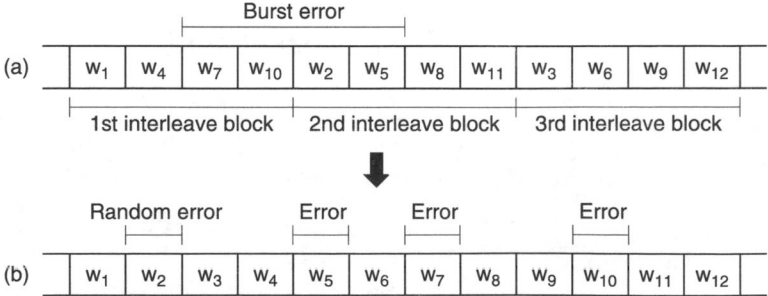

PART TWO
The Compact Disc

8 Overview of the compact disc medium

It is the compact disc (CD) that has introduced most people to digital audio reproduction. Table 8.1 is a comparison of the LP and CD systems, showing that CD is far superior to LP in each aspect of dynamic range, distortion, frequency response, and wow and flutter specifications. In particular, CD exhibits a remarkably wide dynamic range (90 dB) throughout the entire audible frequency spectrum. In contrast, the dynamic range of the LP is 70 dB at best. Harmonic distortion of CD reproduction is less than 0.01%, which is less than one-hundredth of that of LP. Wow and flutter are simply too minute to be measured in a CD system. This is because, in playback, digital data are first stored in an RAM and then released in perfect, uniform sequence determined by a reference clock of quartz precision.

With a mechanical system like that of the LP, the stylus must be in physical contact with the disc. Therefore, both the stylus and the disc will eventually wear out, causing serious deterioration of sound quality. With the CD's optical system, however, lack of contact between the disc and the pick-up means that there is no sonic deterioration no matter how many times the disc is played.

Mechanical (and, for that matter, variable capacitance) systems are easily affected by dust and scratches, as signals are impressed directly on the disc surface. A compact disc, however, is covered with a protective layer (the laser optical pick-up is focused underneath this) so that the effect of dust and scratches is minimized.

Table 8.1 System comparison between CD and LP

	CD system	Conventional LP player
Specifications		
Frequency response	20 Hz–20 kHz ± 0.5 dB	30 Hz–20 kHz ± 3 dB
Dynamic range	More than 90 dB	70 dB (at 1 kHz)
S/N	90 dB (with MSB)	60 dB
Harmonic distortion	Less than 0.01%	1–2%
Separation	More than 90 dB	25–30 dB
Wow and flutter	Quartz precision	0.03%
Dimensions		
Disc	12 cm (diameter)	30 cm (diameter)
Playing time (on one side)	60 minutes (maximum 74 minutes)	20–25 minutes
Operation/reliability		
Durability disc	Semi-permanent	High-frequency response is degraded after being played several tens of times
Durability stylus	Over 5000 hours	500–600 hours
Operation	– Quick and easy access due to micro computer control	– Needs stylus pressure adjustment
	– A variety of programmed play possible	– Easily affected by external vibration
	– Increased resistivity to external vibration	
Maintenance	Dust, scratches and fingerprints are made almost insignificant	Dust and scratches cause noise

Furthermore, a powerful error-correction system, which can correct even large burst errors, makes the effect of even severe disc damage insignificant in practice.

Figure 8.1 CD player.

Main parameters

Main parameters of the CD compared to the LP are shown in Table 8.2. Figure 8.2 compares CD and LP disc sizes. Figure 8.3 compares track pitch and groove dimensions of a CD with an LP; 60 tracks of the CD would fit into one track of an LP.

Optical discs

Figure 8.4 gives an overview of the optical discs available. The CD single is the digital equivalent of a 45 rpm single. It can contain about 20 minutes of music and is fully compatible with any CD player. A CD Video (CDV) contains 20 minutes of digital audio which can be played back on an ordinary CD player and 6 minutes of video with digital audio. To play back the video part you need a Video Disc Player or a Multi Disc Player (MDP).

Multi Disc Players are capable of playing both Compact Discs and Laser Discs (LDs). Optical discs containing video signals can be distinguished from discs containing only digital audio by their colour. CDVs and LDs have a gold shine, while CDs and CD singles have a silver shine.

Recording and read-out system on a CD

The data on a compact disc are recorded on the master by using a laser beam photographically to produce pits in the disc surface, in a clockwise spiral track starting at the centre of the disc. The length of the pits and the distance between them from the recorded information are as shown in Figure 8.5.

Table 8.2 Parameter comparison between CD and LP

	CD	LP
Disc diameter	120 mm	305 mm
	568–228 rpm (at 1.4 m s^{-1})	33⅓ rpm
	486–196 rpm (at 1.2 m s^{-1})	
Playing time (maximum)	74 min	32 min (one side)
No. of tracks	20,625	1060 maximum
Track spacing	1.6 μm	85 μm
Lead-in diameter	46 mm	302 mm
Lead-out diameter	116 mm	121 mm
Total track length	5300 mm	705 m maximum
Linear velocity	1.2 or 1.4 m s^{-1}	528–211 mm s^{-1}

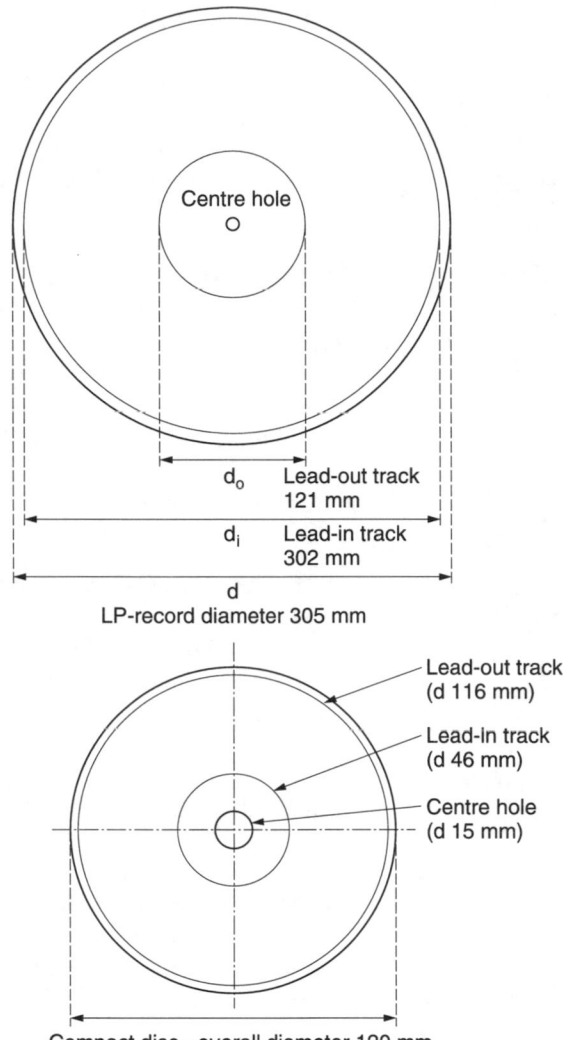

Figure 8.2 A comparison of CD and LP sizes.

d_o Lead-out track 121 mm

d_i Lead-in track 302 mm

d LP-record diameter 305 mm

Lead-out track (d 116 mm)

Lead-in track (d 46 mm)

Centre hole (d 15 mm)

Compact disc - overall diameter 120 mm

In fact, on the user disc, the pits are actually bumps. These can be identified by focusing a laser beam onto the disc surface: if there is no bump on the surface, most of the light that falls on the surface (which is highly reflective) will return in the same direction. If there is a bump present, however, the light will be scattered and only a small portion will return in the original direction (Figure 8.6). The disc has a 1-mm-thick protective transparent layer over the signal layer, i.e., the pits. More important, the spot size of the laser beam is about 1 mm in diameter at the surface of the disc, but is as small as 1.7 μm across at the signal layer. This means that a dust particle or a scratch on the disc

Figure 8.3 Track comparison between CD and LP.

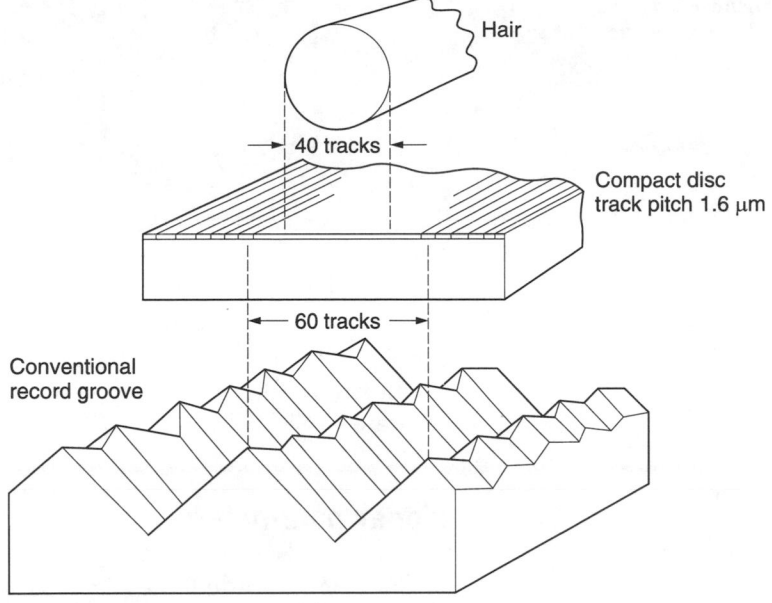

Hair

40 tracks

Compact disc track pitch 1.6 μm

60 tracks

Conventional record groove

Figure 8.4 Optical discs.

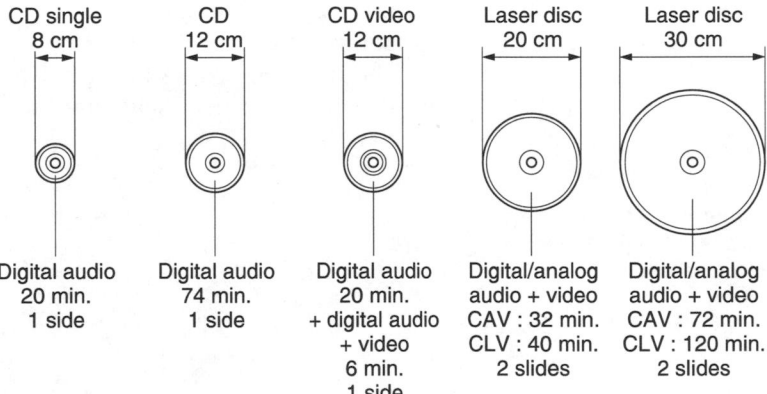

CD single 8 cm	CD 12 cm	CD video 12 cm	Laser disc 20 cm	Laser disc 30 cm
Digital audio 20 min. 1 side	Digital audio 74 min. 1 side	Digital audio 20 min. + digital audio + video 6 min. 1 side	Digital/analog audio + video CAV : 32 min. CLV : 40 min. 2 slides	Digital/analog audio + video CAV : 72 min. CLV : 120 min. 2 slides

Figure 8.5 Pits on a CD, viewed from the label side.

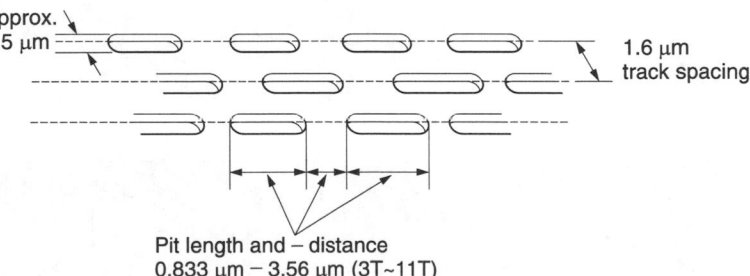

Approx. 0.5 μm

1.6 μm track spacing

Pit length and – distance 0.833 μm – 3.56 μm (3T~11T)

Figure 8.6 CD laser beam reflection: (a) from disc surface; (b) from pit.

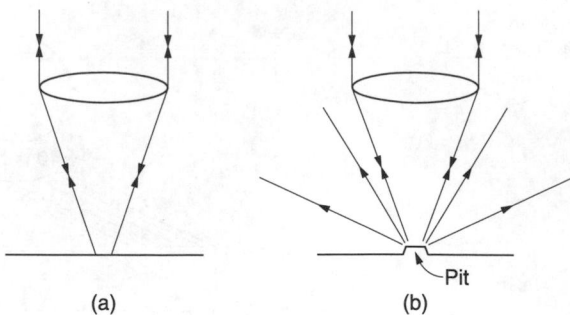

surface is literally out of focus to the sensing mechanism. Figure 8.7 illustrates this. Obviously, control of focus must be extremely accurate.

Signal parameters

Before being recorded, the digital audio signal (which is a 16-bit signal) must be extended with several additional items of data. These include:

* error correction data;
* control data (time, titles, lyrics, graphics and information about the recording format or emphasis);
* synchronization signals, used to detect the beginning of each data block;

Figure 8.7 How a dust particle on the disc surface is out of focus.

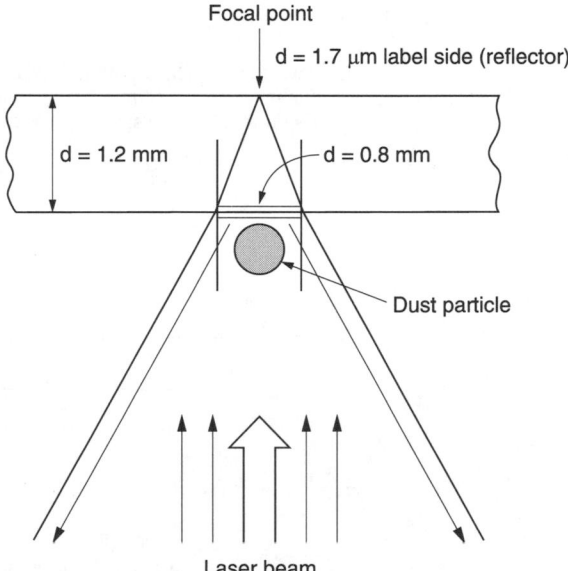

- merging bits: added between each data symbol to reduce the DC component of the output signal.

Audio signal

The audio signal normally consists of two channels of audio, quantized with a 16-bit linear quantization system at a sampling frequency of 44.1 kHz. During recording, pre-emphasis (slight boost of the higher frequencies) may be applied. Pre-emphasis standards agreed for the compact disc format are 50 and 15 µs (or 3183 and 10 610 Hz).

Consequently, the player must in this case apply a similar de-emphasis to the decoded signal to obtain a flat frequency response (Figure 8.8).

A specific control code recorded along with the audio signal on the compact disc is used to inform the player whether pre-emphasis is used, and so the player switches in the corresponding de-emphasis circuit to suit.

Alternatively, audio information on the CD may comprise four music channels instead of two; this is also identified by a control code to allow automatic switching of players equipped with a four-channel playback facility. Although, on launching CD, there were no immediate plans for four-channel discs or players, the possibility for later distribution was already provided in the standard.

Figure 8.8 CD pre-emphasis and de-emphasis characteristics.

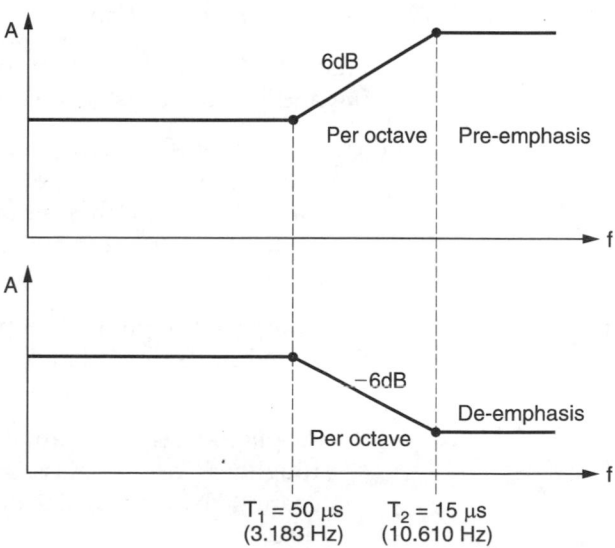

Additional information on the CD

Before the start of the music programme, a 'lead-in' signal is recorded on the CD. When a CD is inserted, most players immediately read this lead-in signal, which contains a 'table of contents' (TOC). The TOC contains information on the contents of the disc, such as the starting point of each selection or track, number of selections, duration of each selection. This information can be displayed on the player's control panel, and/or used during programme search operation.

At the end of the programme, a lead-out signal is similarly recorded which informs the player that playback is complete.

Furthermore, music start flags between selections inform the player that a new selection follows.

Selections recorded on the disc can be numbered from 1 through 99. In each track, up to 99 indexes can be given, which may separate specific sections of the selection. Playing time is also encoded on the disc in minutes, seconds and 1/75ths of a second; before each selection, this time is counted down.

There is further space available to encode other information, such as titles, performer names, lyrics and even graphic information, which may all be displayed, for instance, on a TC screen during playback.

Compact disc production

Compact disc 'cutting'

Figure 8.9 is a block diagram comparing CD digital audio recording and playback systems with analog LP systems.

The two systems are quite similar and, in fact, overlap can occur at record production stage. However, where LP masters are mechanically cut, CD masters are 'cut' in an electro-optical photographic process: no 'cutting' of the disc surface actually takes place.

The compact disc production process follows seven main stages, illustrated in Figure 8.10:

1   A glass plate is polished for optimum smoothness.
2   A photo-resistive coating is applied to its surface. The roughness of the glass surface and the thickness of the coating determines the depth of the pits on the compact disc.

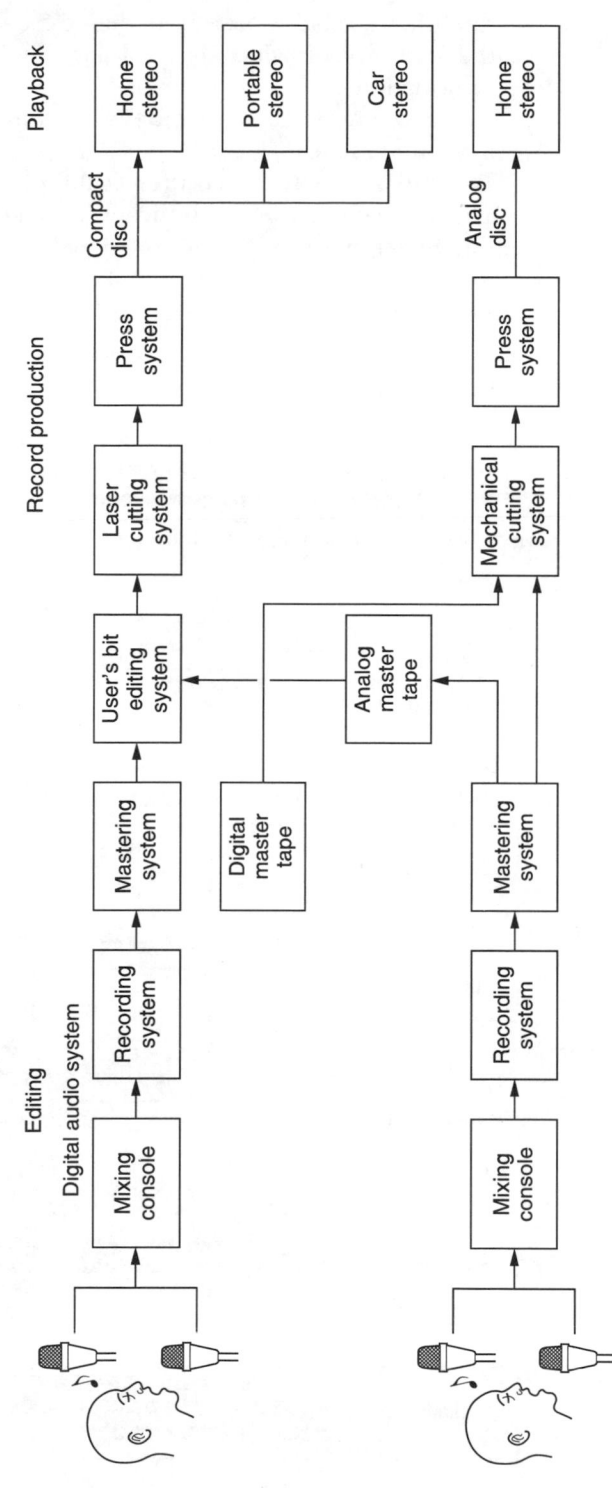

Figure 8.9 A comparison of CD and LP recording and playback systems.

3   The photo-resistive coating is then exposed to a laser beam, the intensity of which is modulated with digitized audio information.
4   The photo-resistive layer is developed and the pits of information are revealed.
5   The surface is silvered to protect the pits.
6   The surface is plated with nickel to make a metal master.
7   The metal master is then used to make mother plates. These mothers are in turn used to make further metal masters, or stampers.

Compact disc stamping

The stamping process, although named after the analogous stage in LP record production, is, in fact, an injection moulding, compression moulding or polymerization process, producing plastic discs (Figure 8.11). The signal surface of each disc is then coated

Figure 8.10 Stages in the 'cutting' of a compact disc.

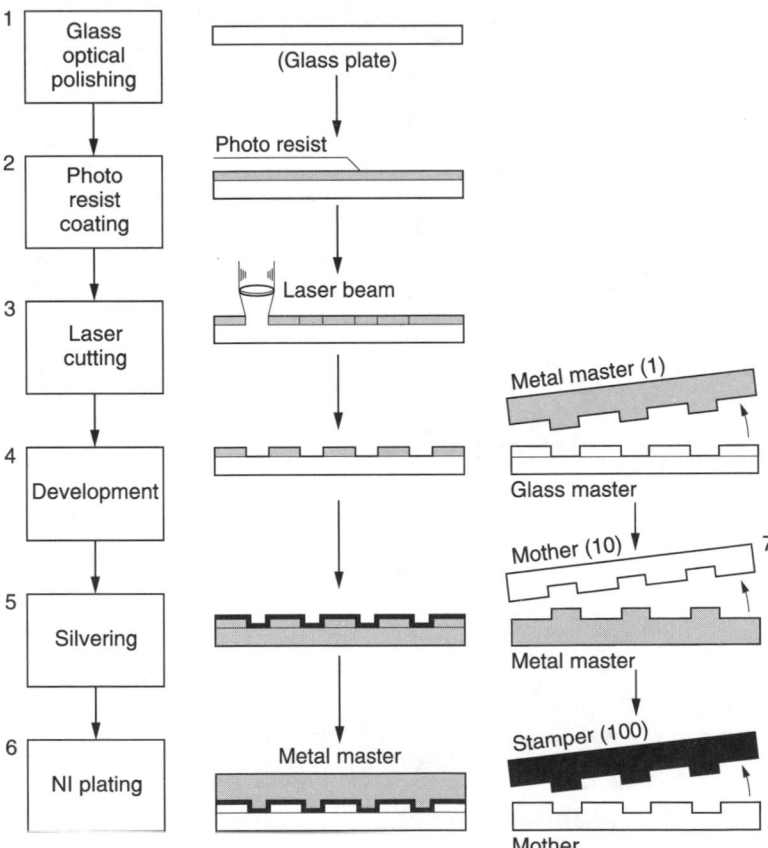

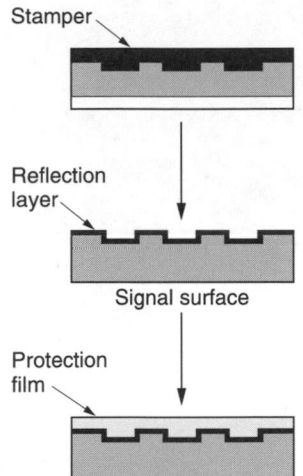

Stamper

Reflection layer

Signal surface

Protection film

Figure 8.11 Stages in 'stamping' a compact disc.

with a reflective material (vaporized aluminium) to enable optical read-out, and further protected with a transparent plastic layer which also supports the disc label.

9 Compact disc encoding

A substantial amount of information is added to the audio data before the compact disc is recorded. Figure 9.1 illustrates the encoding process and shows the various information to be recorded.

There are usually two audio channels with 16-bit coding, sampled at 44.1 kHz. So, the bit rate, after combining both channels, is:

$$44.1 \times 16 \times 2 = 1.4112 \times 10^6 \text{ bit s}^{-1}$$

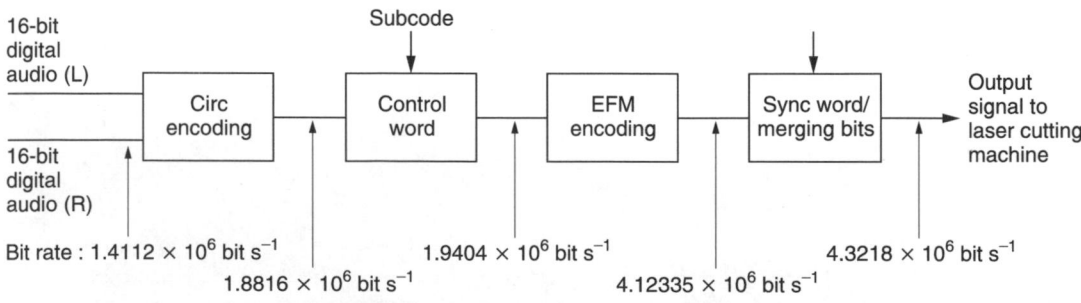

Figure 9.1 The encoding process in compact disc production.

CIRC encoding

Most of the errors which occur on a medium such as CD are random. However, from time to time burst errors may occur due to fingerprints, dust or scratches on the disc surface. To cope with both random and burst errors, Sony and Philips developed the cross-interleave Reed–Solomon error-correction code (CIRC). CIRC is a very powerful combination of several error correction techniques.

It is useful to be able to measure an error-correcting system's ability to correct errors, and as far as the compact disc medium is concerned it is the maximum length of a burst error which is critical. Also, the greater the number of errors received, the greater the probability of some errors being uncorrectable. The number of errors received is defined as the bit error rate (BER). An important system specification, therefore, is the number of data samples per unit time, called the sample interpolation rate, which have to be interpolated (rather than corrected) for given BER values. The lower this rate is, the better the system. Then, if burst errors cannot be corrected, an important specification is the maximum length of a burst error which can be interpolated. Finally, it is important to know the number of undetected errors, resulting in audible clicks. Any specification of an error-correcting system must take all these factors into account. Table 9.1 is a list of all relevant specifications of the CIRC system used in CD.

The CIRC principle is as follows (refer to Figure 9.2):

- The audio signal is sampled (digitized) at the A/D converter and these 16-bit samples are split up into two 8-bit words called symbols.

Table 9.1 Specification of CIRC system in the compact disc

Aspect	Specification
Maximum correctable burst error length	4000 bits (i.e., 2.5 mm on the disc surface)
Maximum interpolatable burst error length	12 300 bits (i.e., 7.7 mm on the disc surface)
Sample interpolation rate	One sample every 10 hours at a BER of 10^{-4} 1000 samples every minute at a BER of 10^{-3}
Undetected error samples	Less than one every 750 hours at a BER of 10^{-3} Negligible at a BER of 10^{-4}
Code rate	On average, four bits are recorded for every three data bits

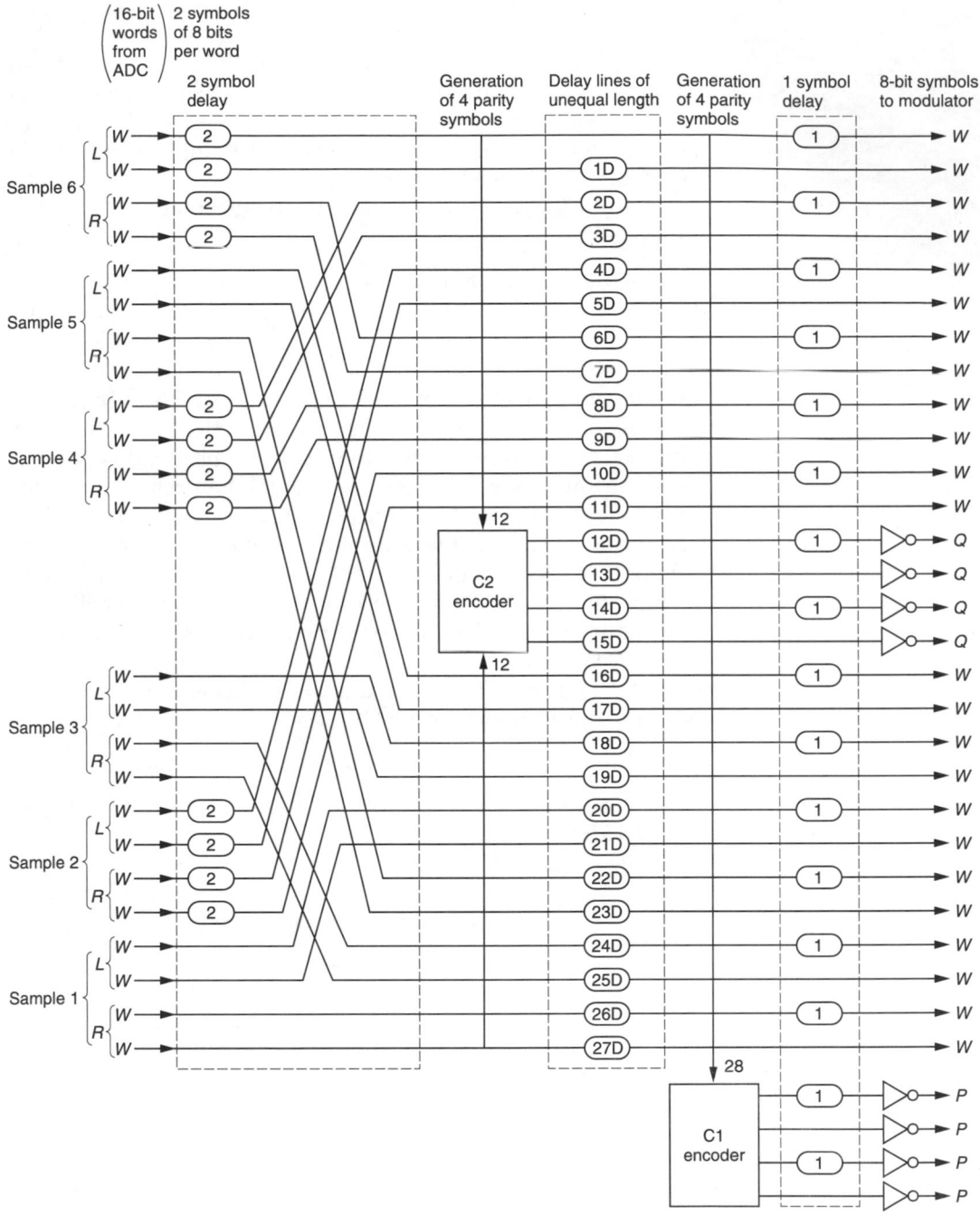

Figure 9.2 CIRC encoder.

- Six of the 16-bit samples from each channel, i.e., 24 8-bit symbols, are applied to the CIRC encoder and stored in an RAM memory.
- The first operation in the CIRC encoder is called scrambling. The scrambling operation consists of a two-symbol delay for the even samples and a mixing up of the connections to the C2 encoder.
- The 24 scrambled symbols are then applied to the C2 encoder, which generates four 8-bit parity symbols called Q words. The C2 encoder inserts the Q words between the 24 incoming symbols, so that at the output of the C2 encoder 28 symbols result.
- Between the C2 and the C1 decoders there are 28 8-bit delay lines with unequal delays. Due to the different delays, the sequence of the symbols is changed completely, according to a determined pattern.
- The C1 encoder generates further four 8-bit parity symbols known as P words, resulting in a total of 32 8-bit symbols.
- After the C1 encoder, the even words are subjected to a one-symbol delay, and all P and Q control words are inverted. The resultant sequenced 32 8-bit symbols are called a **frame**; this is a CIRC-encoded signal that is applied to the EFM modulator. On playback, the CIRC decoding circuit restores the original 16-bit samples, which are then applied to the D/A converter.

The C2 encoder outputs 28 8-bit symbols for 24 symbols at its input: it is therefore called a (24, 28) encoder. The C1 encoder outputs 32 symbols for 28 symbols input: it is a (28, 32) encoder.

The bit rate at the output of the CIRC encoder is:

$$1.4112 \times \frac{32}{24} = 1.8816 \times 10^6 \, \text{bit s}^{-1}$$

The control word

One 8-bit control word is added to every 32-symbol block of data from the encoder. The compact disc standard defines eight additional channels of information or subcodes that can be added to the music information; these subcodes are called P, Q, R, S, T, U, V and W. At the time of writing, only the P and Q subcodes are commonly used:

- The P subcode is a simple music track separator flag that is normally 0 (during music and in the lead-in track), but is 1 at the start of each selection. It can be used for simple search systems. In the lead-out track, it switches between 0 and 1 in a 2 Hz rhythm to indicate the end of the disc.

• The Q subcode is used for more sophisticated control purposes; it contains data such as track number and time.

The other subcodes carry information relating to possible enhancements, such as text and graphics, but will not be discussed here.

Each subcode word is 98 bits long and, as each bit of the control word corresponds to each subcode (i.e., P, Q, R, S, T, U, V, W), a total of 98 complete data blocks or frames must be read from the disc to read each subcode word. This is illustrated in Figure 9.3.

After addition of the control word, the new data rate becomes:

$$1.811600 \times \frac{33}{32} = 1.9404 \times 10^6 \text{ bit s}^{-1}$$

The Q subcode and its usage

Figure 9.4a illustrates the structure of the 98-bit Q subcode word. The R, S, T, U, V and W subcode words are similar. The first 2 bits are synchronizing bits, S0 and S1. They are necessary to allow the decoder to distinguish the control word in a block from the audio information, and always contain the same data.

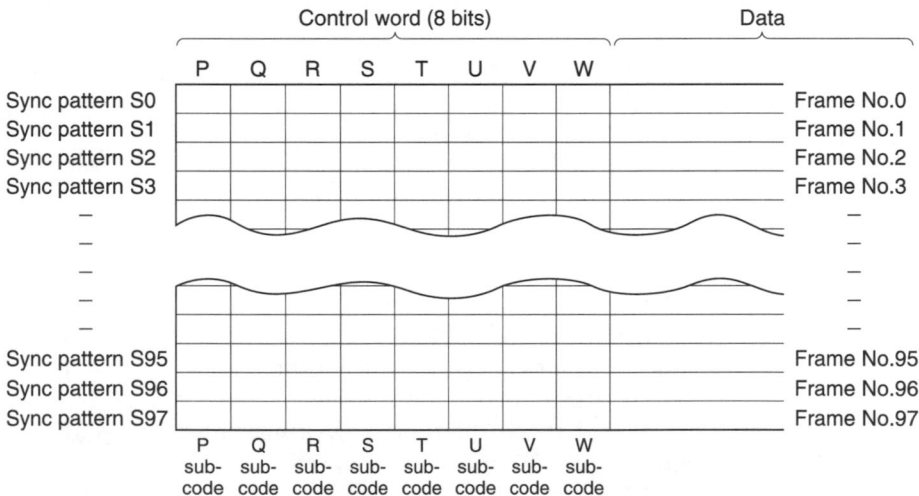

Figure 9.3 How one of each of the six subcode bits are present in every frame of information. A total of 98 frames must therefore be read to read all six subcode words.

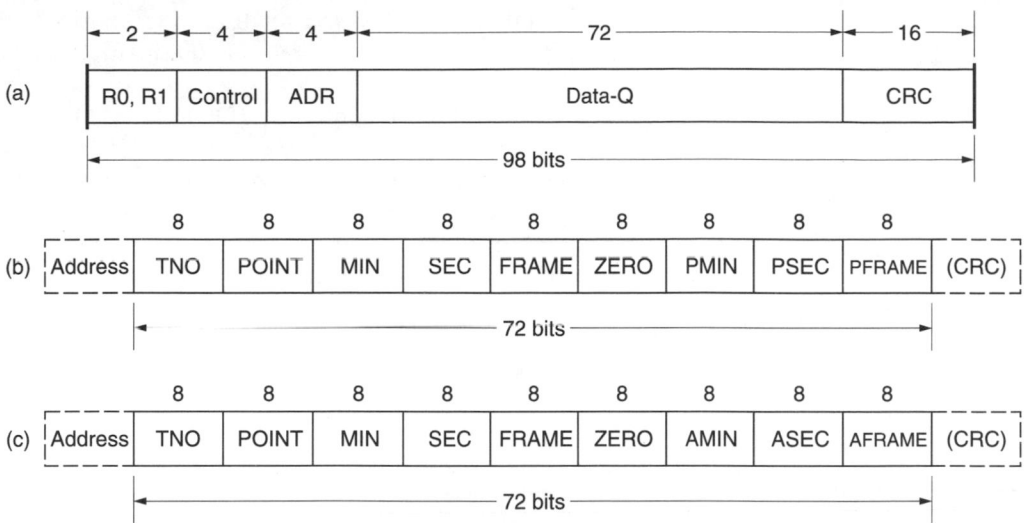

Figure 9.4 Formats of data in the Q subcode: (a) overall format; (b) mode 1 data format in the lead-in track; (c) mode 1 data format in music and lead-out tracks.

The next 4 bits are control bits, indicating the number of channels and pre-emphasis used, as follows:

0000 two audio channels/no pre-emphasis
1000 four audio channels/no pre-emphasis
0001 two audio channels/with pre-emphasis
1001 four audio channels/with pre-emphasis

Four address bits indicate the mode of the subsequent data to follow. For the subcode, three modes are defined.

At the end of the subcode word, a 16-bit CRCC error-correction code, calculated on control, address and data information, is inserted. The CRCC uses the polynomial $P(x) = x^{16} + x^{12} + x^5 + 1$.

The three modes of data in Q subcode words are used to carry various information.

Mode 1 (address = 0001)

This is the most important mode, and the only one which is of use during normal playback. At least nine of 20 consecutive subcode words must carry data in mode 1 format. Two different situations are possible, depending whether the subcode is in the lead-in track or not.

When in the lead-in track, the data are in the format illustrated in Figure 9.4b. The 72-bit section comprises nine 8-bit parts:

- **TNO** – containing information relating to track number: two digits in BCD form (i.e., 2×4 bits). Is 00 during lead-in.
- **POINT/PMIN/PSEC/PFRAME** – containing information relating to the table of contents (TOC). They are repeated three times.

  POINT indicates the successive track numbers, while PMIN, PSEC and PFRAME indicate the starting time of that track. Furthermore, if POINT = A0, PMIN gives the physical track number of the first piece of music (PSEC and PFRAME are zero); if POINT = A1, PMIN indicates the last track on the disc, and if POINT = A2, the starting point of the lead-out track is given in PMIN, PSEC and PFRAME.

  Table 9.2 shows the encoding of the TOC on a disc which contains six pieces of music.
- **ZERO** – 8 bits, all zero.

Table 9.2 Table of contents (TOC) information, on a compact disc with six pieces of music

Frame number	POINT	PMIN, PSEC, PFRAME
n	01	00, 02, 32
$n + 1$	01	00, 02, 32
$n + 2$	01	00, 02, 32
$n + 3$	02	10, 15, 12
$n + 4$	02	10, 15, 12
$n + 5$	02	10, 15, 12
$n + 6$	03	16, 28, 63
$n + 7$	03	16, 28, 63
$n + 8$	03	16, 28, 63
$n + 9$	04	16, 28, 63
$n + 10$	04	16, 28, 63
$n + 11$	04	16, 28, 63
$n + 12$	05	16, 28, 63
$n + 13$	05	16, 28, 63
$n + 14$	05	16, 28, 63
$n + 15$	06	49, 10, 33
$n + 16$	06	49, 10, 33
$n + 17$	06	49, 10, 33
$n + 18$	A0	01, 00, 00
$n + 19$	A0	01, 00, 00
$n + 20$	A0	01, 00, 00
$n + 21$	A1	06, 00, 00
$n + 22$	A1	06, 00, 00
$n + 23$	A1	06, 00, 00
$n + 24$	A2	52, 48, 41
$n + 25$	A2	52, 48, 41
$n + 26$	A2	52, 48, 41
$n + 27$	01	00, 02, 32
$n + 28$	01	00, 02, 32
.	.	.
.	.	.
.	.	.

In music and lead-out tracks, data are in the format illustrated in Figure 9.4c. The 72-bit section now comprises:

- **TNO** – current track number: two digits in BCD form (01 to 99).
- **POINT** – index number within a track: two digits in BCD form (01 to 99).
  If POINT = 00 it indicates a pause in between tracks.
- **MIN/SEC/FRAME** – indicates running time within a track: each part consists of digits in BCD form. There are 75 frames in a second (00 to 74). Time is counted down during a pause, with a value zero at the end of the pause. During lead-in and lead-out tracks, the time increases.
- **AMIN/ASEC/AFRAME** – indicates the running time of the disc in the same format as above. At the start of the programme area, it is set to zero.
- **ZERO** – 8 bits, all zero.

Figure 9.5 shows a timing diagram of P subcode and Q subcode status during complete reading of a disc containing four selections (of which selections three and four fade out and in consecutively without an actual pause).

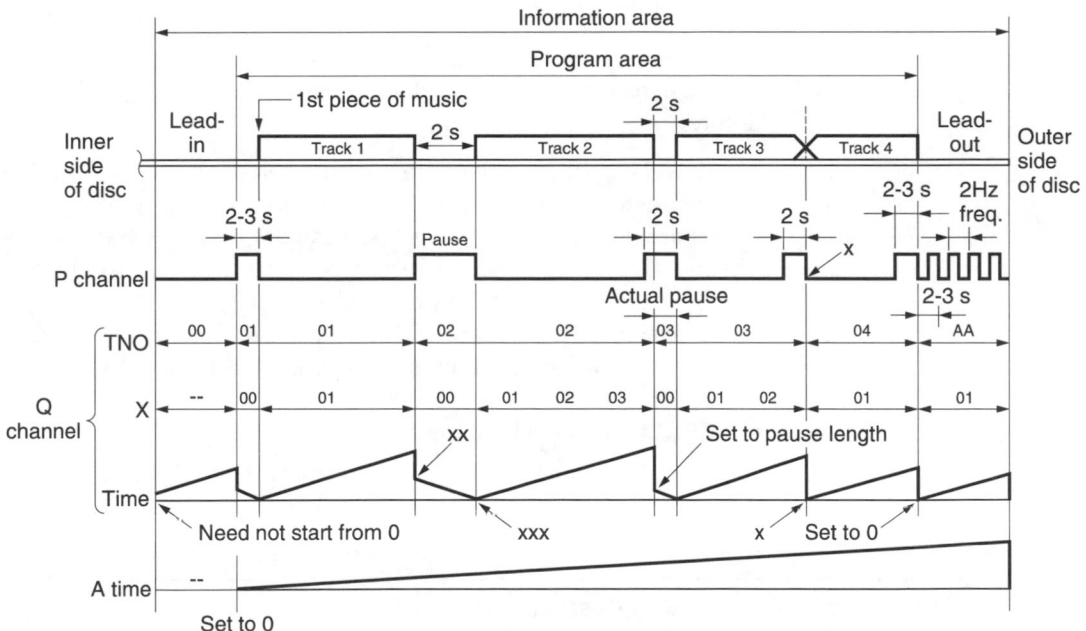

Figure 9.5 Timing diagram of P and Q subcodes.

Mode 2 (address = 0010)

If mode 2 data are present, at least one of 100 successive subcode words must contain it. It is of importance only to the manufacturer of the disc, containing the disc catalogue number. The 98-bit, Q subcode word in mode 2 is shown in Figure 9.6. The structure is similar to that of mode 1, with the following differences:

- **N1 to N13** – catalogue number of the disc expressed in 13 digits of BCD, according to the UPC/EAN standard for bar coding. The catalogue number is constant for any one disc. If no catalogue number is present, N1 to N13 are all zero, or mode 2 subcode words may not even appear.
- **ZERO** – these 12 bits are zero.

Mode 3 (address = 0111)

Like mode 2 data, if mode 3 is present, at least one of 100 successive subcode words will contain it.

Mode 3 is used to assign each selection with a unique number, according to the 12-character International Standard Recording Code (ISRC), defined in DIN-31-621.

If no ISRC number is assigned, mode 3 subcode words are not present. During lead-in and lead-out tracks, mode 3 subcode words are not used, and the ISRC number must only change immediately after the track number (TNO) has been changed.

The 98-bit, Q subcode word in mode 3 is shown in Figure 9.7. The structure is similar to that of mode 1, with the following differences:

- **I1 to I12** – the 12 characters of the selection's ISRC number. Characters I1 and I2 give the code corresponding to country. Characters I3 to I5 give a code for the owner. Characters I6 and I7 give the year of recording. Characters I8 to I12 give the recording's serial number.
  Characters I1 to I5 are coded in a 6-bit format according to Table 9.3, while characters I6 to I12 are 4-bit BCD numbers.
- **00** – these 2 bits are zero.
- **ZERO** – these 4 bits are zero.

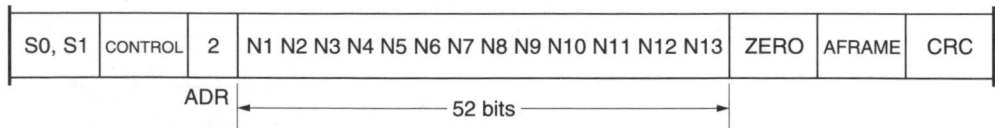

Figure 9.6 Q subcode format with mode 2 data.

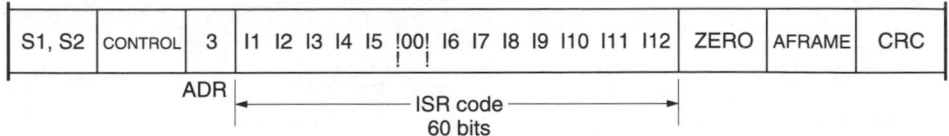

Figure 9.7 Q subcode format with mode 3 data.

Table 9.3 Format of characters I1 to I5 in the ISRC code

Character	Binary	Octal	Character	Binary	Octal
0	000000	00	I	011001	31
1	000001	01	J	011010	32
2	000010	02	K	011011	33
3	000011	03	L	011100	34
4	000100	04	M	011101	35
5	000101	05	N	011110	36
6	000110	06	O	011111	37
7	000111	07	P	100000	40
8	001000	10	Q	100001	41
9	001001	11	R	100010	42
A	010001	21	S	100011	43
B	010010	22	T	100100	44
C	010011	23	U	100101	45
D	010100	24	V	100110	46
E	010101	25	W	100111	47
F	010110	26	X	101000	50
G	010111	27	Y	101001	51
H	011000	30	Z	101010	52

EFM encoding

EFM, or eight-to-fourteen modulation, is a technique which converts each 8-bit symbol into a 14-bit symbol, with the purpose of aiding the recording and playback procedure by reducing required bandwidth, reducing the signal's DC content and adding extra synchronization information. A timing diagram of signals at this stage of CD encoding is given in Figure 9.8.

The procedure is to use 14-bit codewords to represent all possible combinations of the 8-bit code. An 8-bit code represents 256 (i.e., 2^8) possible combinations, as shown in Table 9.4. A 14-bit code, on the other hand, represents 16 384 (i.e., 2^{14}) different combinations, as shown in Table 9.5. Of the 16 384 14-bit codewords, only 256 are selected, having combinations which aid processing of the signal.

For instance, by choosing codewords which give low numbers of individual bit inversions (i.e., 1 to 0, or 0 to 1) between consecutive bits, the bandwidth is reduced. Similarly, by choosing

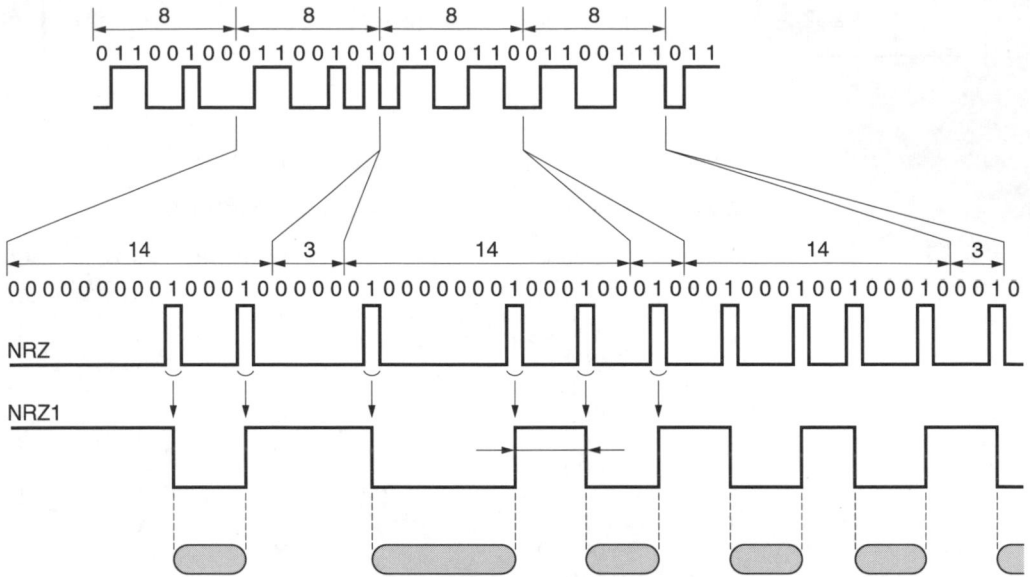

Figure 9.8 Timing diagram of EFM encoding and merging bits.

codewords with only limited numbers of consecutive bits with the same logic level, overall DC content is reduced.

Table 9.4 An 8-bit code

MSB	7SB	6SB	5SB	4SB	3SB	2SB	LSB	N^0
0	0	0	0	0	0	0	0	0
0	0	0	0	0	0	0	1	1
0	0	0	0	0	0	1	0	2
–	–	–	–	–	–	–	–	–
–	–	–	–	–	–	–	–	–
–	–	–	–	–	–	–	–	–
1	1	1	1	1	1	1	0	254
1	1	1	1	1	1	1	1	255

Table 9.5 A 14-bit code

MSB	13SB	12SB	11SB	10SB	9SB	8SB	7SB	6SB	5SB	4SB	3SB	2SB	LSB	N^0
0	0	0	0	0	0	0	0	0	0	0	0	0	0	0
0	0	0	0	0	0	0	0	0	0	0	0	0	1	1
–	–	–	–	–	–	–	–	–	–	–	–	–	–	–
–	–	–	–	–	–	–	–	–	–	–	–	–	–	–
–	–	–	–	–	–	–	–	–	–	–	–	–	–	–
1	1	1	1	1	1	1	1	1	1	1	1	1	0	16382
1	1	1	1	1	1	1	1	1	1	1	1	1	1	16383

A ROM-based look-up table, say, can then be used to assign all 256 combinations of the 8-bit code to the 256 chosen combinations within the 14-bit code. Some examples are listed in Table 9.6.

In addition to EFM modulation, three extra bits, known as merging bits, are added to each 14-bit symbol, with the purpose of further lowering DC content of the signal. Exact values of the merging bits depend on the adjacent symbols.

Finally, the data bits are changed from NRZ into NRZI (non-return to zero inverted) format, by converting each positive-going pulse of the NRZ signal into a single transition. The resultant signal has a minimum length of $3T$ (i.e., three clock periods) and a maximum of $11T$ (i.e., 11 clock periods), as shown in Figure 9.9.

The bit rate is now:

$$1.9404 \times \frac{17}{8} = 4.12335 \times 10^6 \, \text{bit s}^{-1}$$

The sync word

To the signal, comprising 33 symbols of 17 bits (i.e., a total of 561 bits), a sync word and its three merging bits are added, giving 588 bits in total (Figure 9.10). Sync words have two main functions: (1) they indicate the start of each frame; (2) sync word frequency is used to control the player's motor speed.

The 588-bit-long signal block is known as an information frame.

Final bit rate

The final bit rate, recorded on the CD, consequently becomes:

$$4.12335 \times \frac{588}{561} = 4.3218 \times 10^6 \, \text{bit s}^{-1}$$

Table 9.6 Examples of 8-bit to 14-bit encoding

8-bit word	14-bit word
0 0 0 0 0 0 1 1	0 0 1 0 0 1 0 0 0 0 0 0 0 0
0 1 0 0 1 1 1 0	0 1 0 0 0 0 0 1 0 0 1 0 0 0
1 0 1 0 1 0 1 0	1 0 0 1 0 0 0 1 0 0 0 1 0 0
1 1 1 1 0 0 1 0	0 0 0 0 0 0 1 0 0 0 1 0 0 1

Figure 9.9 Minimum and maximum pit length.

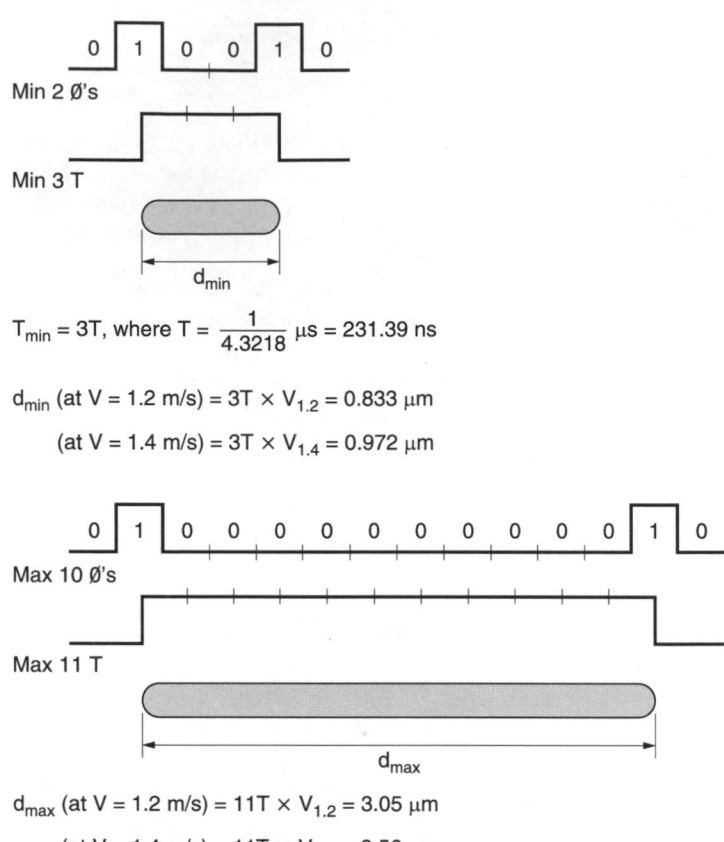

$T_{min} = 3T$, where $T = \dfrac{1}{4.3218}$ μs = 231.39 ns

d_{min} (at $V = 1.2$ m/s) $= 3T \times V_{1.2} = 0.833$ μm

$\quad$ (at $V = 1.4$ m/s) $= 3T \times V_{1.4} = 0.972$ μm

d_{max} (at $V = 1.2$ m/s) $= 11T \times V_{1.2} = 3.05$ μm

$\quad$ (at $V = 1.4$ m/s) $= 11T \times V_{1.4} = 3.56$ μm

Figure 9.10 Adding the sync word.

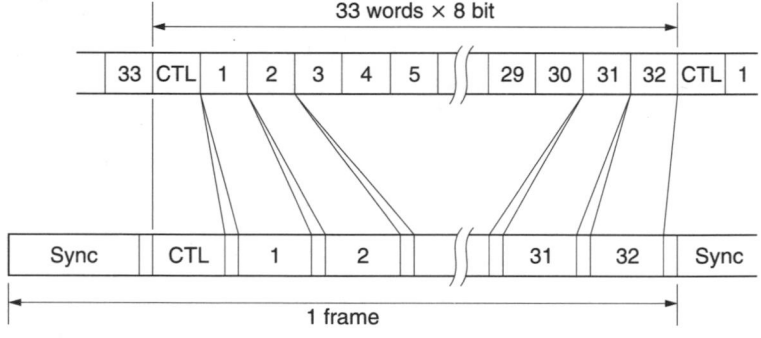

The frame frequency f_{frame} is:

$$\frac{4.3218}{588} = 7350 \text{ Hz}$$

And, as subcodes are in blocks of 98 frames, the subcode frequency f_{sc} is:

$$= \frac{7350}{98} = 75 \text{ Hz}$$

Playing time is calculated by counting blocks of subcode (i.e., 75 blocks = 1 second). A 60-minute-long CD consequently contains:

$60 \times 60 \times 7350 = 26\,460\,000$ frames

As each frame comprises $33 \times 8 = 264$ bits of information, a 1-hour-long CD actually contains $6\,985\,440\,000$ bits of information, or $873\,180\,000$ bytes! Of this, the subcode area contains some 25.8 kbytes (200 Mbits). This gigantic data storage capacity of the CD medium is also used for more general purposes on the CD-ROM (compact disc read-only memory), which is derived directly from the audio CD.

10 Opto-electronics and the optical block

As the compact disc player uses a laser beam to read the disc, we will sketch some basic principles of opto-electronics – the technological marriage of the fields of optics and electronics. The principles are remarkably diverse, involving such topics as the nature of optical radiation, the interaction of light with matter, radiometry, photometry, and the characteristics of various sources and sensors.

The optical spectrum

By convention, electromagnetic radiation is specified according to its wavelength (λ). The frequency of a specific electromagnetic wavelength is given by:

$$f = \frac{c}{\lambda}$$

where f is frequency in Hz, c is the velocity of light (3×10^8 m s^{-1}) and λ is wavelength in m.

The optical portion of the electromagnetic spectrum extends from 10 to 10^6 nm and is divided into three major categories: ultraviolet (UV), visible and infrared (IR).

UV is those wavelengths falling below the visible spectrum and above X-rays. UV is classified according to its wavelength as extreme or shortwave UV (10–200 nm), far (200–300 nm) and near or long-wave UV (300–370 nm).

Visible is those wavelengths between 370 and 750 nm, which can be perceived by the human eye. Visible light is classified according to the various colours its wavelengths elicit in the mind of a standard observer. The major colour categories are: violet (370–455 nm), blue (456–492 nm), green (493–577 nm), yellow (578–597 nm), orange (598–622 nm) and red (623–750 nm).

IR is those wavelengths above the visible spectrum and below microwaves. IR is classified according to its wavelength as near (750–1500 nm), middle (1600–6000 nm), far (6100–40 000 nm) and far-far (41 000–10^6 nm).

Interaction of optical waves with matter

An optical wave may interact with matter by being reflected, refracted, absorbed or transmitted. The interaction normally involves two or more of these effects.

Reflection

Some of the optical radiation impinging upon any surface is reflected away from the surface. The amount of reflection varies

Figure 10.1 Optical block of a compact disc player.

according to the properties of the surface and the wavelength, and in real circumstances may range from more than 98% to less than 1% (a lampblack body). Reflection from a surface may be diffuse, specular or a mixture of both.

A diffuse reflector has a surface which is rough when compared to the wavelength of the impinging radiation (Figure 10.2).

A specular, sometimes called regular, reflector, on the other hand, has a surface which is smooth when compared to the wavelength of the impinging radiation. A perfect specular reflector will thus reflect an incident beam without altering the divergency of the beam.

A narrow beam of optical radiation impinging upon a specular reflector obeys two rules, illustrated in Figure 10.3.

1 The angle of reflection is equal to the angle of incidence.
2 The incident ray and the reflected ray lie in the same plane as a normal line extending perpendicularly from the surface.

Absorption

Some of the optical radiation impinging upon any substance is absorbed by the substance. The amount of absorption varies according to the properties of the substance and the wavelength,

Figure 10.2 Reflection of light from a diffuse reflector.

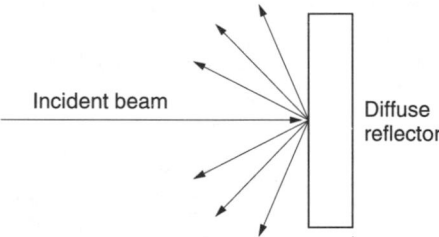

Figure 10.3 Reflection of light from a specular reflector.

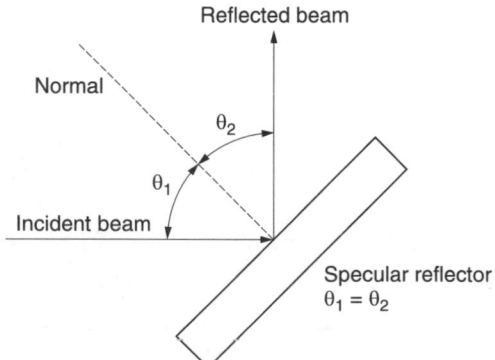

and in real circumstances may range from less than 1% to more than 98%.

Transmission

Some of the optical radiation impinging upon a substance is transmitted into the substance. The penetration depth may be shallow (transmission = 0) or deep (transmission more than 75%).

The reflection (ρ), the absorption (α) and the transmission (σ) are related in the expression:

$$\rho + \alpha + \sigma = 1$$

Refraction

A ray of optical radiation passing from one medium to another is bent at the interface of the two media if the angle of incidence is not 90°.

The index of refraction n is the sine of the angle of incidence divided by the sine of the angle of refraction, as illustrated in Figure 10.4. Refractive index varies with wavelength and ranges from 1.0003 to 2.7.

Optical components

Optical components are used both to manipulate and control optical radiation, and to provide optical access to various sources and sensors.

Glass is the most common optical material at visible and near-IR wavelengths, but other wavelengths require more exotic materials, such as calcium aluminate glass (for middle IR) and lithium fluoride (for UV).

Figure 10.4 The index of refraction.

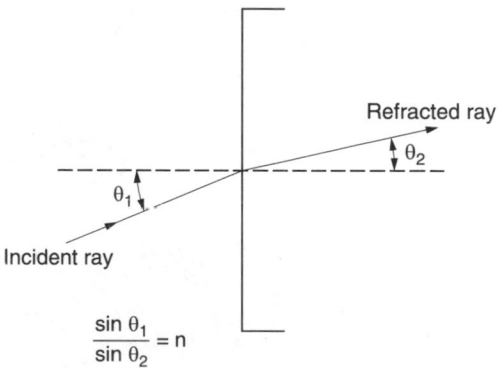

$$\frac{\sin\theta_1}{\sin\theta_2} = n$$

The thin simple lens

Figure 10.5 shows how an optical ray passes through a this simple lens.

A lens may be either positive (converging) or negative (diverging). The focal point of a lens is that point at which the image of an infinitely distant point source is reproduced.

Both the source and the focal point lie on the lens axis. The focal length (f) is the distance between the lens and the focal point. The f/number of a lens defines its light-collecting ability and is given by:

$$f / number = \frac{f}{D}$$

where f is the focal length and D is the diameter of the lens.

A small f/number denotes a large lens diameter for a specified focal length and a higher light-collecting ability than a large f/number.

Numerical aperture (NA) is a measure of the acceptance angle of a lens and is given by:

$$NA = n \sin \theta$$

where n is the refractive index of the objector image medium (for air, $n = 1$) and θ is half the maximum acceptance angle (shown in Figure 10.6). The relation of the focal length (f) to the distances between the lens and the object being imaged (s) and the lens and the focused image (s') is given by the Gaussian form of the thin lens equation:

$$1/s + 1/s' = 1/f$$

Figure 10.5 How light is refracted through a thin simple lens.

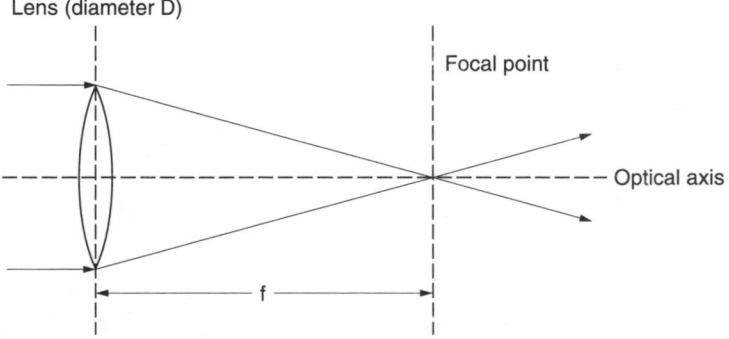

The combined focal length for two thin lenses in contact or close proximity and having the same optical axis is given by:

$$1/f = 1/f_1 + 1/f_2$$

The relationship between the focal length (f) and the refractive index (n) is:

$$1/f = (n - 1)(1/r_1 - 1/r_2)$$

where r_1 is the radius of the left lens surface and r_2 is the radius of the right lens surface.

The cylindrical lens

A cylindrical lens is a section of a cylinder and therefore magnifies in only one plane. An optical beam which enters the cylindrical lens shown in Figure 10.7 is focused only in the horizontal plane. As a result the cross-section of the beam after passing through the lens is elliptic, with the degree varying according to the distance from the lens. By detecting the elliptic degree, a useful measure of whether or not a beam is focused on a surface can be made.

The prism

A prism is an optically transparent body used to refract, disperse or reflect an optical beam. The simplest prism is the right-angle prism, shown in Figure 10.8.

An optical ray perpendicularly striking one of the shorter faces of the prism is totally internally reflected at the hypotenuse, undergoes a 90° deviation and emerges from the second shorter face.

Figure 10.6 Acceptance angle of a lens.

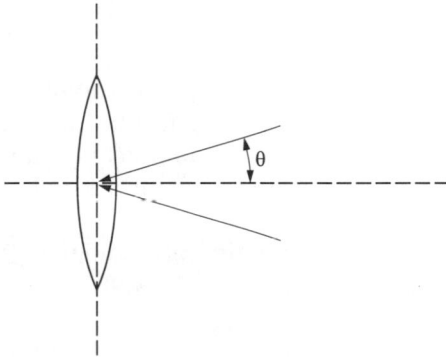

Figure 10.7 Operation of a cylindrical lens.

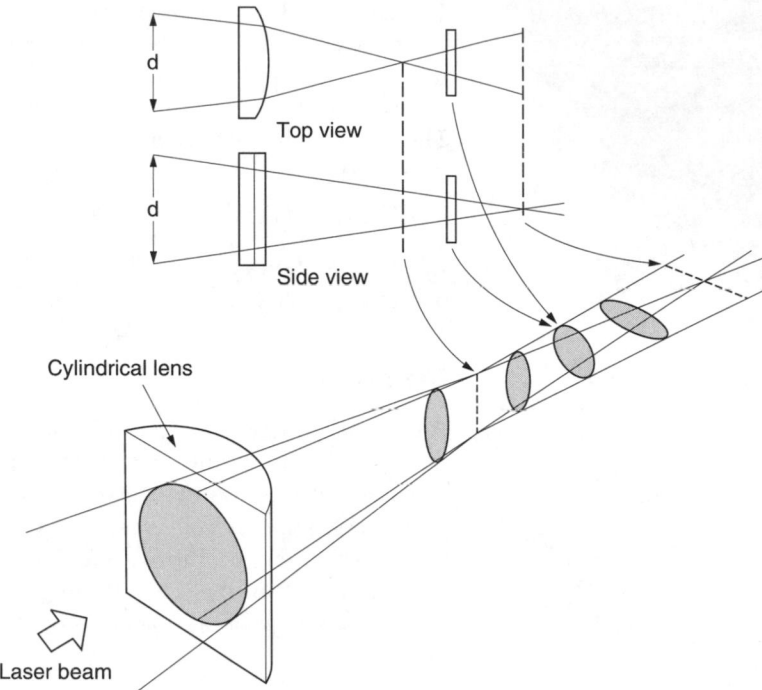

Figure 10.8 Right-angle prism.

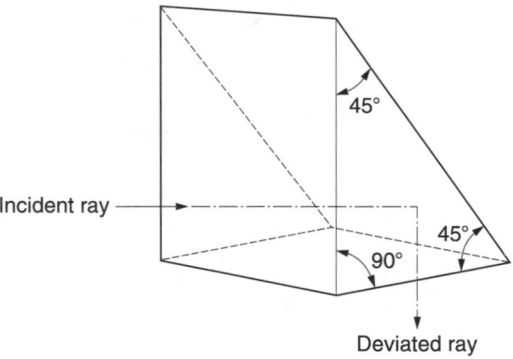

Operation of a totally internally reflecting prism is dependent upon the fact that a ray impinging upon the surface of a material having a refractive index (n) smaller than the refractive index (n') of the medium in which the ray is propagating will be totally internally reflected when the angle of incidence is greater than a certain critical angle (θ_c) given by:

$$\sin \theta_c = \frac{n'}{n}$$

The collimator

The combination of two simple lenses is commonly used to increase the diameter of a beam while reducing its divergence. Such a collimator is shown in Figure 10.9.

Diffraction gratings

Diffraction gratings are used to split optical beams and usually comprise a thin plane parallel plate, one surface of which is coated with a partially reflecting film of thin metal with thin slits. Figure 10.10 shows a diffraction grating used with a prism.

The slits are spaced only a few wavelengths apart. When the beam passes through the grating it diffracts at different angles, and appears as a bright main beam with successively less intensive side beams, as shown in Figure 10.11.

Figure 10.9 Principle of a collimator.

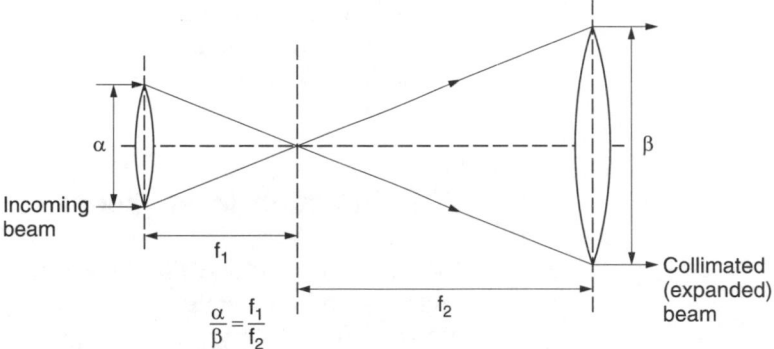

$$\frac{\alpha}{\beta} = \frac{f_1}{f_2}$$

Figure 10.10 A diffraction grating used with a prism (part of a CD player optical pick-up).

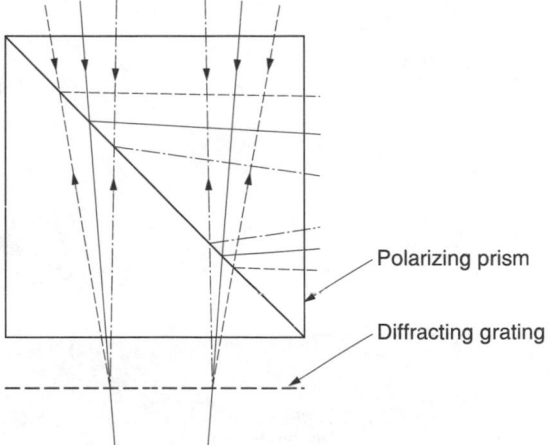

Polarizing prism

Diffracting grating

Figure 10.11 Light passing through a diffraction grating.

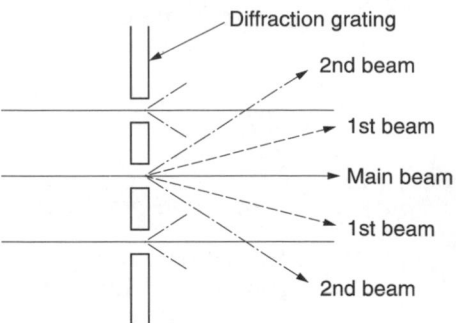

1st beams are used for tracking servo.

Quarter wave plate

When linear polarized light is passed through an anisotropic crystal (Figure 10.12), the polarization plane will be contorted during the traverse of the crystal. The thickness d of the crystal required to obtain a contortion of more than 180° is equal to one wavelength of the light. In order to obtain a contortion of more than 45°, only one-quarter of d is therefore required, and a crystal with this thickness is used in an optical pick-up in a CD player, and is called a 1/4 wave plate, quarter wave plate or QWP.

The injection laser diode (ILD)

The basic operation of any laser consists of pumping the atoms into a stimulated state, from which electrons can escape and fall to the lower energy state by giving up a photon of the appro-

Figure 10.12 Light passing through an anisotropic crystal becomes rotated in polarization.

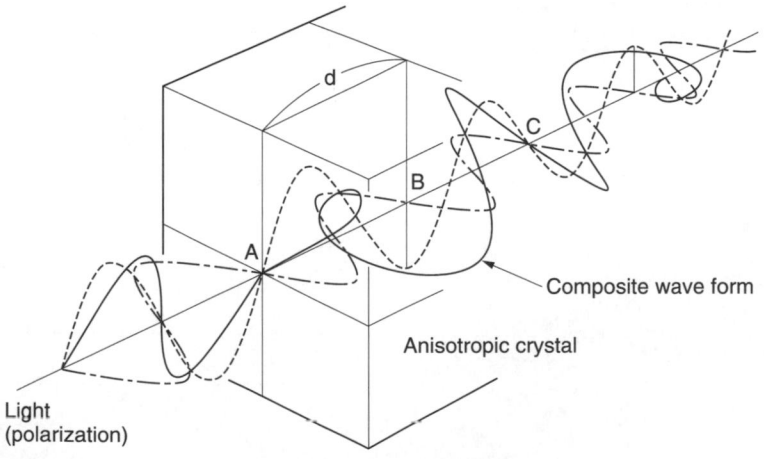

priate energy. In a solid-state laser, the input energy populates some of the usually unpopulated bands. When a photon of energy equal to the band gap of the crystal passes through, it stimulates other excited photons to fall in step with it. Thus, a small priming signal will emerge with other coherent photons.

The laser can be operated in a continuous wave (CW) oscillator mode if the ends of the laser path are optically flat, parallel and reflective, so forming an optical resonant cavity. A spontaneously produced photon will rattle back and forth between the ends, acquiring companions due to stimulated emission.

This stimulated emission can be like an avalanche, completely draining the high-energy states in a rapid burst of energy. On the other hand, certain types of lasers can be produced which will operate in an equilibrium condition, giving up photons at just the input energy pumping rate. The injection laser diode (ILD) is such a device.

The material in an injection laser diode is heavily doped so that, under forward bias, the region near the junction has a very high concentration of holes and electrons. This produces the inversion condition with a large population of electrons in a high-energy band and a large population of holes in the low-energy band. In this state, the stimulated emission of photons can overcome photon absorption and a net light flux will result.

In operation, the forward current must reach some threshold, beyond which the laser operates with a single 'thread' of light, and the output is relatively stable but low. As the current is increased, light output increases rapidly.

The ILD characteristic is highly temperature-sensitive. A small current variation or a modest temperature change can cause the output to rise so rapidly that it destroys the device. A photo-diode, monitoring the light output, is commonly used in a feed-back loop to overcome this problem.

Like the LED, the ILD must be driven from a current source, rather than a voltage source, to prevent thermal runaway. With a voltage source, as the device junction begins to warm the forward voltage drop decreases, which tends to increase the current, in turn decreasing the forward voltage drop, and so on until the current tends towards infinity and the device is destroyed.

In addition, the ILD has a tendency to deteriorate with operation. Deterioration is greatly accelerated when operating the ILD outside of its optimum limits. A typical ILD characteristic is shown in Figure 10.13.

Figure 10.13 Characteristic of a typical ILD.

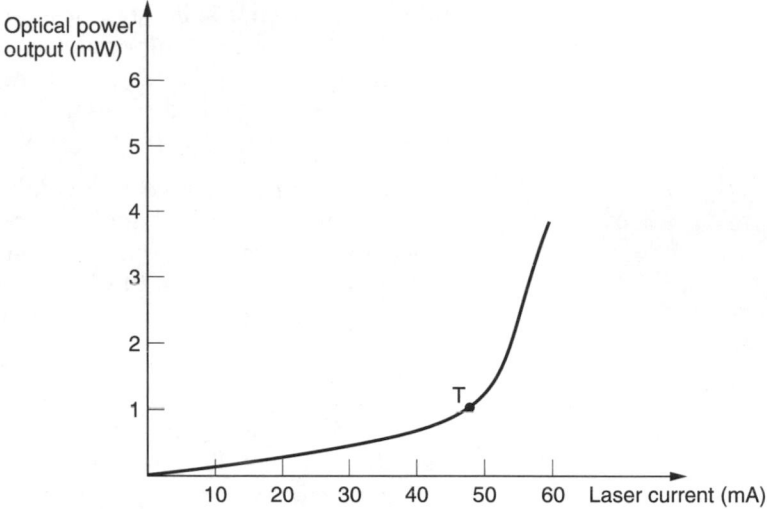

The ILD with a double heterostructure (DH)

A very narrow P-N junction layer of GaAs semiconductor is sandwiched between layers of AlGaAs.

The properties of the outer semiconductor layers confine the electrons and holes to the junction layer, leading to an inverted population at a low input current.

Quantum aspects of the laser

A laser (light amplification by stimulated emission of radiation) is a maser (microwave amplification by stimulated emission of radiation) operating at optical frequencies.

Because the operations of masers and lasers are dependent upon quantum processes and interactions, they are known as quantum electronic devices. The laser, for example, is a quantum amplifier for optical frequencies, whereas the maser is a quantum amplifier for microwave frequencies.

Masers and lasers utilize a solid or gaseous active medium, and their operations are dependent on Planck's law:

$$\Delta E = E_2 - E_1 = hf$$

An atom can make discontinuous jumps or transitions from one allowed energy level (E_2) to another (E_1), accompanied by either the emission or absorption of a photon of electromagnetic radiation at frequency f.

The energy difference E is commonly expressed in electronvolts (abbreviation: eV) and:

$1\ eV = 1.6 \times 10^{-19}\ J$

The constant h is Planck's constant and equals $6.6262 \times 10^{-34}\ J\ s$. The frequency f (in hertz) of the radiation and the associated wavelength λ (in metres) are related by the expression:

$$\lambda = \frac{c}{f}$$

where c (the velocity of light) is $3 \times 10^8\ m\ s^{-1}$.

Example:

$$\Delta E = E_2 - E_1 = 1.6\ \ eV$$
$$= 1.6 \times 1.6 \times 10^{-19}\ J$$
$$= 2.5632 \times 10^{-19}\ J$$

However, ΔE also equals:

$$hf = h\frac{c}{\lambda}$$

So:

$$\lambda = \frac{hc}{\Delta E}$$
$$= \frac{6.6262 \times 10^{-34} \times 3 \times 10^8}{2.5632 \times 10^{-19}}\ m$$
$$= 775 \times 10^{-19}\ m$$
$$= 775\ nm$$

TOP: T-type optical pick-up

The schematic of an optical pick-up of the three-beam type used in CD players is shown in Figure 10.14. This diagram shows the different optical elements composing a pick-up and indicates the laser beam route through the unit.

The laser beam is generated by the laser diode. It passes through the diffraction grating, generating two secondary beams called **side beams**, which are used by a tracking servo circuit to maintain correct tracking of the disc.

The beam enters a polarizing prism (called a beam splitter) and only the vertical polarized light passes. The light beam, still divergent at this stage, is converged into a parallel beam by the collimation lens and passed through the 1/4 wave plate, where

Figure 10.14 A three-beam optical pick-up.

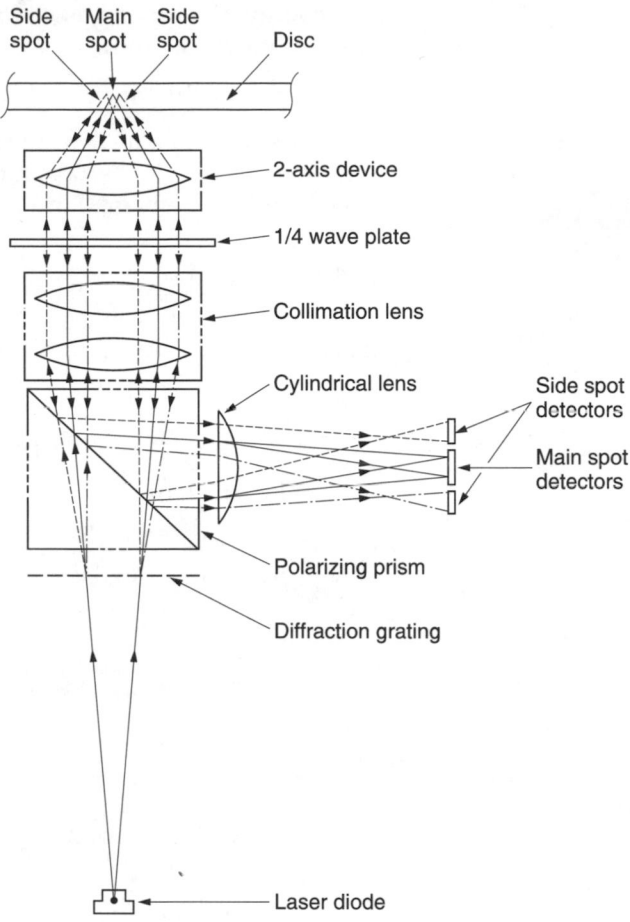

Side spot   Main spot   Side spot

Disc

2-axis device

1/4 wave plate

Collimation lens

Cylindrical lens

Side spot detectors

Main spot detectors

Polarizing prism

Diffraction grating

Laser diode

the beam's polarization plane is contorted by 45°. The laser beam is then focused by a simple lens onto the pit surface of the disc. The simple lens is part of a servo-controlled mechanism known as a two-axis device.

The beam is reflected by the disc mirrored surface, converged by the two-axis device lens into a parallel beam, and re-applied to the 1/4 wave plate. Again, the polarization plane of the light is contorted by 45°, so the total amount of contortion becomes 90°, i.e., the vertically polarized laser beam has been twisted to become horizontally polarized.

After passing through the collimation lens, the laser beam is converged onto the slating surface of the polarizing prism. Now polarized horizontally, the beam is reflected internally by the slanted surface of the prism towards the detector part of the pick-up device.

In the detector part, the cylindrical lens focuses the laser beam in only one plane, onto six photo-detectors in a format shown in Figure 10.15, i.e., four main spot detectors (A, B, C and D) and two side spot detectors (E and F), enabling read-out of the pit information from the disc.

The T-type optical pick-up (TOP) is shown in detail in Figure 10.16.

FOP: flat-type optical pick-up

Later models use a flat-type optical pick-up (FOP), shown in Figure 10.17. In the FOP, a non-polarizing prism is used and no 1/4 wave plate. The non-polarizing prism is a half mirror which

Figure 10.15 Relative positions of the six photo-detectors of a CD optical pick-up.

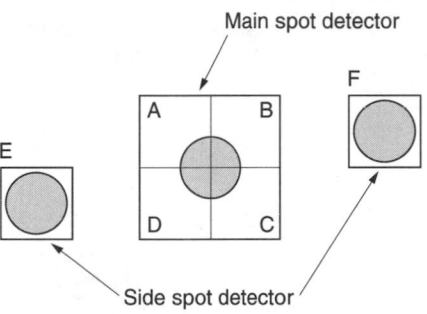

Figure 10.16 T-type optical pick-up (TOP).

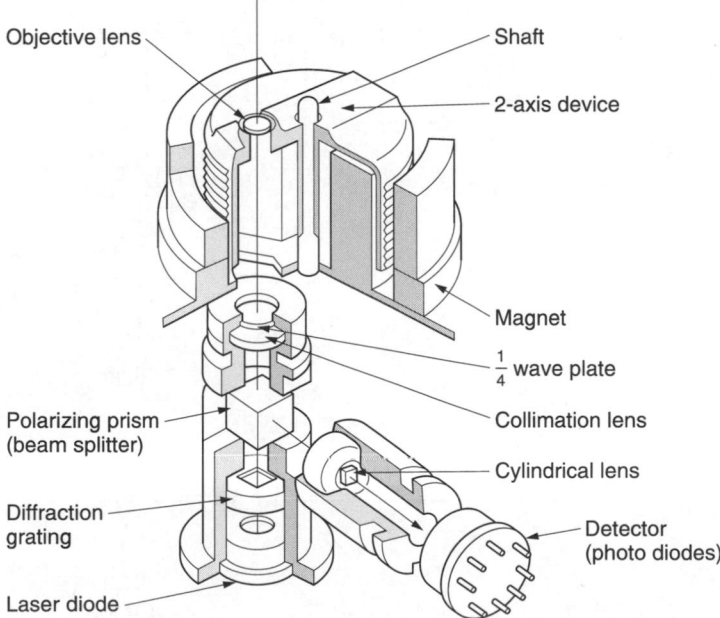

Figure 10.17 Flat-type
optical pick-up (FOP).

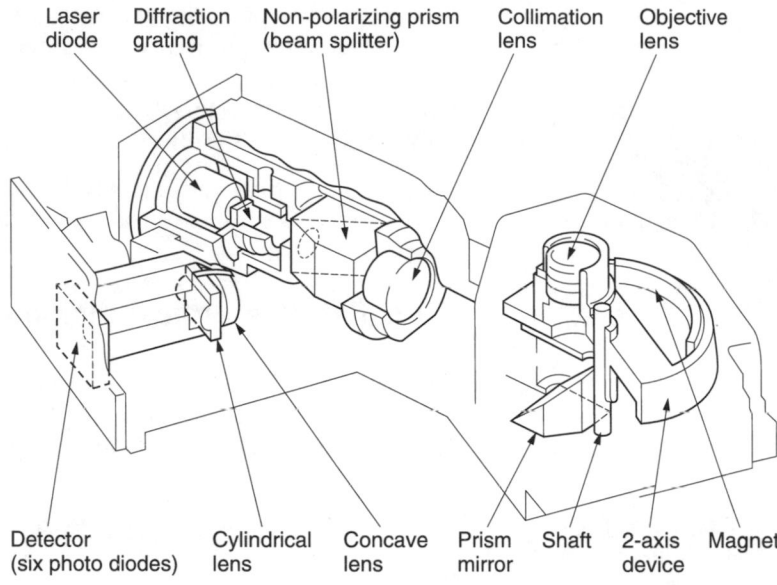

Laser diode   Diffraction grating   Non-polarizing prism (beam splitter)   Collimation lens   Objective lens

Detector (six photo diodes)   Cylindrical lens   Concave lens   Prism mirror   Shaft   2-axis device   Magnet

Figure 10.18 Beam splitting
in the FOP.

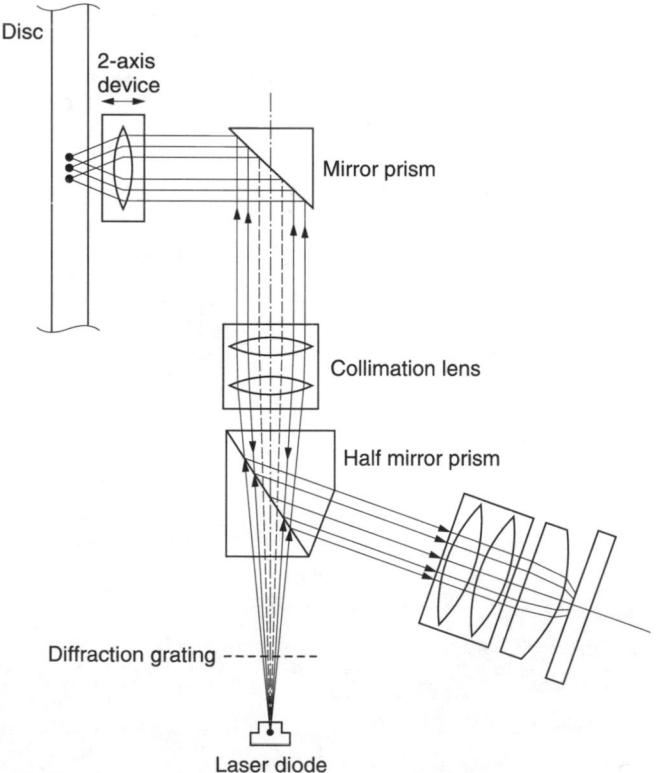

Disc

2-axis device

Mirror prism

Collimation lens

Half mirror prism

Diffraction grating

Laser diode

reflects half of the incident light and lets pass the other half (Figure 10.18).

This means that half of the light passes through the prism and returns after being reflected by the mirror. This secondary beam, 50% of the original, is reflected again for 50% and passed through for 50%, so that the resulting light beam intensity is $1/2 \times 1/2 = 1/4$ of the original beam (Figure 10.19).

The new principle enables the elimination of the influence of double refraction (i.e. the change of the deflection angle when reflected by the disc surface), caused by discs with mirror impurities.

Two-axis device

Optical pick-ups contain an actuator for objective lens position control. The compact disc player, due to the absence of any physical contact between the disc and the pick-up device, has to contain auto-focus and auto-tracking functions.

These functions are performed by the focus and tracking servo circuits via the two-axis device, enabling a movement of the objective lens in two axes: vertically for focus correction and horizontally for track following. Figure 10.20 shows such a two-axis device construction.

The principle of operation is that of the moving coil in a magnetic field. Two coils, the focus coil and the tracking coil, are suspended between magnets (Figure 10.21), creating two magnetic fields. A current through either coil, due to the magnetic field, will cause the coil to be subjected to a force, moving the coil in the corresponding direction.

Figure 10.19 Light distribution in the FOP.

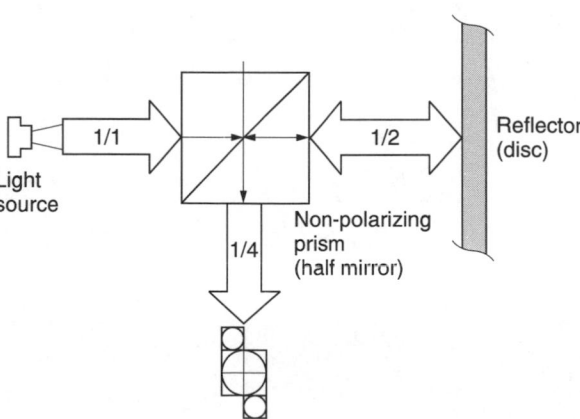

Figure 10.20 Construction of a two-axis device, used to focus the laser beam onto the surface of a compact disc.

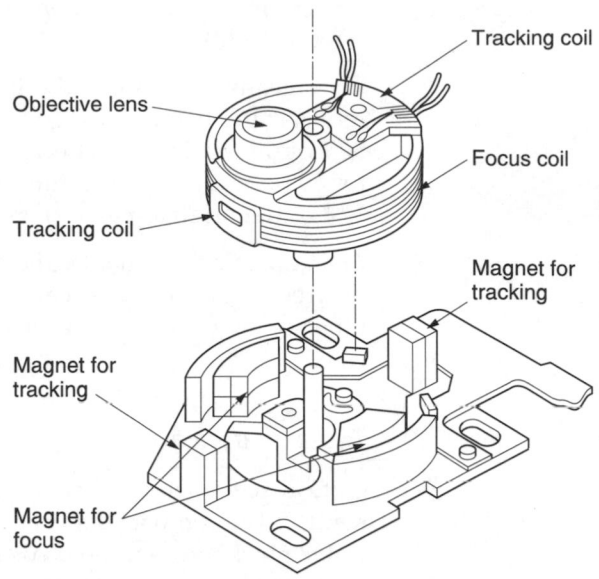

Figure 10.21 Operating principle of a two-axis device.

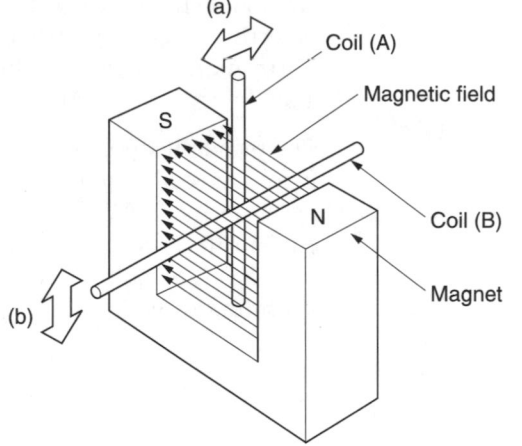

11 The servo circuits in CD players

A feedback control circuit is one in which the value of a controlled variable signal is compared with a reference value. The difference between these two signals generates an actuating error signal which may then be applied to the control elements in the control system. This principle is shown in Figure 11.1. The amplified actuating error signal is said to be fed back to the system, thus tending to reduce the difference. Supplementary power for signal amplification is available in such systems.

The two most common types of feedback control systems are regulators and servo circuits. Fundamentally, both systems

Figure 11.1 Feedback control system: block diagram.

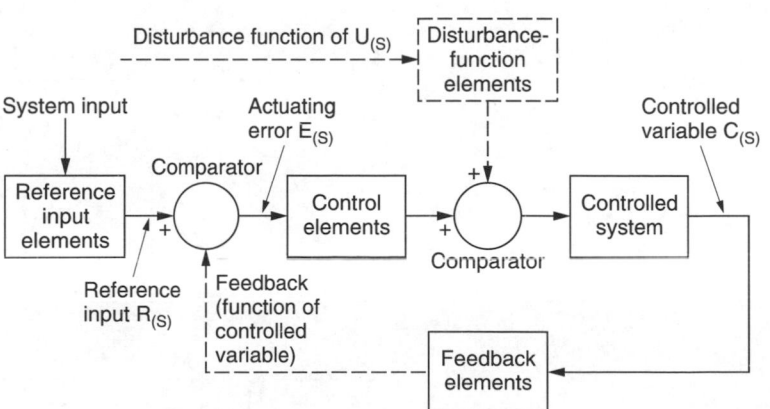

are similar, but the choice of systems depends on the nature of reference inputs, the disturbance to which the control is subjected and the number of integrating elements in the control.

Regulators are designed primarily to maintain the controlled variable or system output very nearly equal to a desired value in the presence of output disturbances. Generally, a regulator does not contain any integrating elements. An example of a regulator is shown in Figure 11.2, a stabilized power supply with series regulator.

The non-inverting input of a comparator is connected to a reference voltage (V_{ref}), and a fraction of the output voltage V_{out} is fed back to the comparator's inverting input. The closed-loop gain G of this circuit is given by:

$$G = \frac{R_1 + R_2}{R_2}$$

and the output voltage by:

$$V_{out} = G \cdot V_{ref}$$

A servo circuit, on the other hand, is a feedback control system in which the controlled variable is mechanical, usually a displacement or a velocity. Ordinarily in a servo circuit, the reference input is the signal of primary importance; load disturbances, while they may be present, are of secondary importance. Generally, one or more integrating elements are contained in the forward transfer function of the servo circuit.

An example of a servo circuit is shown in Figure 11.3, where a motor is driven at a constant speed. This circuit is a phase-locked system consisting of a phase-frequency detector, an amplifier with a filter, a motor and an encoder. The latter is a device which

Figure 11.2 A voltage regulator.

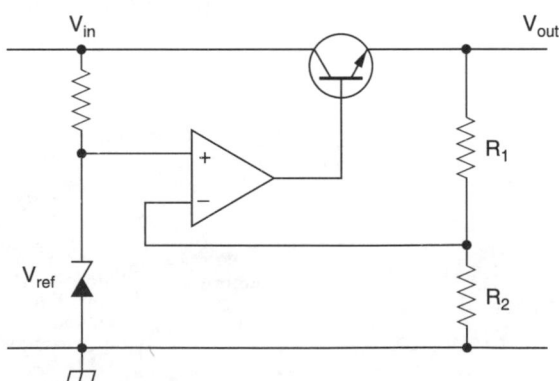

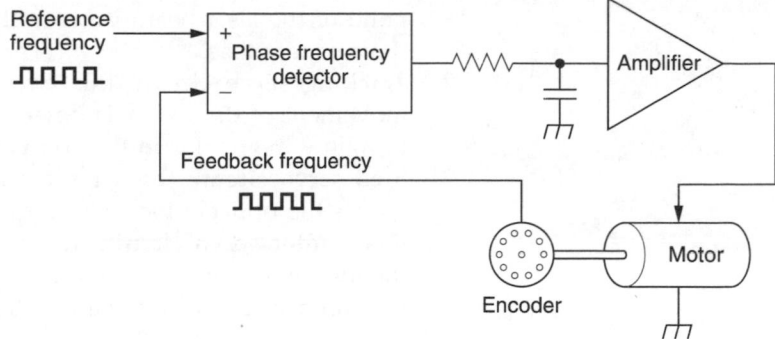

Figure 11.3 Possible motor speed control servo circuit.

emits a number of pulses per revolution of the motor shaft. Therefore, the frequency of the encoder signal is directly proportional to the motor speed.

The objective of the system is to synchronize the feedback frequency with the reference frequency. This is done by comparing the two signals and correcting the motor velocity according to any difference in frequency or phase.

Summary of the servo circuits in a CD player

For this explanation, the Sony CDP-101 CD player is used as an example. It uses four distinct servo circuits, as shown in Figure 11.4. These are:

1   **Focus servo circuit**: this servo circuit controls vertical movement of the two-axis device and guarantees that the focal

Figure 11.4 Servo circuits in the CDP-101 compact disc player.

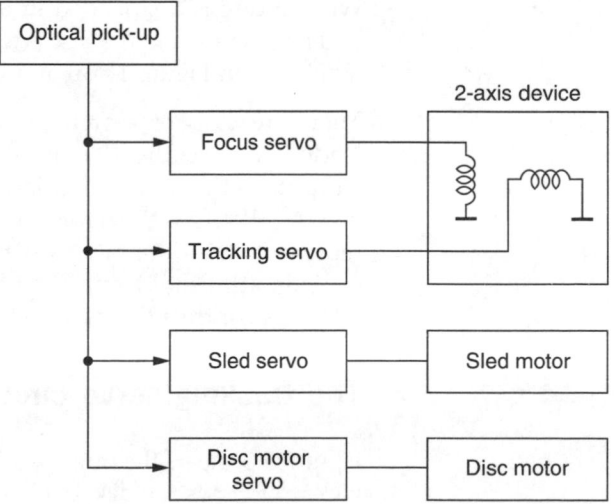

151

point of the laser beam is precisely on the mirror surface of the compact disc.

2   **Tracking servo circuit**: this circuit controls the horizontal movement of the two-axis device and forces the laser beam to follow the tracks on the compact disc.

3   **Sled servo circuit**: this circuit drives the sled motor which moves the optical block across the compact disc.

4   **Disc motor servo circuit**: this circuit controls the speed of the disc motor, guaranteeing that the optical pick-up follows the compact disc track at a constant linear velocity.

The optical pick-up is the source of the feedback signals for all four servo circuits.

The focus servo circuit

Detection of the correct focal points

The reflected laser beam is directed to the main spot detector (Figure 11.5a), an array of four photodiodes, labelled A, B, C and D. When the focus is OK, the beam falls equally on the four diodes, and the focus error signal, $(A + C) - (B + D)$, is zero.

On the other hand, when the beam is out of focus (Figure 11.5b, c), an error signal is generated, because the beam passes through a cylindrical lens, which makes the beam elliptic in shape (Figure 11.6). The resultant focus error signal from the main spot detector, $(A + C) - (B + D)$, is therefore not zero.

The focus search circuit

When a disc is first loaded in the player, the distance between the two-axis device and the disc is too large: the focus error signal is zero (as shown in Figure 11.7a) and the focus servo circuit is inactive.

Therefore, a focus search circuit is used which, after the disc is loaded, moves the two-axis device slowly closer to the disc. Outputs of the four photodiodes are combined in a different way (i.e., $A + B + C + D$) to form a radio frequency (RF) signal, which represents the data bits read from the disc. When the RF signal exceeds a threshold level (Figure 11.7b), the focus servo is enabled and now controls the two-axis device for a zero focus error signal.

The tracking servo circuit

Figure 11.8 shows the three possible tracking situations as the optical pick-up follows the disc track. In Figure 11.8a and b, the main spot

Figure 11.5 Detection of correct focus (a) arrangement of optical pick-up: focus is correct (b) and focus is not correct (c).

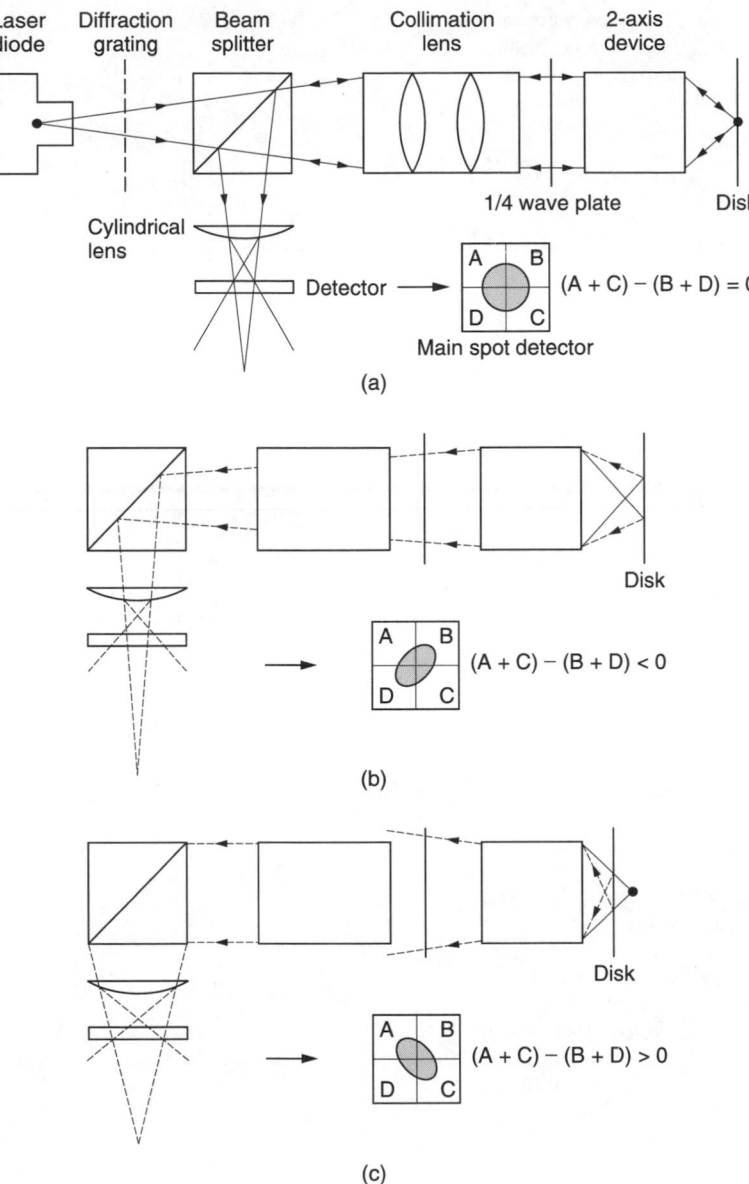

detector is not correctly tracked, and so one or other of the side spot detectors gives a large output signal as the pit is traversed. In Figure 11.8c, on the other hand, the main spot detector is correctly tracked, and both side spot detectors give small output signals.

Side spot detectors consist of two photo-diodes (E and F) and generate a tracking error signal:

Figure 11.6 How elliptical beams are produced by the cylindrical lens, when the optical pick-up is out of focus.

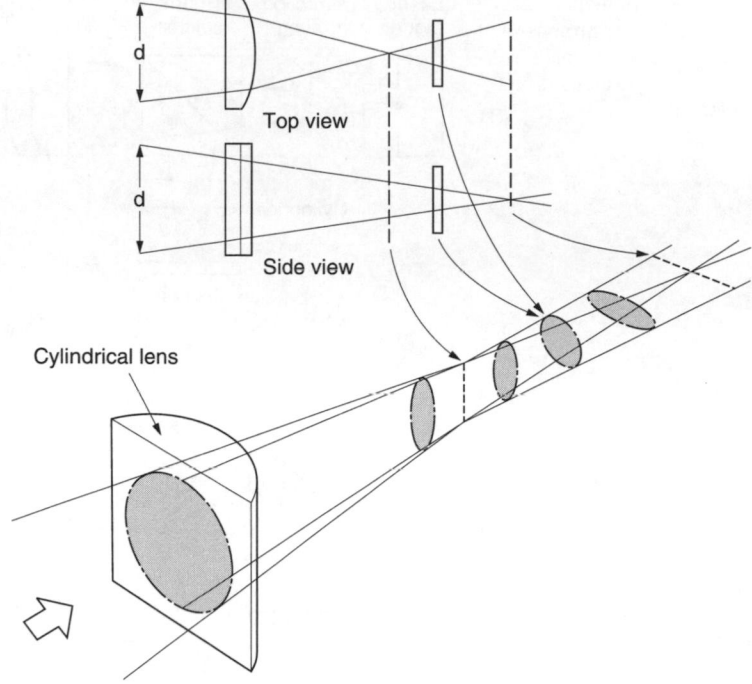

Top view

Side view

Cylindrical lens

Figure 11.7 Signals within the focus servo circuit.
(a) A focus error (FE) signal detects when the pick-up is in focus by means of combining the photo-detector outputs as (A + C) – (B + D): zero voltage means focus has been obtained.
(b) A radio frequency (RF) signal, obtained by combining the photo-detector outputs as (A + B + C + D), must exceed a threshold level before the focus servo is activated.

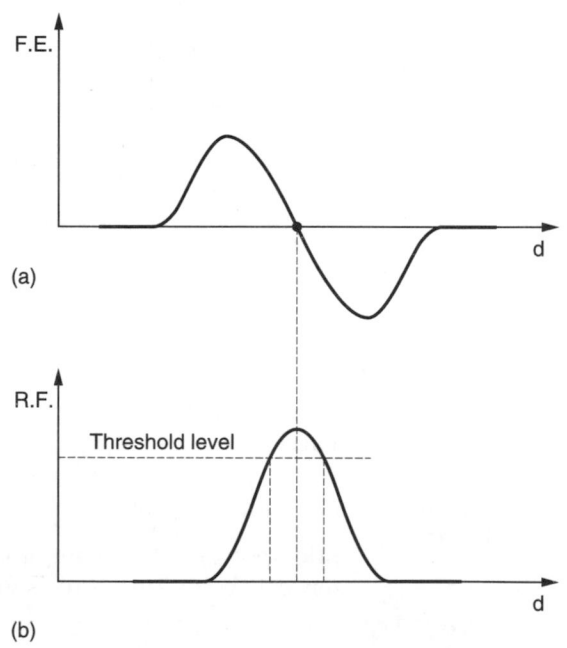

F.E.

d

(a)

R.F.

Threshold level

d

(b)

Figure 11.8 Three possible tracking situations: (a, b) mis-tracking; (c) correct tracking.

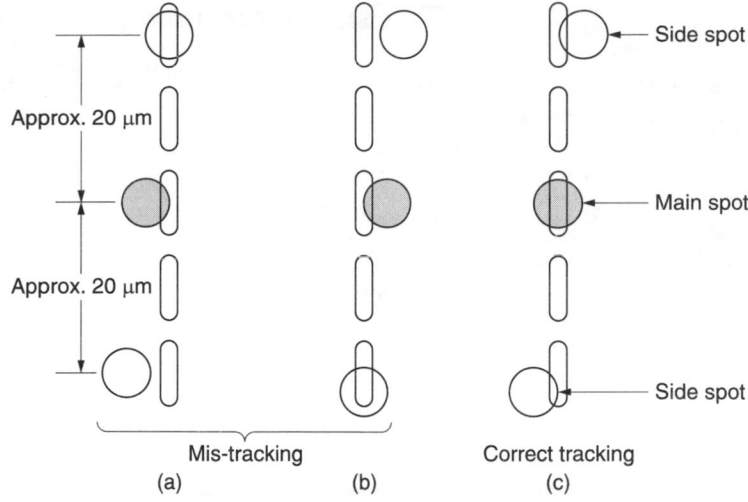

Side spot

Approx. 20 µm

Main spot

Approx. 20 µm

Side spot

Mis-tracking

Correct tracking

(a)   (b)   (c)

$$TE = E - F$$

The tracking servo acts in such a way that the tracking error signal is as small as possible, i.e., the main spot detector is exactly on the pits of the track. Tracking error, focus error and resultant focus signals are shown, derived from the optical pick-up's photodiode detectors, in Figure 11.9.

Figure 11.9 How the various error signals are obtained from the photo-detectors of the CD optical pick-up.

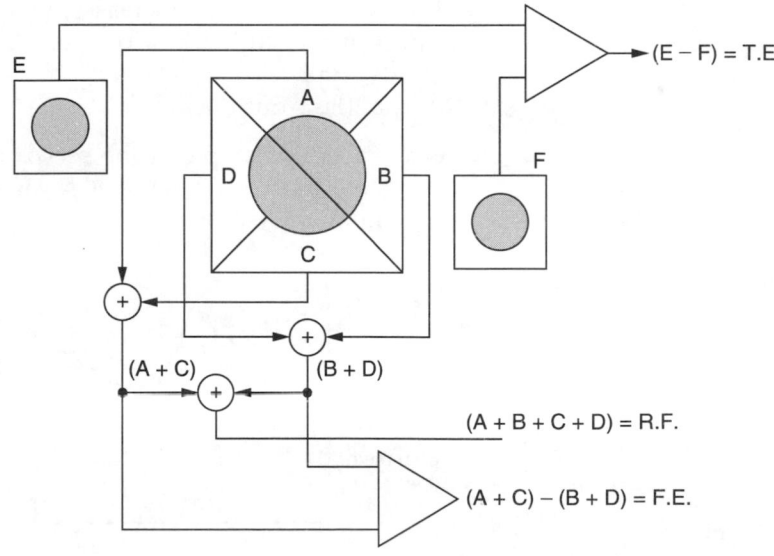

$(E - F) = T.E.$

$(A + B + C + D) = R.F.$

$(A + C) - (B + D) = F.E.$

The sled servo motor

The two-axis device allows horizontal movement over a limited number of tracks, giving a measure of fine tracking control. Another servo circuit, called the sled servo circuit, is used to move the complete optical unit across the disc for coarse tracking control. It uses the same tracking error signal as the tracking servo of the two-axis device.

However, the output of the tracking servo circuit is linearly related to the tracking error signal, whereas the output of the sled servo circuit has a built-in hysteresis: Only when the TE signal exceeds a fixed threshold level does the sled servo drive the sled motor. Tracking error and sled motor drive signals are shown in Figure 11.10.

The disc motor servo circuit

In Chapter 2, we saw how each frame of information on the disc starts with a sync word. One of the functions of the sync words is to control the disc motor.

The sync word frequency is compared against a fixed frequency (derived from a crystal oscillator) in a phase comparator and the motor is driven according to any frequency or phase difference.

As the length of the tracks increases linearly from the inner (lead-in) track to the outer (lead-out) track, the number of frames per track increases in the same manner. This means that the frequency of the sync words also increases, which causes the motor speed to decrease, resulting in a constant linear velocity. The angular velocity typically decreases from 500 rpm (lead-in track) to 200 rpm (lead-out track).

As a practical example of the servo circuit ICs, the CXA1082 is one of the most used servo ICs. Figure 11.11 presents a block diagram of a CXA1082.

Figure 11.10 Tracking error and sled motor drive signals within the CD player.

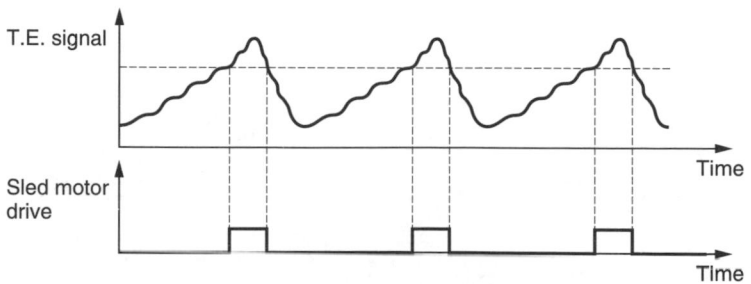

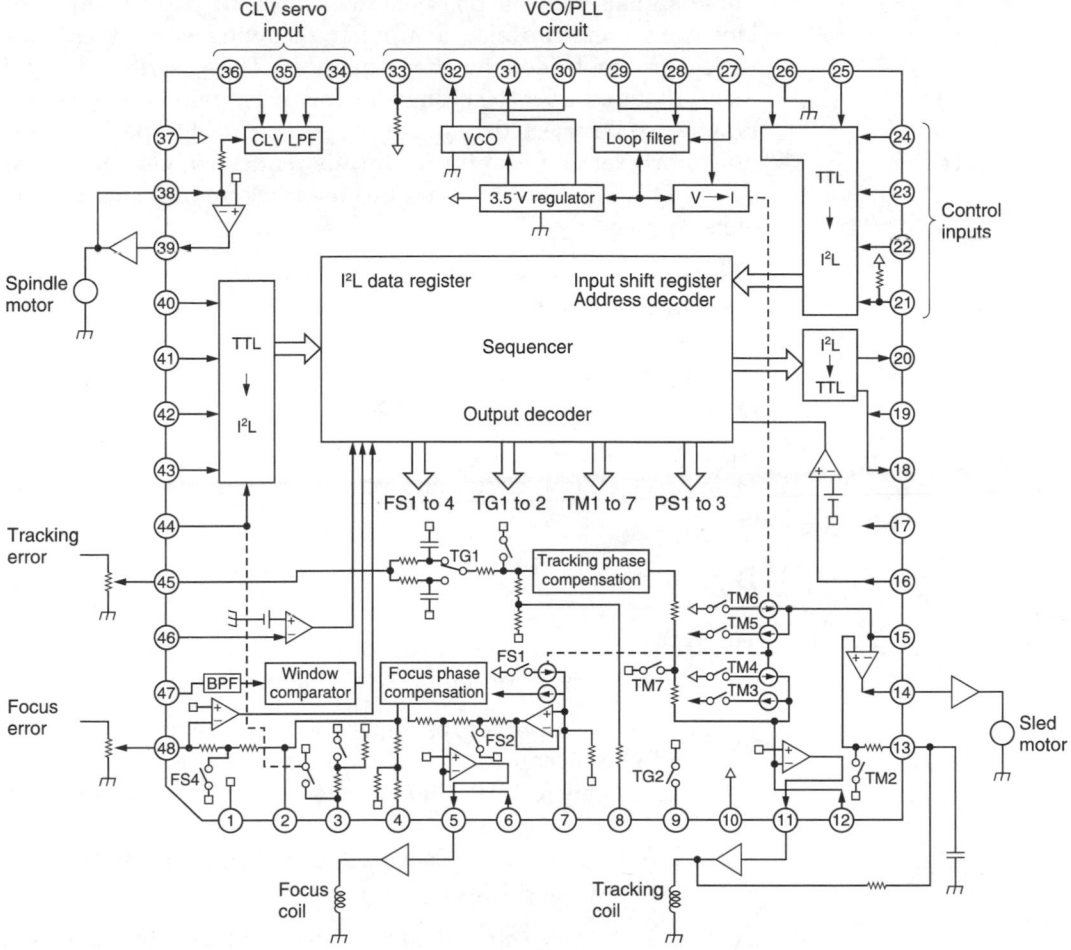

Figure 11.11 CXA1082 block diagram.

This IC has two defined parts: a digital control part and an analog servo control part.

The TTL-I²L registers are inputs/outputs to control the servo IC. In this way the servo IC will be instructed by the system control. Main inputs/outputs are to/from the system controller, which will trigger the execution of most operations. Besides these main instructions, there are a number of dedicated inputs/outputs, for specific functions.

The analog part consists mainly of the focus and the tracking control lines. Focus error and tracking error are input from the RF part of the CD player.

These signals will be phase-compensated to correct any error. However, there are also a number of switches on these lines which are controlled by the sequencer. These switches enable manipulation of servo settings; in this way, gain can be set and power can be injected to drive the servo during specific operations; for example, if a track jump is requested, closing of TM3 and TM4 will inject the current to the tracking and sled coils for either direction.

The same circuits are used for initial control; thus, when the set is starting up it will try to focus the laser on the disc. In order to do this, the focus coil is driven up/down a number of times. At that moment the servo loop is disabled by FS4, and the up/down cycles are driven by FS1 and FS2.

Apart from these focus/tracking gain circuits, this IC also comprises a VCO/PLL circuit and part of the spindle servo circuit.

Digital servo

The latest development in servo circuitry is the use of digital servos. The benefits of digital servo circuits are listed below.

*   Playability improved, as adjustments which used to be semi-fixed now become automatic and continuous.
*   Adjustment-free; all adjustments are automatic, performed by the system itself.
*   Use of digital signal processors (DSPs), which improves the possibilities and future features.

There are five servo circuits in a CD player: focus, tracking, sled, spindle and RF-PLL. Spindle and RF-PLL were digitized similar to drum motor/capstan motor circuits used in VCR and DAT sets. Focus, tracking and sled servos were digitized using DSP and D/A technology. Figure 11.12 shows the main servo circuits.

Spindle and RF-PLL use input from a reference crystal (RFCK) and from the RF retrieved from the disc. Focus, tracking and sled servos use the A/D–DSP–D/A set-up shown here. Figure 11.13 shows a block diagram of a digital servo.

*   The A/D circuit will make 8-bit conversions of input signals at a rate of 88.2 kHz for focus and tracking error, and at a rate of 345 Hz for sled error.
*   The DSP circuit will carry out (digital) phase compensation; detection of specific signals (focus OK, defect and mirror) is also carried out based upon the input signals. Also, similar to the analog servo circuits, if specific actions need to be

Figure 11.12 Main servo circuits.

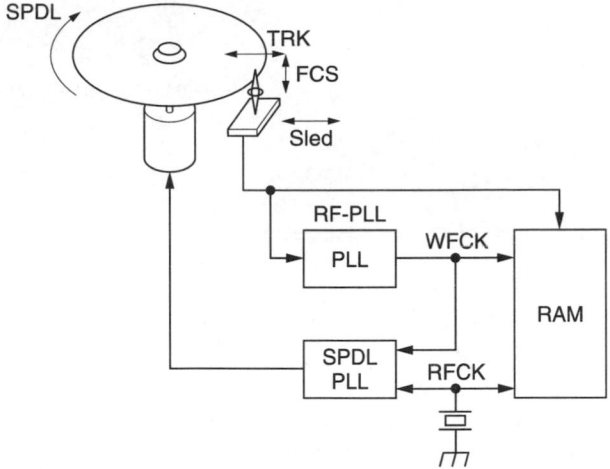

taken, such as track jumps, etc. this can be implemented easily by the DSP stages.

- The D/A circuit will convert the error output signal to a 7-bit pulse width modulation (PWM) signal at 88.2 kHz for the focus and tracking coils, and a 7-bit PWM signal at 345 Hz for the sled motor. The PWM signals will of course be fed to operational amplifiers where they are converted to a correct analog driving signal for coils and motor.

The CXD-2501 digital servo IC block diagram in Figure 11.14 shows the full integration of tracking, sled and focus servo into one IC.

Figure 11.13 Digital servo block diagram.

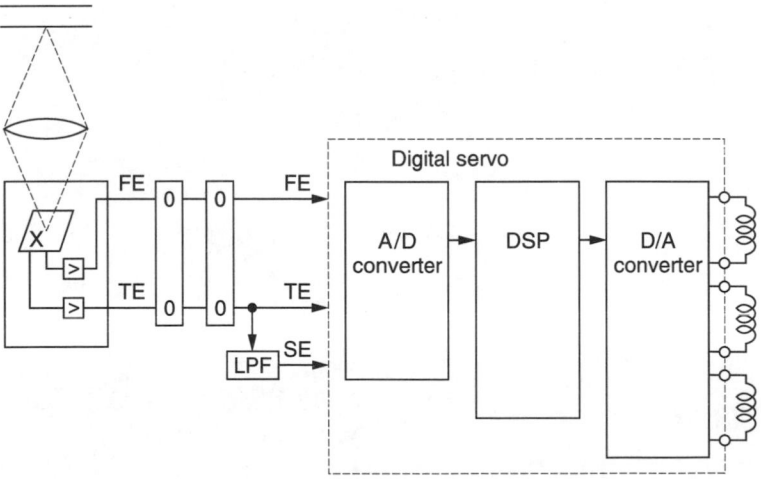

Figure 11.14 CXD2501
block diagram.

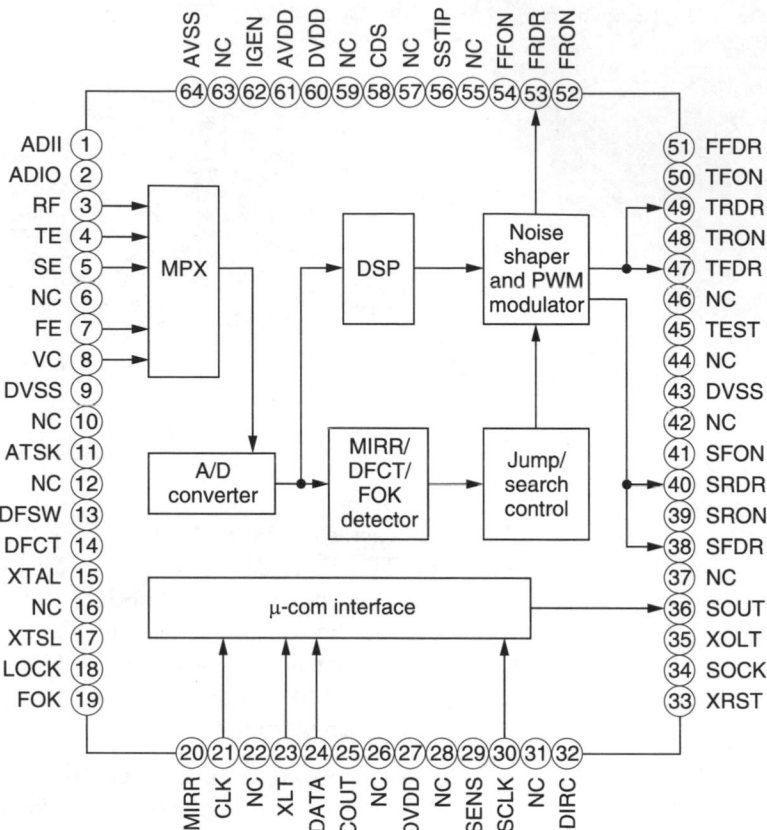

12 Signal processing

RF amplification

The signal that is read from the compact disc and contains the data information is the RF signal, that consists of the sum of signals (A + B + C + D) from the main spot detector. At this stage, the signal is a weak current signal and requires current-to-voltage conversion, amplification and waveform shaping. The CDP-101 RF amplifier circuit is shown in Figure 12.1. IC402 is a current-to-voltage converter and amplifier stage, while IC403 is an offset amplifier, correcting the offset voltage of the RF signal and delivering the amplified RF signal.

Figure 12.2 shows the waveshaper circuit used in the CDP-101. Because the RF signal from the RF amplifier is a heterogeneous signal due to disc irregularities, the waveshaper circuit detects correct zero-cross points of the eye pattern and transforms the signal into a square wave signal. After waveshaping by IC404, the signal is integrated in the feedback loop through a low-pass filter circuit to obtain a DC voltage applied to the input, so as to obtain correct slicing of the eye pattern signal.

Figure 12.3 shows a timing diagram of the RF signal before and after waveform shaping.

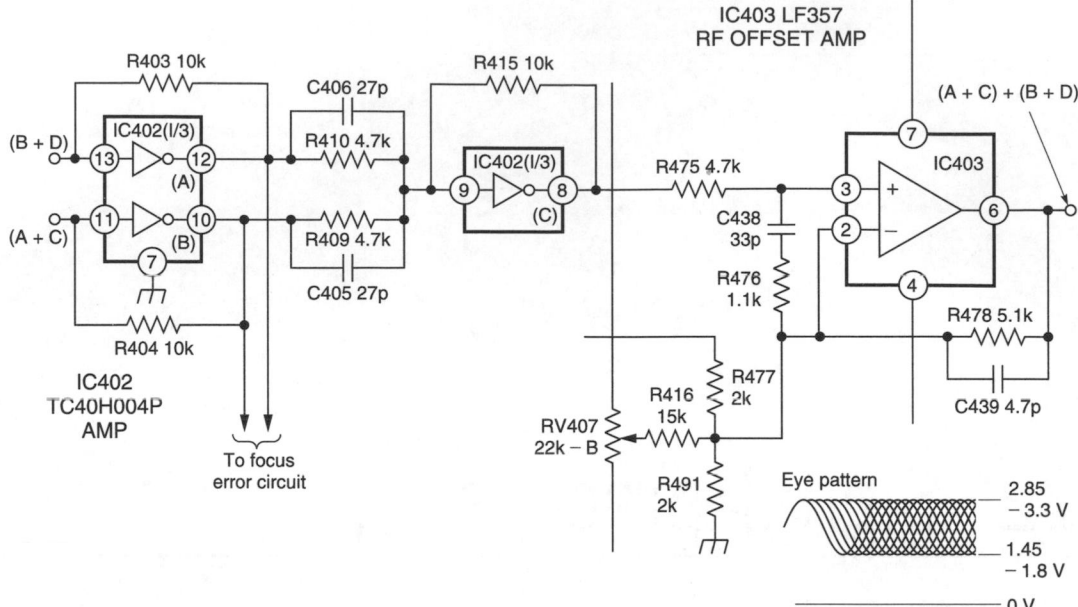

Figure 12.1 RF amplification circuit of the CDP-101 compact disc player.

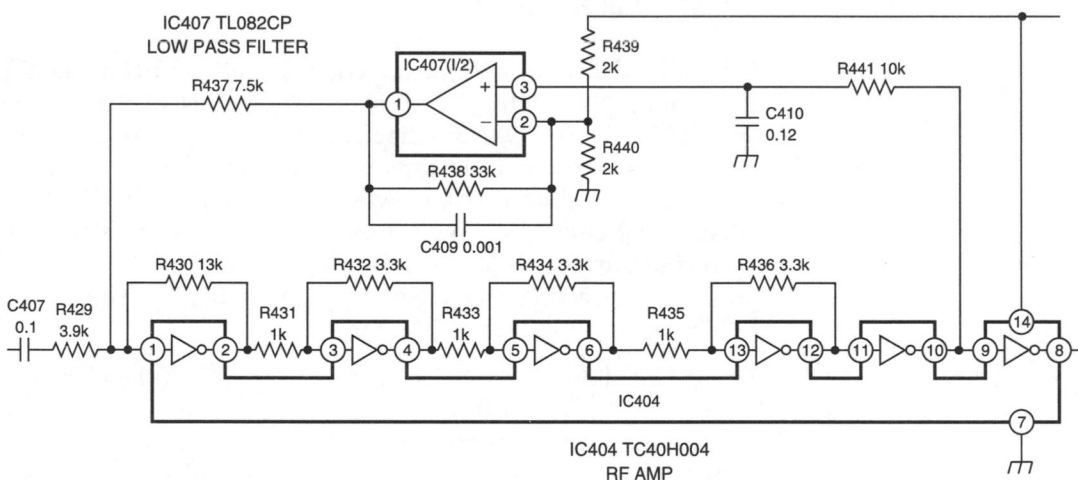

Figure 12.2 Waveshaping circuit used in the CDP-101.

Signal decoding

The block diagram in Figure 12.4 represents the basic circuit blocks of a compact disc player. After waveshaping, the RF signal is applied to a phase-locked loop (PLL) circuit in order to detect the clock information from the signal which, in turn, synchronizes the data.

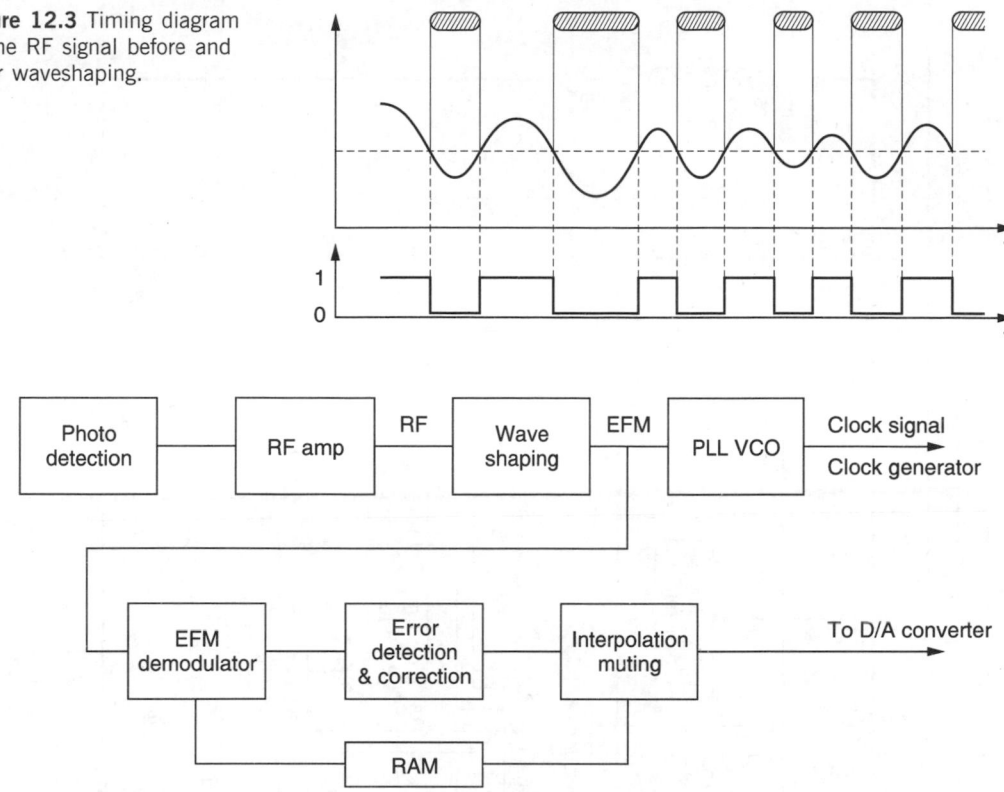

Figure 12.3 Timing diagram of the RF signal before and after waveshaping.

Figure 12.4 Signal decoding within the compact disc player.

Also, the RF signal is applied to an EFM demodulator and associated error stages in order to obtain a demodulated signal. In the CDP-101, a single integrated circuit, the CX7933, performs EFM demodulation; a block diagram is shown in Figure 12.5.

This block diagram shows frame sync detection, fourteen-to-eight demodulation to a parallel 8-bit data output, subcode Q detection and generation by an internal counter and timing generator of the WFCK (write frame clock) and WRFQ (write request) synchronization signals. Figure 12.6 shows the EFM decoding algorithm (for comparison with the encoding scheme in Figure 9.2).

CIRC decoding is performed by a single integrated circuit, the CX7935 (Figure 12.7), on the data stored in the RAM memory. An RAM control IC, the CX7934 (Figure 12.8), is used to control data manipulations between the RAM and the rest of the demodulation stage circuits.

The data are checked, corrected if necessary and de-interleaved during read-out. If non-correctable errors are found, a pointer for

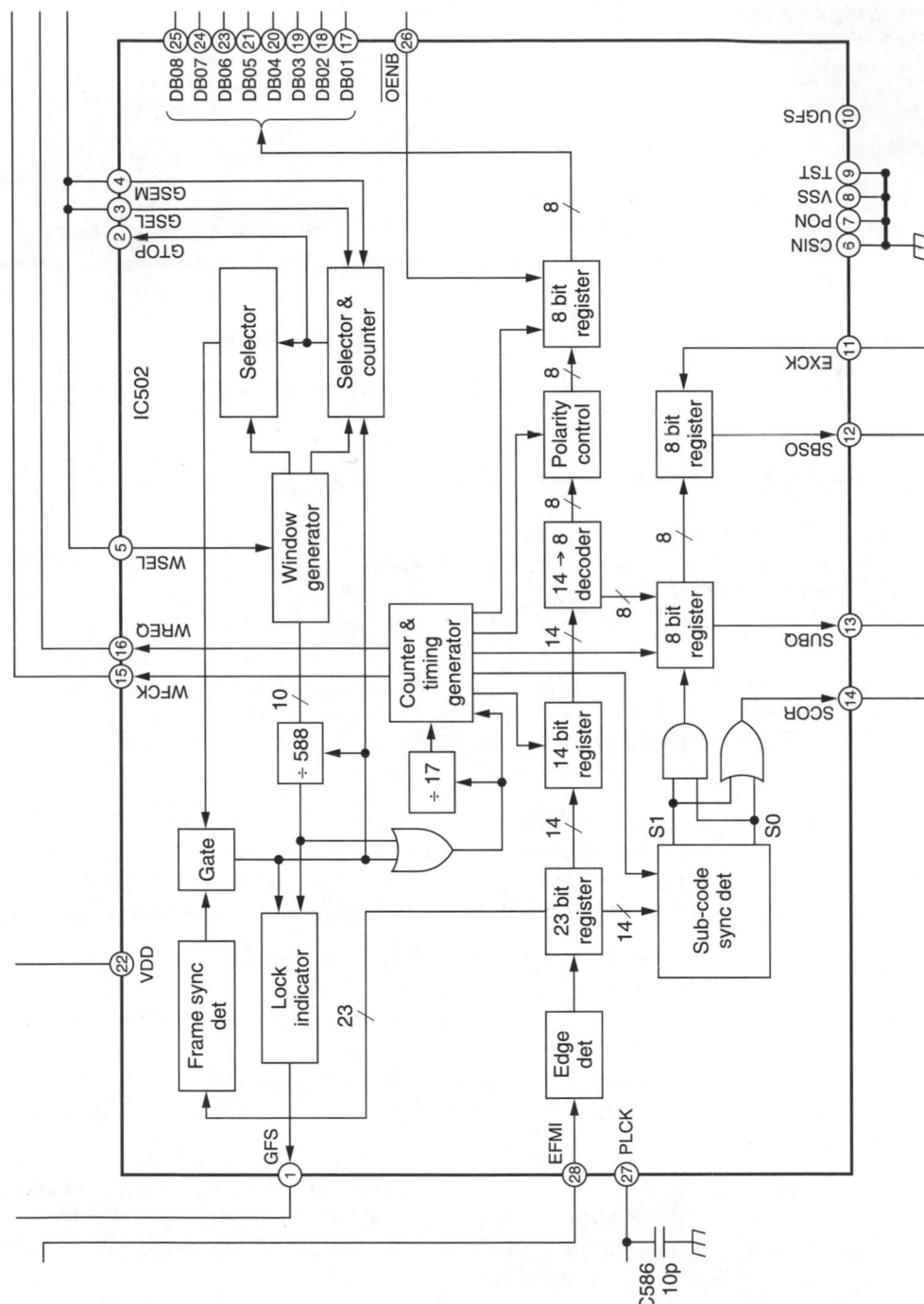

Figure 12.5 Block diagram of the CX7933 integrated circuit, which performs EFM demodulation in the CDP-101.

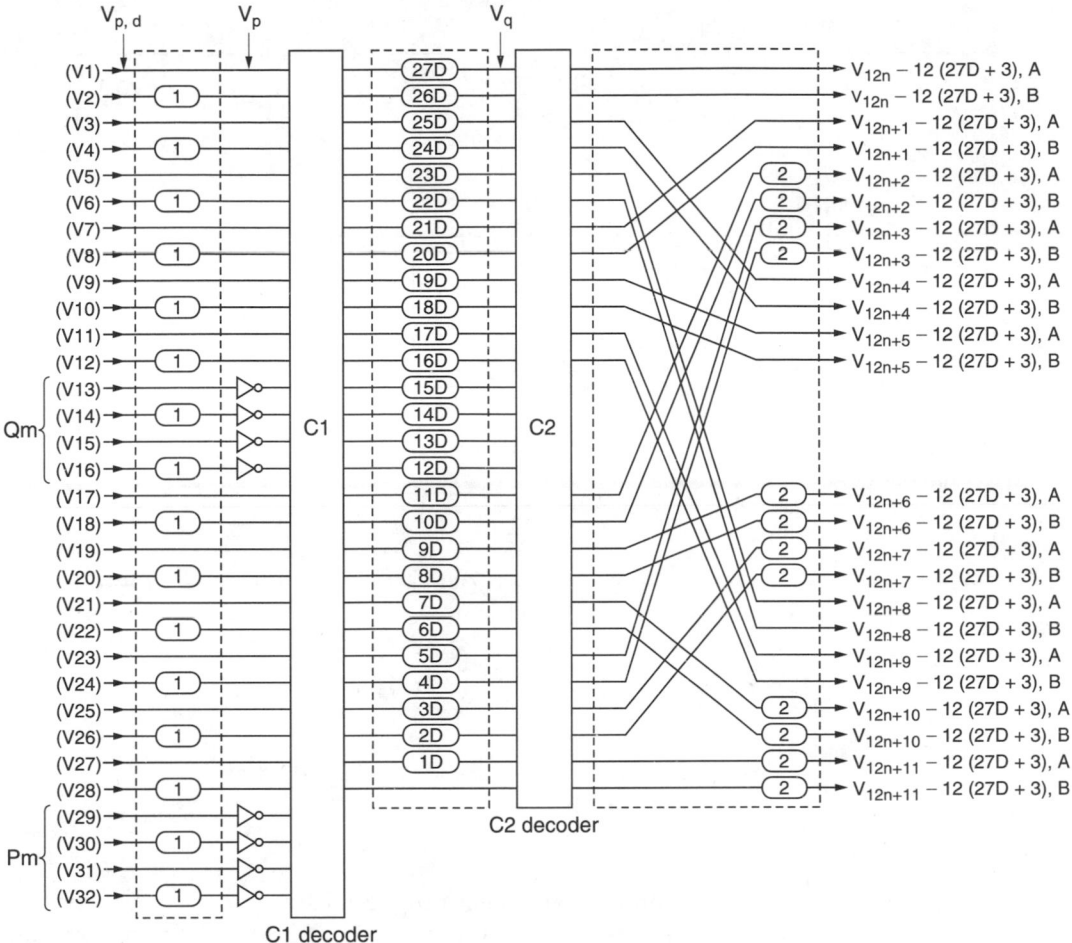

Figure 12.6 CIRC decoding algorithm.

this data word is stored in memory and the circuit corrects the data by interpolation. Figure 12.9 shows the complete signal decoding circuit as used in the CDP-101. In the latest Sony CD players, signal decoding is performed by a single integrated circuit, the CX23035, shown in Figure 12.10.

D/A converter

The D/A converter follows the signal processing and decoding circuits. Figure 12.11 represents the CX20017 integrated circuit D/A converter as used in Sony CD players. The converter is

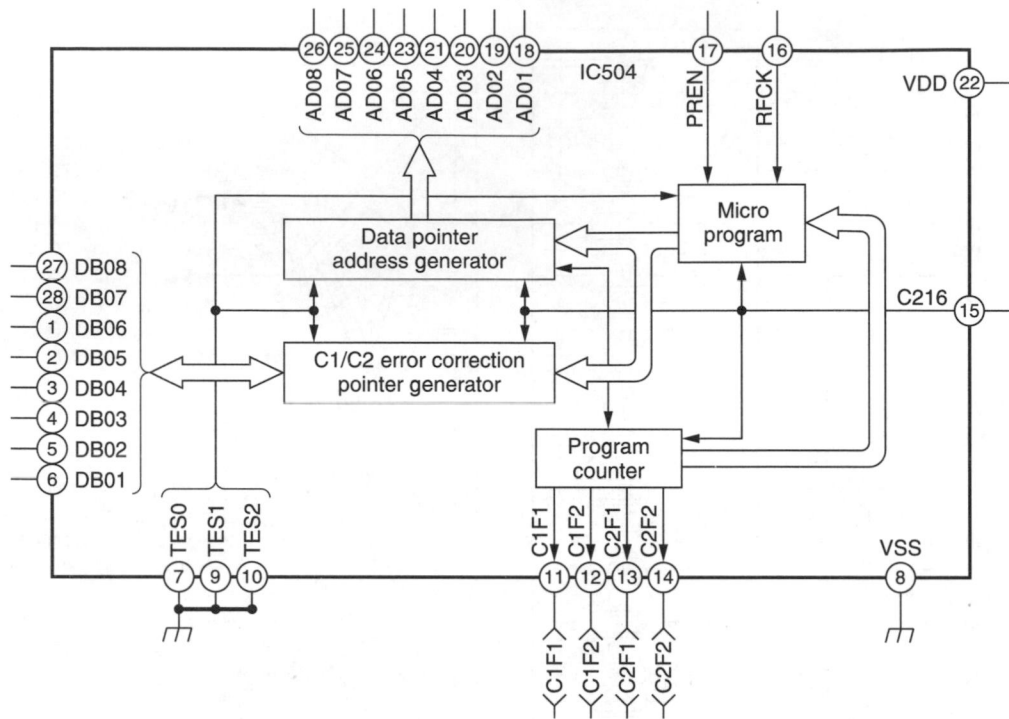

Figure 12.7 Block diagram of the CX7935 CIRC decoder.

formed around an integrating dual-slope A/D converter. Two counters, one for the eight most significant and one for the eight least significant bits, control the two constant-current sources (which have a ratio: $I_o/i_o = 256$) used for charging the integrator capacitor. Conversion is controlled by the LRCK (left/right clock), BCLK (bit clock) and WCLK (word clock) signals.

Figure 12.12 represents the operating principle, where 16-bit data are loaded into the two 8-bit counters by a latch signal. With data in the counters, no carry signals exist and the current switches are closed. The integration capacitor C charges with a total current $I = I_1 + I_2$.

During conversion, each counter counts down to zero, whereupon the carry signals open the current switches, stopping further charging of the capacitor. The final charge across the capacitor, as an analog voltage, represents the 16-bit input.

Figure 12.13 shows a practical application of a D/A converter circuit in a CD player.

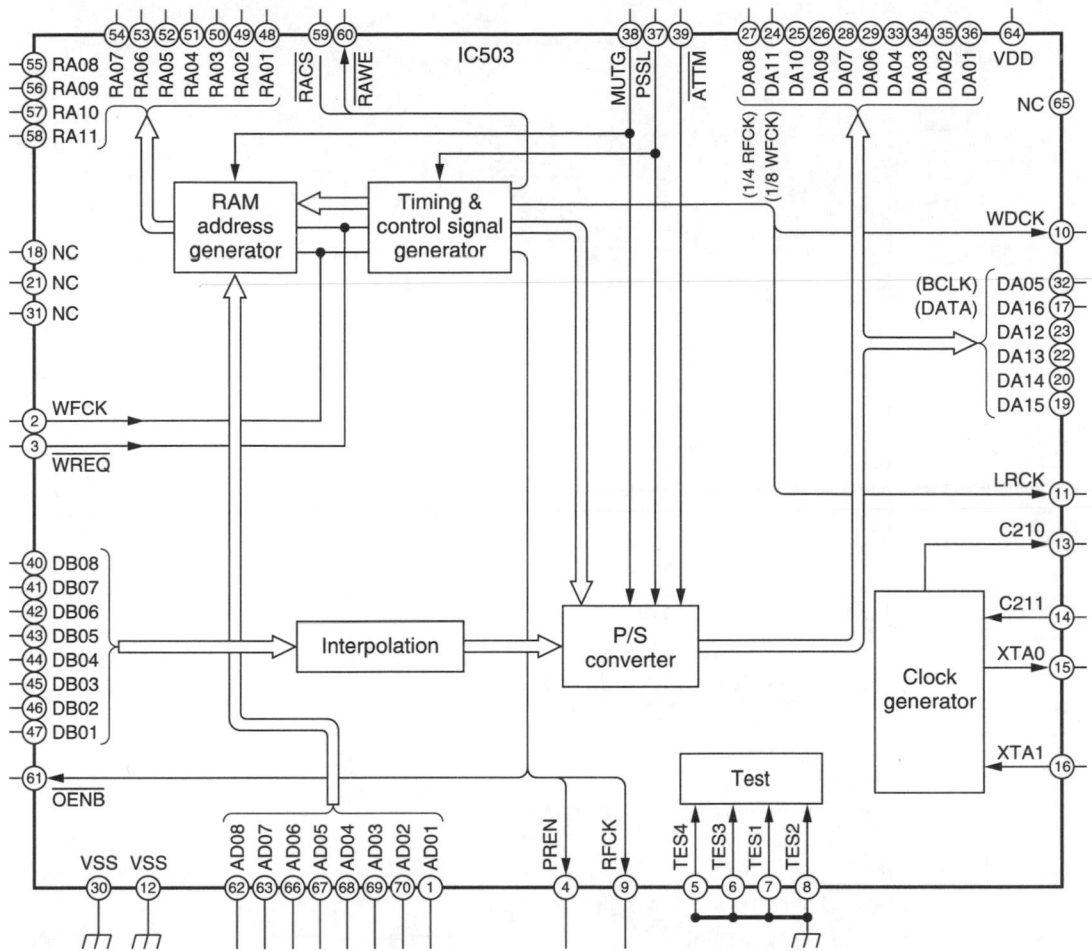

Figure 12.8 Block diagram of the CX7934 RAM controller.

High-accuracy D/A conversion

18-Bit digital filter/eight times oversampling

The CXD-1144 is a digital filter allowing the conversion of 16-bit samples into 18-bit samples with very high precision (Figure 12.14). The remaining ripple in the audible range is reduced to ±0.00001 dB. The attenuation is 120 dB and the echo rejection is about 124 dB. In particular, the reproduction of pulse-shaped tones, such as those of a keyboard or piano, is remarkably improved by enhancing the rising edges of the signal and the high echo rejection.

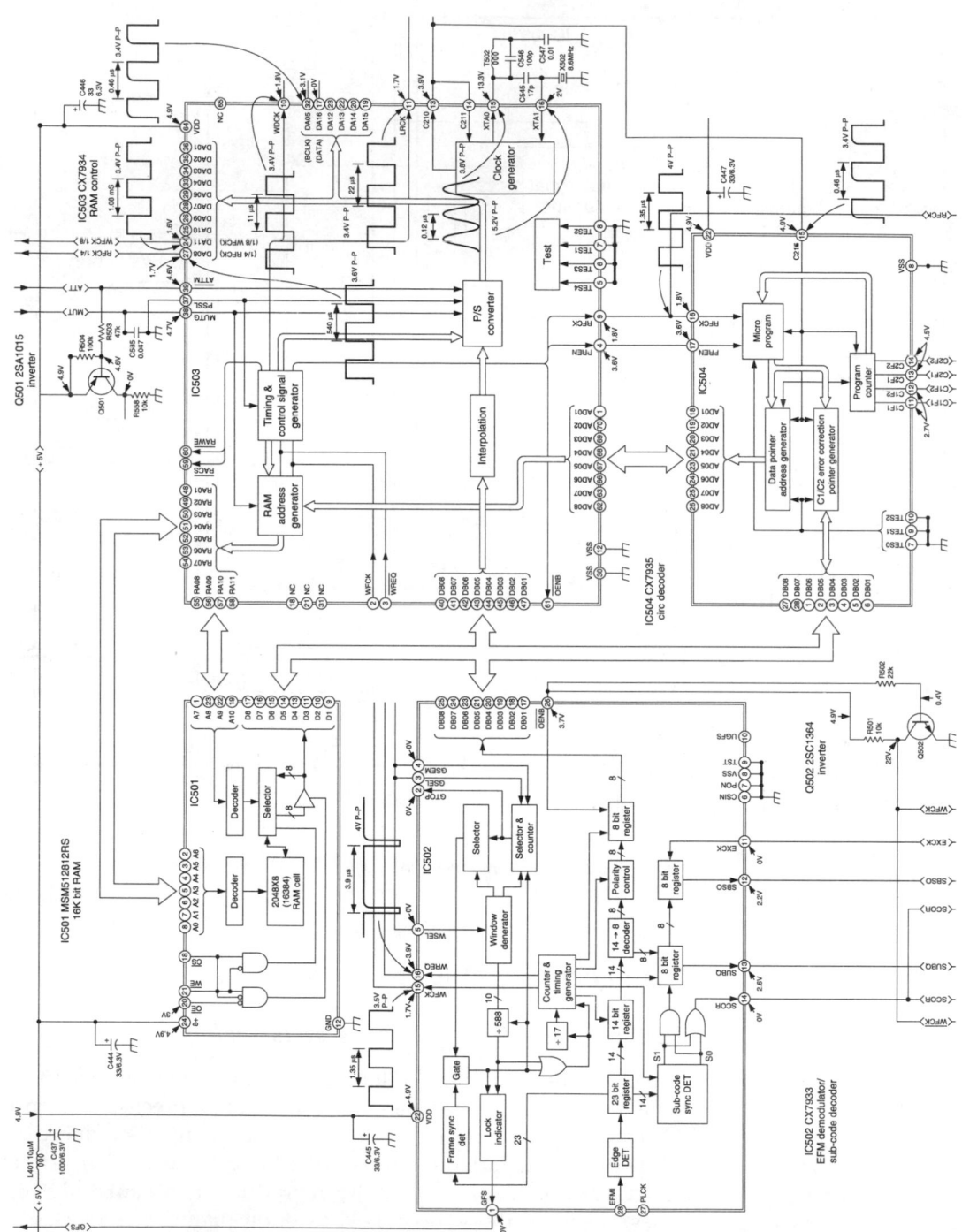

Figure 12.9 Complete signal decoding circuit of the CDP-101.

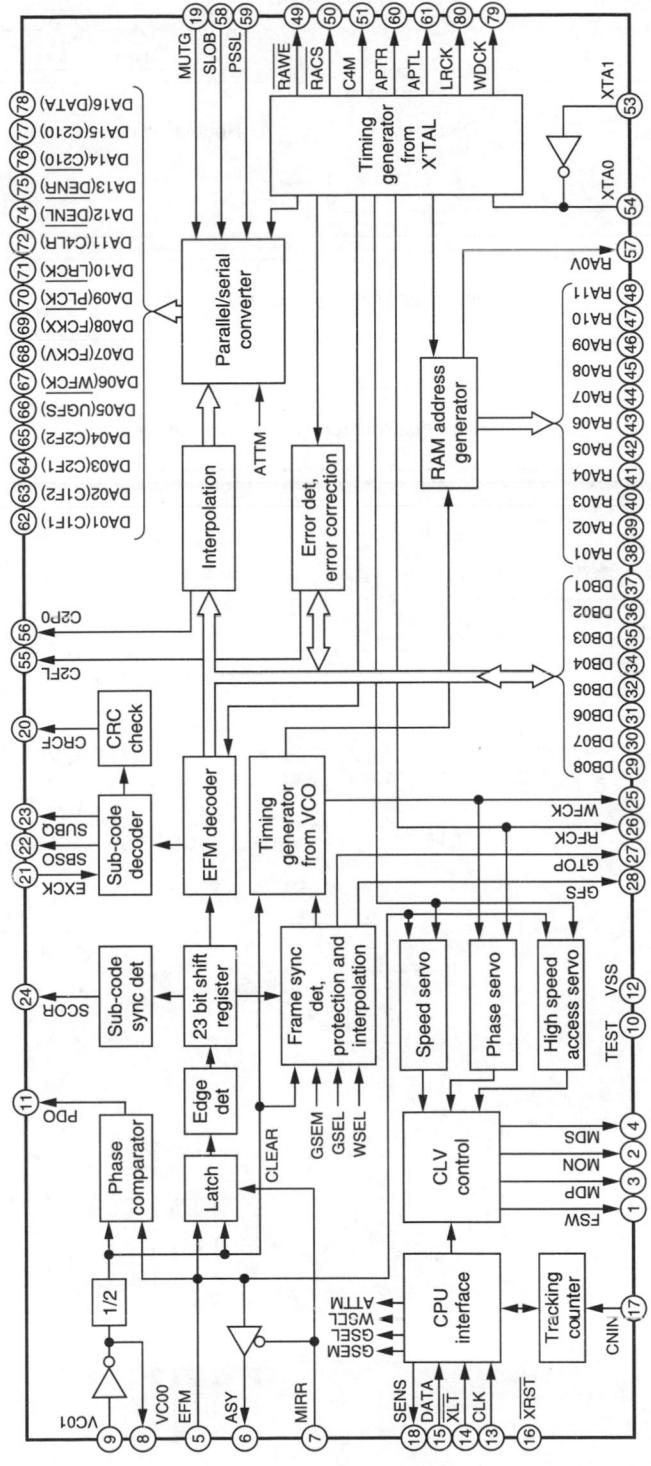

Figure 12.10 Block diagram of the CX23035 integrated circuit, which performs all signal decoding within the latest Sony CD players.

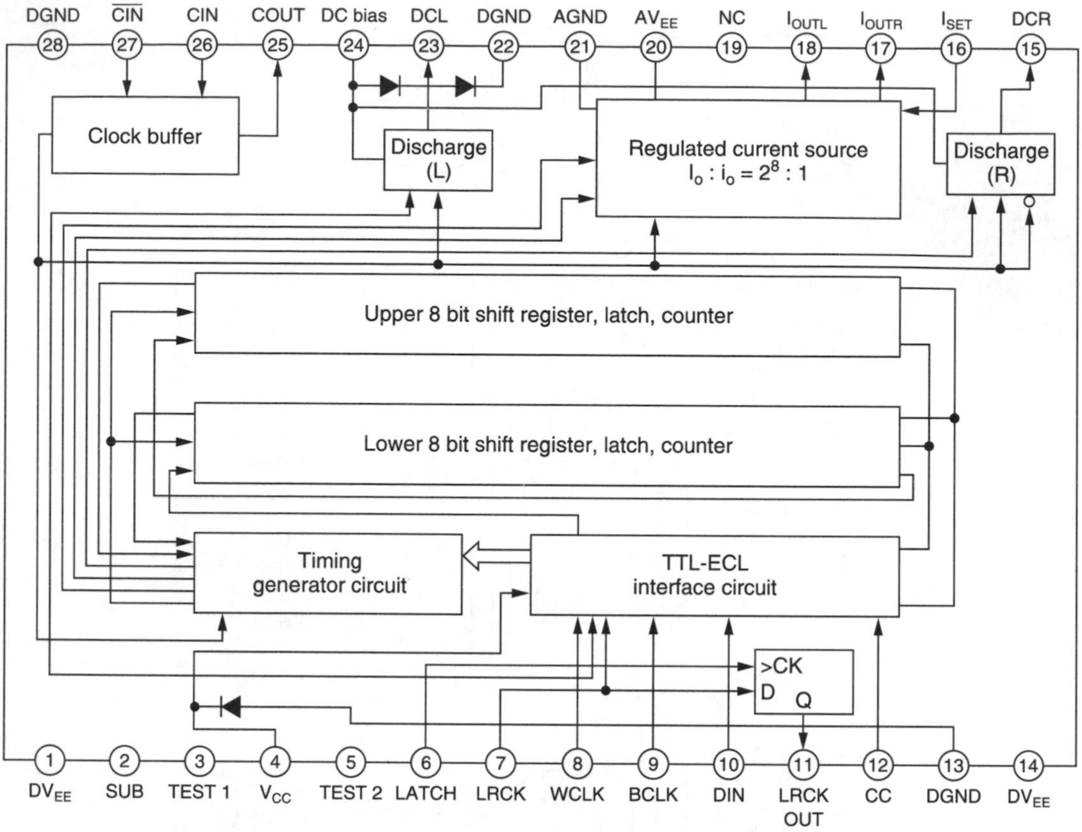

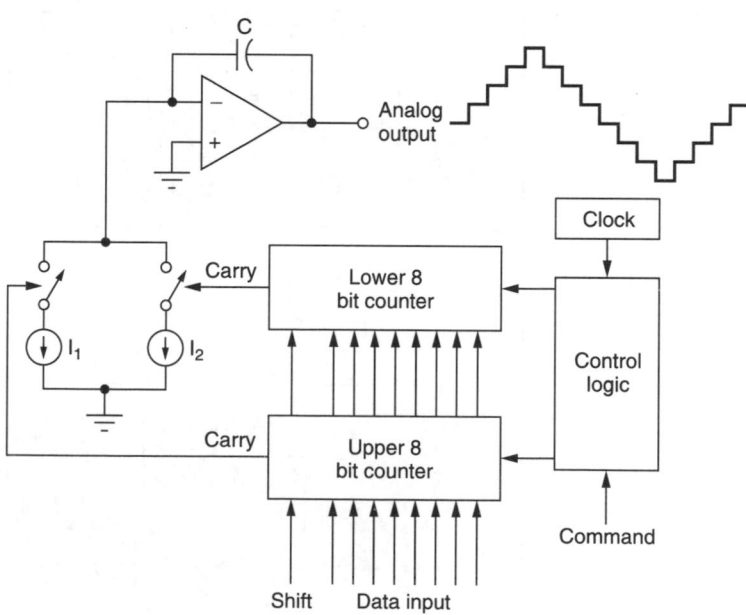

Figure 12.11 Block diagram of the CX20017 D/A converter.

Figure 12.12 Operating principle of the CX20017 D/A converter.

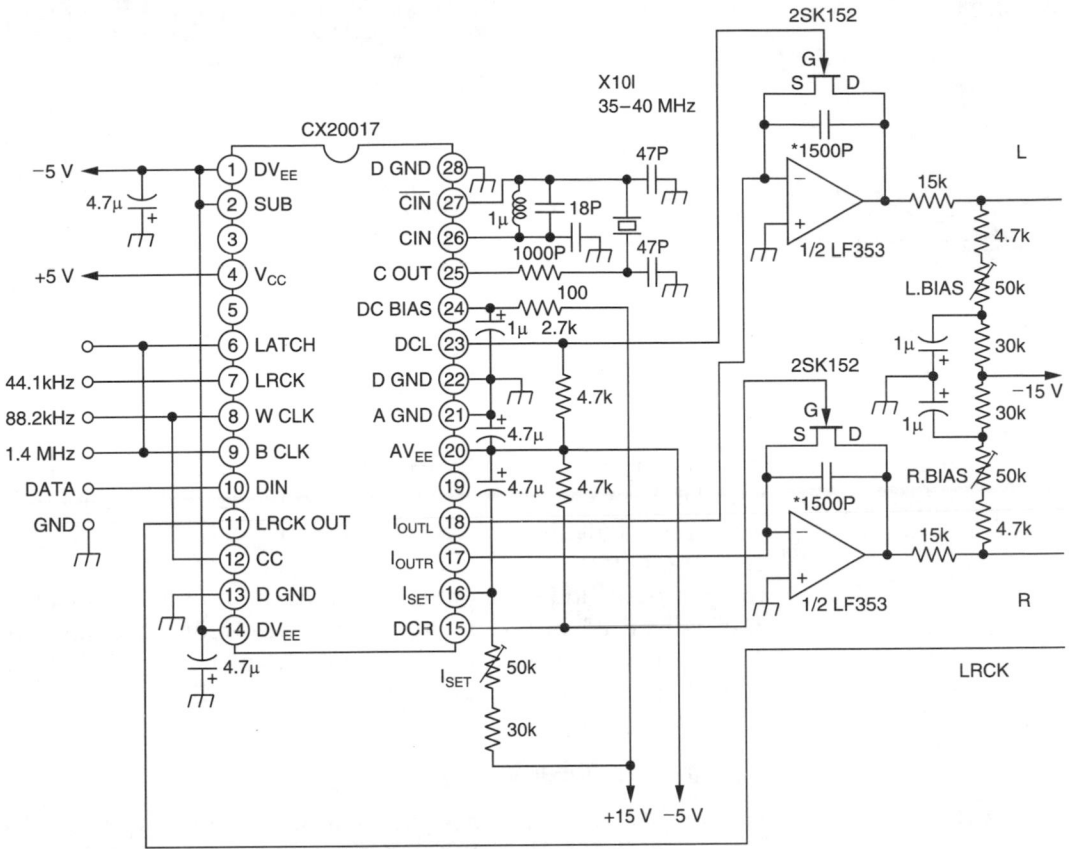

Figure 12.13

The digital filter is also used to create oversampling, calculating the intermediate values of the samples. An oversampling of $8 f_s$ or 352.8 kHz can be achieved. This increase in sampling rate gives an important reduction of quantization noise, which allows a more pure and analytical playback of music. Also, a lower order LPF can be used, improving the group delay and linearity in the audio range.

In order to cope with 18-bit and such a high conversion rate, great care must be taken in the designing of the D/A converter. To reduce the load imposed on the D/A converter, Sony developed an 'overlapped staggered D/A conversion system' (Figure 12.15). The basic idea is to use a digital filter circuit at $8 f_s$ output, combined with two D/A converters for each channel.

The conversion rate is $4 f_s$, so the digital filter output is at $2 \times 4 = 8 f_s$ oversampling. The even and odd samples for each channel

171

Figure 12.14 CXD-1144
digital filter.

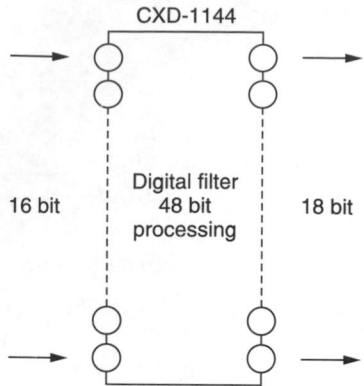

are applied to separate D/A converters. By adding the output of both D/A converters, corresponding to the formula $(L_1 + L_2)/2$, a staircase signal is obtained which corresponds to an $8 f_s$ over-sampled output signal. Since the outputs of the two D/A converters are added continuously, a maximum improvement of 3 dB is realized in quantization noise with reduced distortion. The output current is also doubled, improving the signal-to-noise ratio of the analog noise by a maximum of 6 dB.

High-density linear converter

Figure 12.16 shows a block diagram of a single-bit pulse D/A converter. The digital filter, CXD-1244, uses an internal 45-bit accumulator to perform accurate oversampling needed in the single-bit converter. The pulse D/A converter combines a third-order noise shaper and a Pulse Length Modulation (PLM) converter to produce a train of pulses at its output. By using a low-order LPF, the analog signal is obtained. A digital sync circuit is inserted between the D/A converter and the digital filter to prevent jitter of the digital signal.

Figure 12.15 Overlapped
staggered D/A conversion
system.

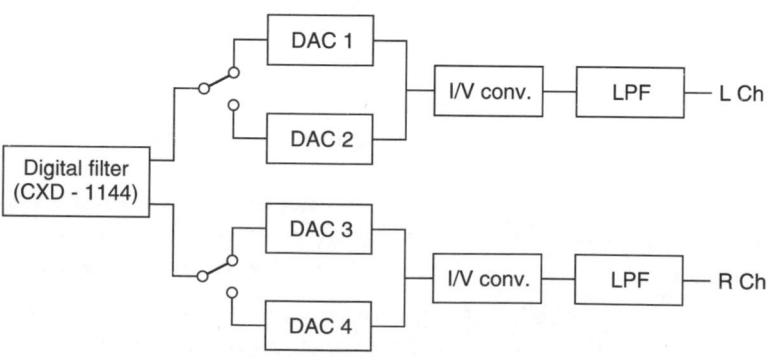

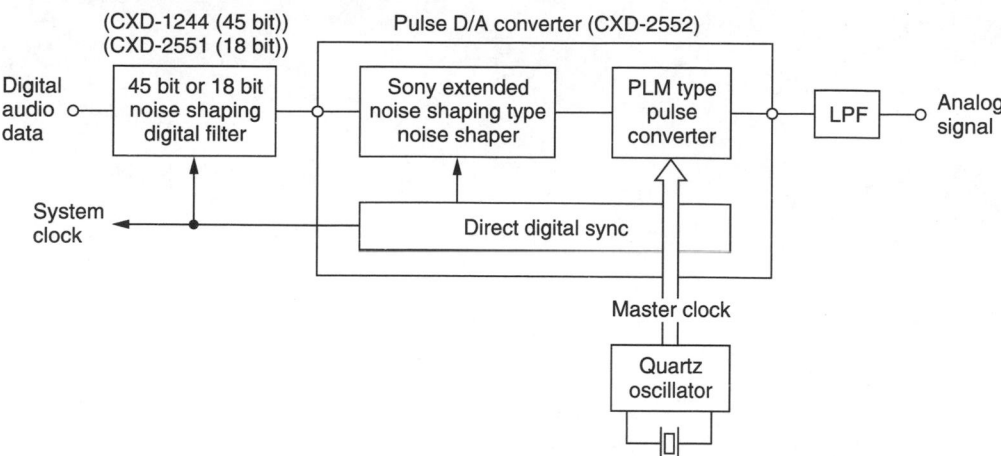

Figure 12.16 Single-bit pulse D/A converter.

Compared with conventional D/A converters, the high-density linear converter provides highly accurate D/A conversion with improved dynamic range and extremely low harmonic distortion.

PART THREE
Digital Audio Recording
Systems

13 Outline

The advantages of digital techniques were first realized in the field of magnetic recording. Analog magnetic recording creates significant deterioration of the original sound: using analog techniques, for example, it is difficult to obtain a flat frequency response at all signal levels. Furthermore, the signal-to-noise ratio is limited to some 70 dB, the sound deteriorates by speed variations of the recorder mechanism, crosstalk and print-through problems arise, and any additional copying deteriorates the characteristics even further. In addition to this, to keep the equipment within close specifications, as required in a professional environment, frequent and costly realignment and maintenance are required.

Digital magnetic recording, on the other hand, solves virtually all of these drawbacks. Recording of digital data, however, presents some specific problems:

- The required bandwidth is increased dramatically compared to the original signal.
- Specific codes must be used for recording (in contrast to the simple data codes mentioned before).
- Error-correction data must be recorded.
- Synchronization of the recorded data stream is necessary to allow for reconstruction of the recorded words.
- In contrast to analog recordings, editing is very complicated and requires complex circuits. For tape-cut editing, common

practice in the analog recording field, a very strong error-correction scheme together with interleaving are needed. Even then, very careful handling is a must; for instance, the tape cannot be touched with bare fingers.

Several different techniques have been developed, outlined in the following chapters.

14 Video PCM formats

Video 8 PCM format

In 1985, a new consumer video format was launched, called Video 8. Sony expected this format to gradually replace the older Betamax and VHS consumer video formats. The Video 8 format uses a much smaller cassette than older video formats, enabling construction of very small video recorders. Almost all major manufacturers in the field of consumer electronics are supporting this new format.

Audio information can be recorded on Video 8 recorders as either an FM signal, along with the picture, or as a PCM signal written in a section of the tape where no picture information is recorded. In some recorders, however, PCM data can be recorded in the video area, instead of the picture signal. By doing this, six channels of high-quality audio can be recorded on a tape.

The specification of the Video 8 PCM standard is summarized in Table 14.1.

A/D–D/A conversion

As only 8 bits per channel are used, audio characteristics would be poor if special measures were not taken. These measures include:

Figure 14.1 Video 8 cassette.

Table 14.1 Specification of Video 8 format

Item	Specification
Number of channels	2 (CH-1 = left, CH-2 = right)
Number of bits	8 bits/word
Quantization	linear
Digital code	2's complement
Modulation	bi-phase (FSK) 2.9 MHz, 5.8 MHz
Sampling frequency	31 250 kHz (PAL)/31 468.53 Hz (NTSC)
Bit transmission rate	5.8×10^6 bit s^{-1}
Video signal	PAL/NTSC
Number of PCM tracks	6

- audio compression and expansion for noise reduction purposes;
- 10-bit sampling;
- non-linear quantization by 10-bit to 8-bit compression and expansion.

These are illustrated, in a block diagram, in Figure 14.2. The characteristics of the noise reduction (NR) system are shown in Figure 14.3, while Figure 14.4 shows the characteristics of the non-linear encoder. The upper limit of the frequency response of the system

Figure 14.2 Block diagram of Video 8 signal processing.

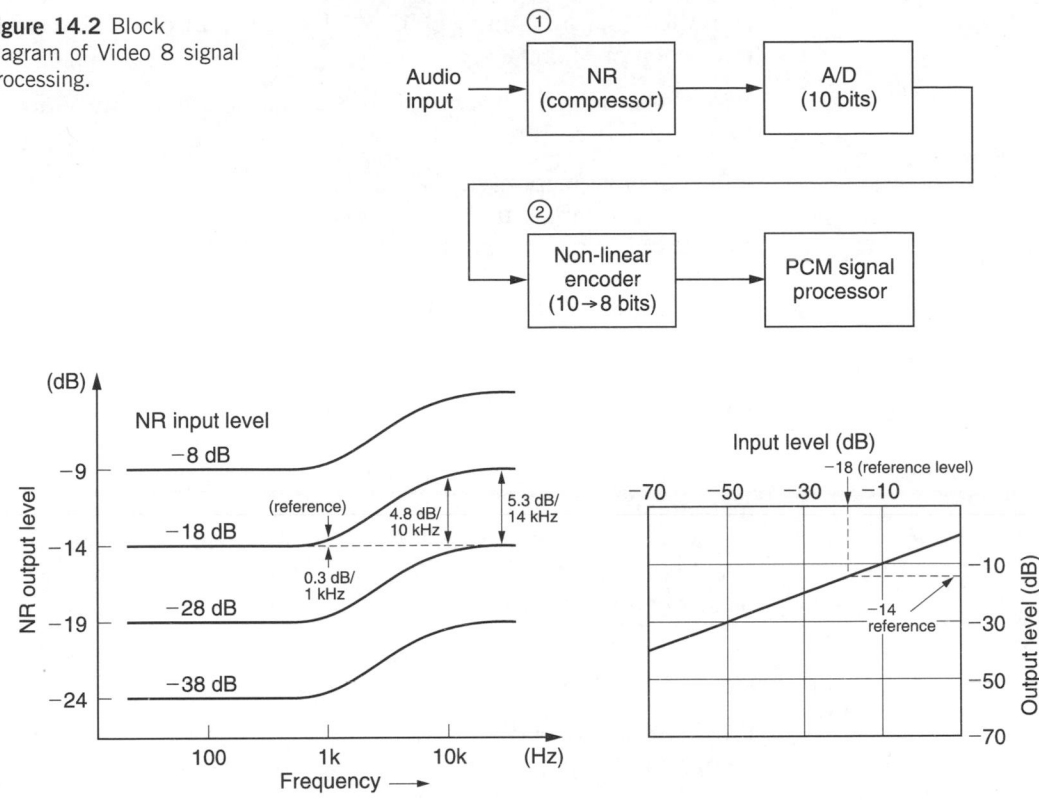

Figure 14.3 Noise reduction system characteristics.

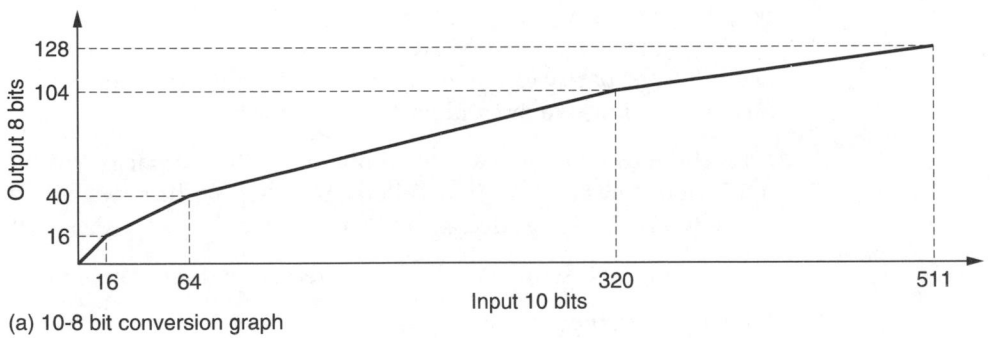

(a) 10-8 bit conversion graph

Y = X	(0 ≤ X < 16)
Y = [X/2] + 8	(16 ≤ X < 64)
Y = [X/4] + 24	(64 ≤ X < 320)
Y = [X/8] + 64	(320 ≤ X ≤ 511)

X: Input absolute value
Y: Output absolute value

(b) Conversion algorithm

Encode law	Input data	Output data
10 bits → 10 bits	0–15	0–15
10 bits → 9 bits	16–63	16–39
10 bits → 8 bits	64–319	40–103
10 bits → 7 bits	320–511	104–127

(c)

Figure 14.4 Non-linear encoder characteristics.

is limited to a maximum of 15 625 Hz, i.e., half of the sampling frequency of 31 250 Hz.

With these measures, typical audio characteristics of Video 8 PCM audio recordings are:

- frequency response: 20–15 000 Hz;
- dynamic range: more than 88 dB;
- sampling frequency: 31.5 kHz;
- quantization: 8-bit non-linear;
- wow and flutter: less than 0.005%.

Description of the format

One track on the section of tape used to record PCM audio information holds 157 blocks of data for a PAL machine and 132 blocks for an NTSC machine. Each block contains eight 8-bit data words, two 8-bit parity words (P and Q), one 16-bit error-detection word, one 8-bit address word and three sync bits.

So, one block comprises 107 bits, and each track comprises 16 799 bits in PAL mode and 14 017 bits in NTSC mode.

Error-correction and detection words are added as shown in Figure 14.5.

The error-correction code adopted for Video 8 PCM is a modified cross-interleaved code (MCIC) in which the code is composed of blocks which are related to the video fields. The version used is called improved MCIC, in which ICIC, the initial value necessary for parity calculation, can be any value and has numerous applications, such as identification words.

As eight audio data words are combined with two parity words, the Video 8 system is often called an 8w-2p coding system. A CRCC word is also added as an error detector.

In encoding, the sequence can be expressed as follows:

P-parity sequence

$$P(n) = Q(n+D) + \sum_{i=2}^{8} W_{i-2}(n+iD)$$

Q-parity sequence

$$Q(n+D) = P(n+d) + \sum_{i=2}^{8} W_{i-2}[n+i(D-d)+d]$$

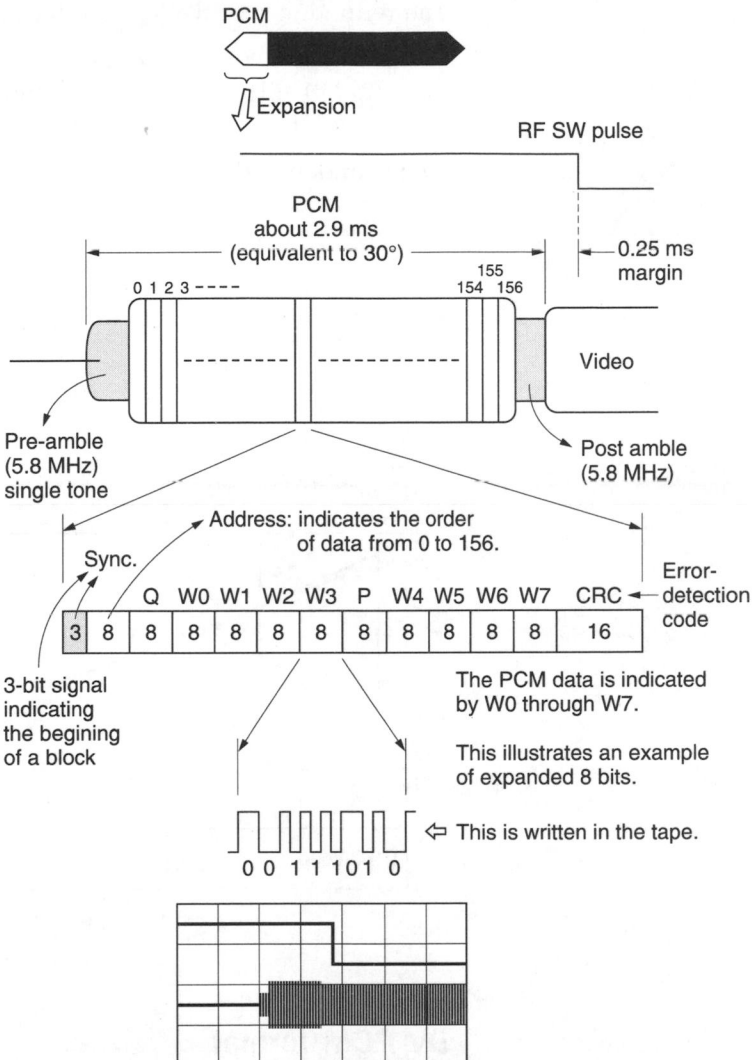

Figure 14.5 Error-correction encoding prior to recording.

where n is the block number ($0 < n < 157$ for PAL recordings, and $0 < n < 132$ for NTSC), D is the delay of the P-parity sequence which converts a burst error into random errors (17 for PAL, 1514 for NTSC) and d is the Q-parity sequence delay behind the P-parity sequence (three for PAL and 32 for NTSC).

The error-detection code is a 16-bit CRCC and its polynomial is given by:

$$g(X) = X^{16} + X^{11} + X^1 + 1$$

In decoding, the pointer method is used, which corrects an erroneous word using a pointer flag.

The redundancy of the Video 8 format is as follows:

there are $8 \times 8 = 64$ audio data bits and
$(2 \times 8) + 16 = 32$ error-correction and detection bits, plus
$1 \times 8 = 8$ address bits

So, redundancy R is:

$$\frac{32+8}{32+8+64} = 38.5\%$$

Words are interleaved onto the PCM section of tape as shown in Figure 14.6.

Figure 14.6 Data interleaving in Video 8 PCM.

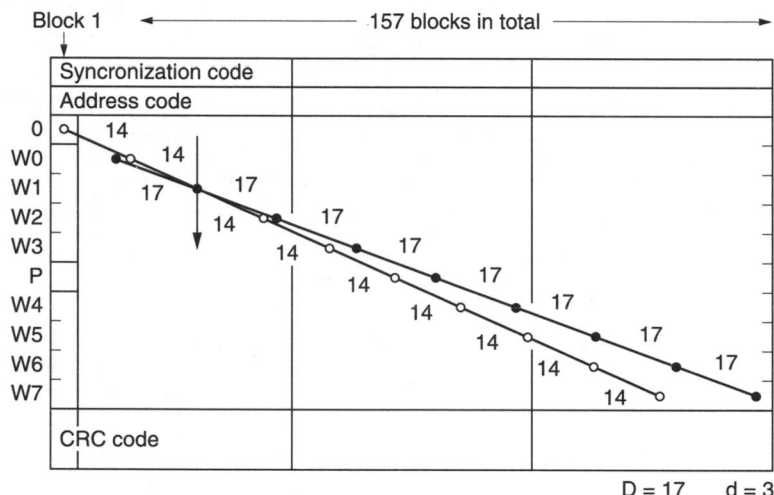

DV-PCM format

As of the mid-1990s, the analog 8 mm and Hi8 video formats were initially complemented and now rather superseded by digital video formats like DV and D8. The audio encoding of these new digital video is a logical evolution from the PCM formats used in their analog predecessors.

Recording format

In the DV recording format, one frame of video signal is recorded on 12 tracks for a PAL system and 10 tracks for NTSC. Each track consists of Insert and Track Information (ITI), Audio, Video and Sub Code.

The audio signal is recorded on two audio blocks, each of which consists of six consecutive tracks for a PAL set. For the NTSC set, each block consists of five consecutive tracks.

Encoding modes are defined in each audio block. They are classified by the type of sampling frequency, quantization and the number of channels in the audio block. Four types of audio mode are available in the DV format (Table 14.2).

Block and byte allocation

In the 48k, 44.1k and 32k modes, one audio channel signal is recorded in an audio block at its respective sampling frequency. The encoded data are represented by samples of 16-bit linear. The 16-bit encoded data are divided into two bytes and form an SD-2ch audio block.

In the 32k-2ch mode, two channels of the audio signal are recorded in an audio block at 32 kHz sampling frequency. The encoded data are expressed by two samples of 12-bit non-linear. Each pair of 12-bit encoded data is divided into three bytes and forms an SD-4ch audio block, as shown in Figure 14.8.

The audio blocks are allocated to channels CH1 and CH2 (Table 14.3). The track positions for CH1 and CH2 are different for PAL and NTSC systems.

Two different kinds of frame mode are available in DV format: the unlock mode and the lock mode. Those two modes are

Figure 14.7 Audio encoding modes.

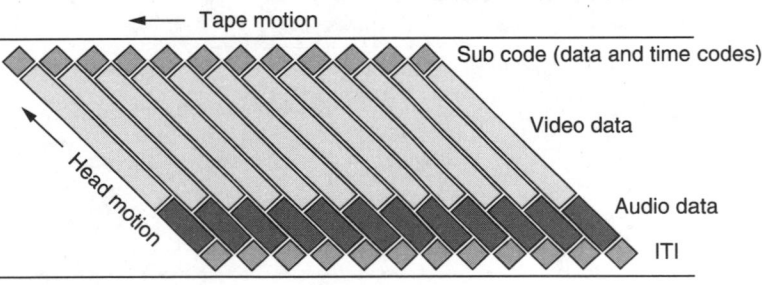

Tape motion

Sub code (data and time codes)

Video data

Head motion

Audio data

ITI

Table 14.2 Audio modes available in DV format

Encode mode	Ch no.	Sampling frequency (kHz)	Quantization
48k mode	2	48	16-bit, linear
44.1k mode	2	44.1	16-bit, linear
32k mode	2	32	16-bit, linear
32k-2ch mode	4	32	12-bit, non-linear

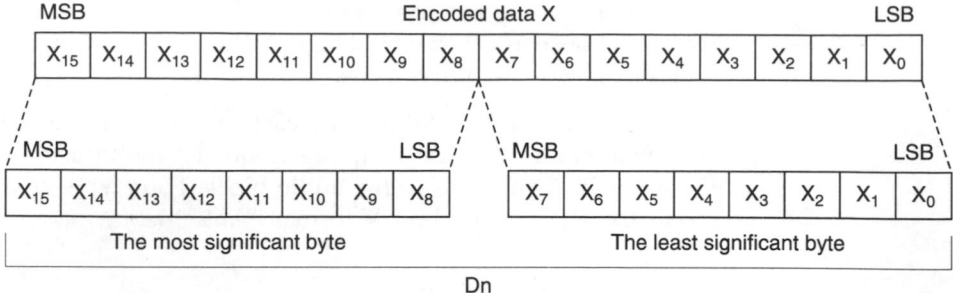

SD-2ch audio block

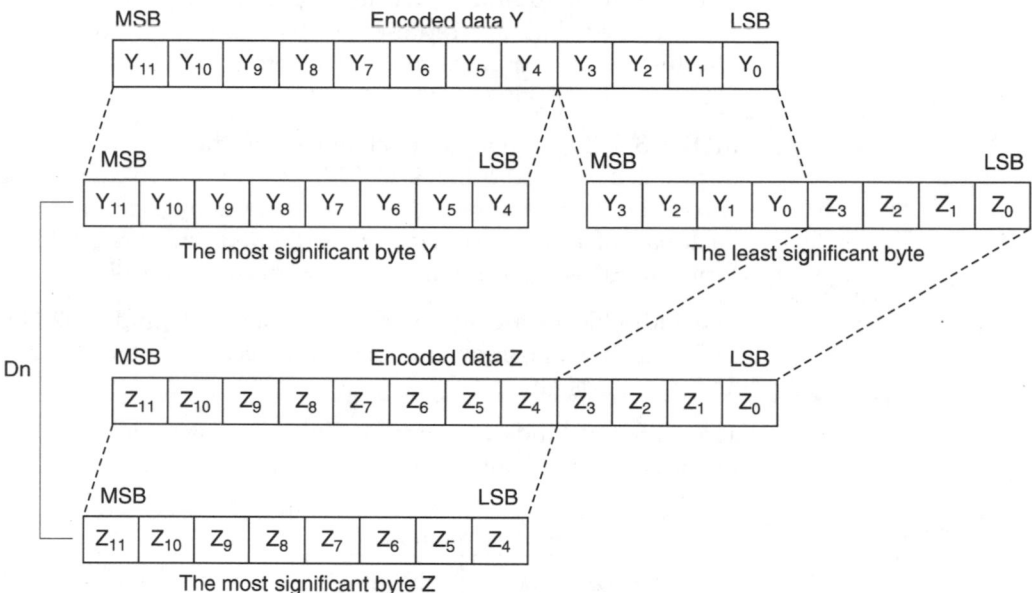

SD-4ch audio block

Figure 14.8 Audio channel allocation.

Table 14.3 Audio channels

		CH1	CH2
Track position	PAL	Tracks of 0–5	Tracks 6–11
	NTSC	Tracks 0–4	Tracks 5–9
Audio block	SD-2ch	48k mode	48k mode
		44.1k mode	44.1k mode
		32k mode	32k mode
	SD-4ch	32k-2ch mode	32k-2ch mode

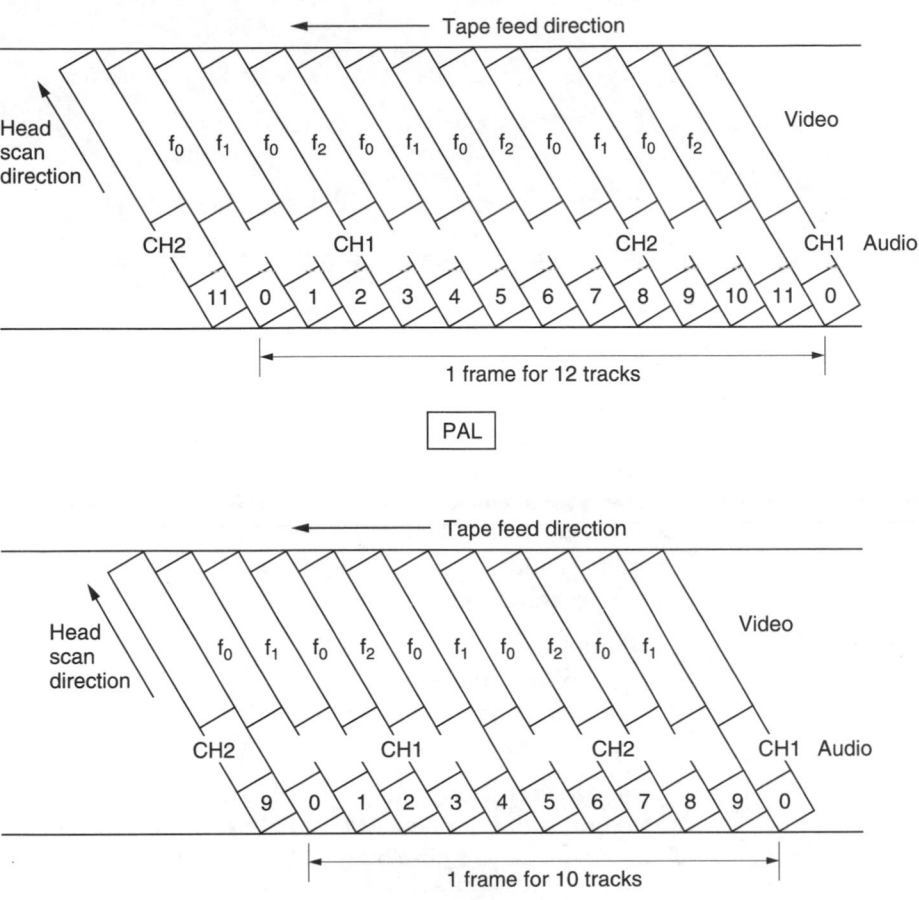

Figure 14.9 Audio frame modes.

differentiated by the synchronization relation between the audio signal and the video signal.

Lock mode

The audio clock used in the sampling process of the audio signal is precisely synchronized to the video clock. The main advantage is that the exact same number of audio samples is recorded per frame. Lock mode is only available for sampling frequencies of 48 and 32 kHz. The number of audio samples per frame is different for PAL and NTSC systems (Table 14.4).

187

Table 14.4 Lock mode

System	Encode mode	Samples per frame
PAL	48k mode	1920
	32k mode	1280
	32k-2ch mode	
NTSC	48k mode	Frame 1: 1600
		Frames 2–5: 1602
	32k mode	Frames 1 and 8: 1066
	32k-2ch mode	Frames 2–7 and 9–15: 1068

Unlock mode

In unlock mode, there is no precise synchronization between audio and video signal. Instead of an exact number of samples per frame as in lock mode, the audio sample clock keeps track of the video clock with a variation of maximum ±25 audio samples per frame. Unlock mode is available for every sampling frequency and the variation of audio samples per frame is as shown in Table 14.5.

A complete frame (Video, Audio and Subcode data) in DV format consists of 175 blocks of 90 bytes. Each block is composed of 2 bytes of synchronization (SYNC), 3 bytes of identification code (ID), a 1-byte header (H), 76 bytes of data and 8 bytes of parity.

The audio sector consists of nine blocks of 76 data bytes, giving a maximum data rate of 12 (10) tracks/frame × 9 blocks/track × 76 bytes/block × 8 bits/byte × 25 (30) frames/s = 1.6416 Mbps for a PAL (NTSC) system. The maximum audio rate of 1.536 Mbps is reached when 48k mode (2ch × 48/1000 × 16) or 32k-2ch mode (4ch × 32/1000 × 12) are used. This leaves 105.6 kbps for audio auxiliary data (AAUX). The audio auxiliary data are the parameters of information of the recorded data and consist of two data types (Tables 14.6 and 14.7).

Table 14.5 Unlock mode

System	Encode mode	Max. samples/ frame	Min. samples/ frame
PAL	48k mode	1944	1896
	44.1k mode	1786	1742
	32k mode	1296	1264
	32k-2ch mode	1296	1264
NTSC	48k mode	1620	1580
	44.1k mode	1489	1452
	32k mode	1080	1053
	32k 2ch mode	1080	1053

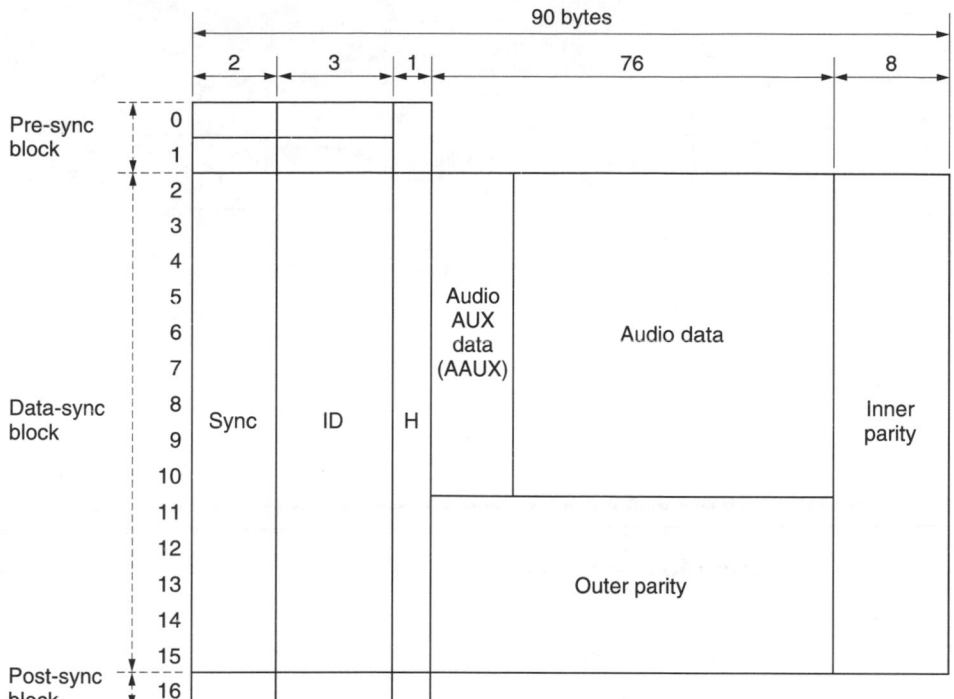

Audio sector format

Figure 14.10 Audio sector format.

Table 14.6 AAUX source pack

Encoding parameters	SMP	Sampling frequency
	QU	Quantization
	CHN	Number of channel in audio block
	AF SIZE	Number of audio samples per frame
	LF	State of the audio processing
Video signal information	50/60	50- or 60-field system
	STYPE	Video signal type
Emphasis state	EF	Emphasis on/off
	TC	Time constant
Audio block contents	AUDIO MODE	

Table 14.7 AAUX source control pack

Copy generation	CGMS	Copy generation management system
Recording starting point	REC ST	Recording start point or not
Recording end point	REC END	Recording end point or not
Recording mode	REC MODE	

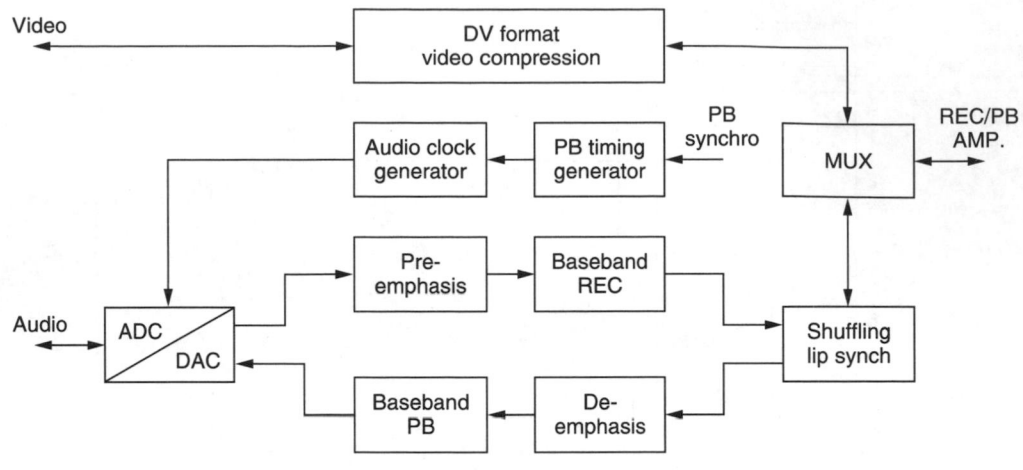

Block diagram of DV signal processing

Figure 14.11 DV signal processing.

15 Digital audio tape (DAT) format

Although digital audio processors have been developed and used for many years, using conventional video recorders to store high-quality audio information, it is inevitable that some form of tape mechanism be required to do the job in a more compact way. Two main formats have been specified.

The first format, known as rotary-head, digital audio tape (R-DAT), is based on the same rotary-head principle as a video recorder, and so has the same limitations in portability. The second format, known as stationary-head, digital audio tape (S-DAT), was developed under the name DCC, mentioned in the short history chapter.

R-DAT

One important difference between standard video recorder and R-DAT techniques is that in a video recorder the recorded signal is continuous; two heads on the drum make contact with the tape for 180° each (i.e., the system is said to have a 180° wrap angle, as shown in Figure 15.2a), or 221° each (a 221° wrap angle, as in Figure 15.2b). In the R-DAT system, where the digital audio signal is time-compressed, meaning that the heads only need to make contact with the tape for a smaller proportion of the time (actually 50%), a smaller wrap angle may be used (90°, as shown in Figure 15.2c).

Figure 15.1 DAT
mechanism.

Figure 15.2

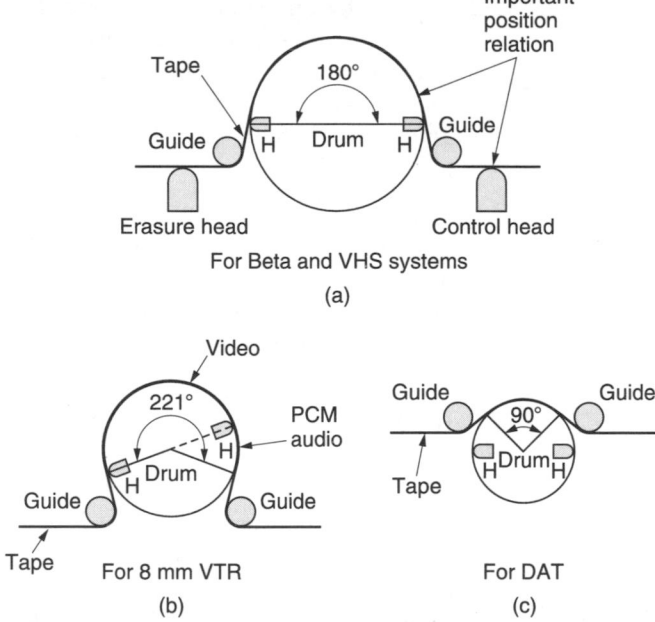

This means only a short length of tape is in contact with the drum
at any one time. Tape damage is consequently reduced, and only
a low tape tension is necessary with resultant increase in head
life.

The R-DAT standard specifies three sampling frequencies:

- 48 kHz; this frequency is mandatory and is used for recording and playback.
- 44.1 kHz; this frequency, which is the same as for CD, is used for playback of pre-recorded tapes only.
- 32 kHz; this frequency is optional and three modes are provided.

32 kHz has been selected as it corresponds with the broadcast standard.

Quantization:

- A 16-bit linear quantization is the standard for all three sampling rates.
- A 12-bit non-linear quantization is provided for special applications such as long play mode at reduced drum speed, 1000 rpm (mode III) and U-channel applications.

Figure 15.3 shows a simplified R-DAT track pattern.

The standard track width is 13.591 μm, the track length is 23.5 mm and the linear tape speed is 8.1 mm s^{-1}. The tape speed of the analog compact cassette (TM) is 47.6 mm s^{-1}. This results in a packing density of 114 Mbit s^{-1} m^{-2} (see Table 15.1).

The R-DAT format specifies a track width of only 13.6 μm, but the head width is about 1.5 times this value, around 20 μm. A procedure known as overwrite recording is used, where one head partially records over the track recorded by the previous head, illustrated in Figure 15.4. This means that as much tape as possible is used – rotating-head recorders without this overwrite record facility must leave a guardband between each track on the

Figure 15.3 Simplified R-DAT track pattern.

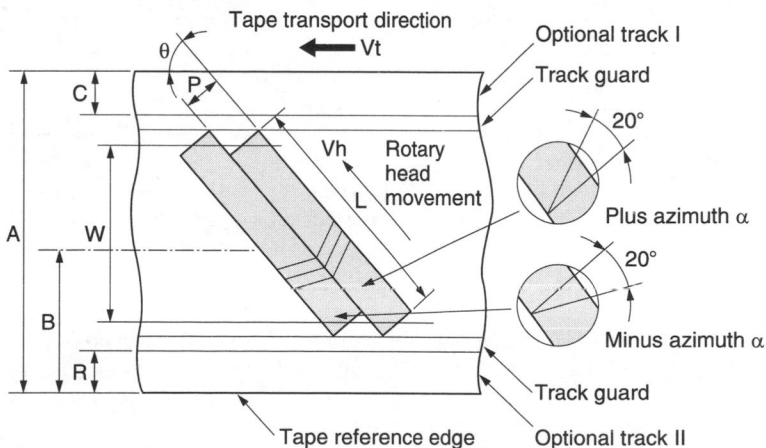

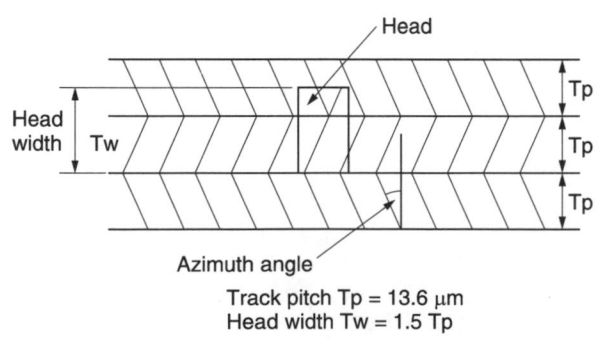

Table 15.1

Mode \ Item	DAT (REC/PB mode)				Pre-recorded tape (PB only)	
	Standard	Option 1	Option 2	Option 3	Normal track	Wide track
Channel number (CH)	2	2	2	4	2	2
Sampling frequency (kHz)	48	32	32	32	44.1	
Quantization bit number (bit)	16 (Linear)	16 (Linear)	12 (Nonlinear)	12 (Nonlinear)	16 (Linear)	16 (Linear)
Linear recording density (kbit in^{-1})	61.0	61.0			61.0	61.1
Surface recording density (Mbit in^{-2})	114	114			114	76
Transmission rate (Mbit s^{-1})	2.46	2.46	1.23	2.46	2.46	
Subcode capacity (kbit s^{-1})	273.1	273.1	136.5	273.1	273.1	
Modulation system	8–10 conversion					
Correction system	Double Reed–Solomon code					
Tracking system	Area sharing ATF					
Cassette size (mm)	73 × 54 × 10.5					
Recording time (min)	120	120	240	120	120	80
Tape width (mm)	3.81					
Tape type	Metal powder					Oxide tape
Tape thickness (μm)	13 ± 1 μm					
Tape speed (mm s^{-1})	8.15	8.15	4.075	8.15	8.15	12.225
Track pitch (μm)	13.591				13.591	20.41
Track angle	6°22′59.5″					6°23′29.4″
Standard drum specifications	30 mm diameter 90° wrap					
Drum rotations (rpm)	2000		1000	2000	2000	
Relative speed (m s^{-1})	3.133		1.567	3.133	3.133	3.129
Head azimuth angle	± 20°					

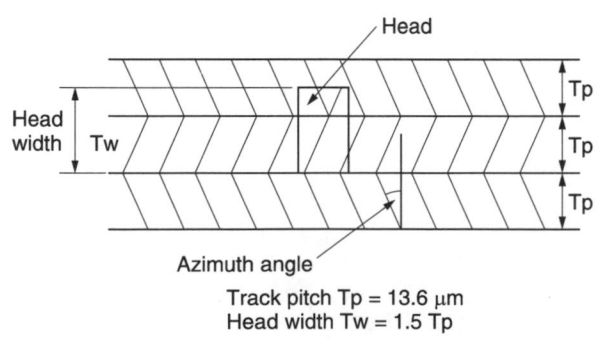

Figure 15.4 Overwrite recording is used to ensure each track is as narrow as possible and no guard-band is required.

Track pitch Tp = 13.6 μm
Head width Tw = 1.5 Tp

Track pattern

tape. Because of this, recorders using overwrite recording techniques are sometimes known as guard-bandless. To prevent crosstalk on playback (as each head is wide enough to pick up all of its own track and half of the next), the heads are set at azimuth angles of ±20°. This enables, as will be explained later, automatic track following (ATF).

These overwrite record and head azimuth techniques are fairly standard approaches to rotating-head video recording, and are used specifically to increase the recording density.

Figure 15.5 shows the R-DAT track format on the tape, while Table 15.2 shows the track contents. Table 15.2 lists each part of a track and gives the recording angle, recording period and number of blocks allocated to each part. Frequencies of these blocks which are not of a digital-data form are also listed.

As specified in the standard, a head drum of 30 mm diameter is applied and rotates at a speed of 2000 rpm. However, in future applications smaller drums with appropriate speeds can be used. At this size and speed, the drum has a resistance to external disturbances similar to that of a gyroscope.

Under these conditions, the 2.46 Mbit s^{-1} signal to be recorded, which includes audio as well as many other types of data, is compressed by a factor of 3 and processed at 7.5 Mbit s^{-1}. This enables the signal to be recorded continuously.

In order to overcome the well-known low-frequency problems of coupling transformers in the record/playback head, an 8/10 modulation channel code converts the 8-bit signals to 10-bit signals.

Figure 15.5 R-DAT tape track format.

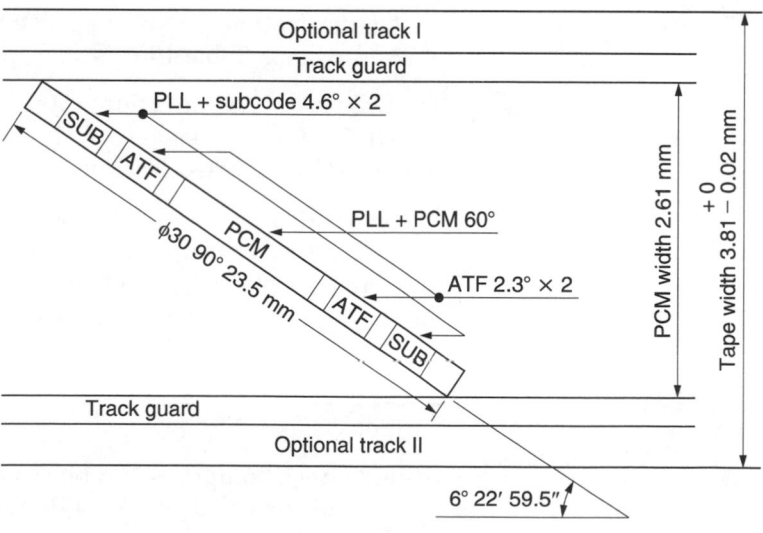

195

Table 15.2

1	2	3 (SUB)	4	5	6	7	8	9 (PCM)	10	11	12	13	14 (SUB)	15	16

		Frequency	Angle* (deg)	Number of blocks	Time (µs)
1	MARGIN	$1/2\,f_{ch}$	5.051	11	420.9
2	PLL (SUB)	$1/2\,f_{ch}$	0.918	2	76.5
3	SUB-1		3.673	8	306.1
4	POST AMBLE	$1/2\,f_{ch}$	0.459	1	38.3
5	IBG	$1/6\,f_{ch}$	1.378	3	114.8
6	ATF		2.296	5	191.3
7	IBG	$1/6\,f_{ch}$	1.378	3	114.8
8	PLL (PCM)	$1/2\,f_{ch}$	0.918	2	76.5
9	PCM		58.776	128	4898.0
10	IBG	$1/6\,f_{ch}$	1.378	3	114.8
11	ATF		2.296	5	191.3
12	IBG	$1/6\,f_{ch}$	1.378	3	114.8
13	PLL (SUB)	$1/2\,f_{ch}$	0.918	2	76.5
14	SUB-2		3.673	8	306.1
15	POST AMBLE	$1/2\,f_{ch}$	0.459	1	38.3
16	MARGIN	$1/2\,f_{ch}$	5.051	11	420.9
	Total		90	196	7500

Recording density 61.0 kbit in^{-1}. f_{ch}=9.408 MHz.
*Values for 30 mm diameter, 90° wrap angle, 2000 rpm cylinder.

This channel coding also gives the benefit of reducing the range of wavelengths to be recorded. The resultant maximum wavelength is only four times the minimum wavelength. This allows overwriting, eliminating the need for a separate erase head.

The track outline is given in Figure 15.5. Each helical track is divided into seven areas, separated by inter-block gaps. As can be seen, each track has one PCM area, containing the modulated digital information (audio data and error codes), and is 128 blocks of 288 bits long. Table 15.2 lists all track parts of a track.

The PCM area is separated from the other areas by an inter-block gap (IBG), three blocks long. At both sides of the PCM area, two ATF areas are inserted, each five blocks long.

Again, an IBG is inserted at both ends of the track, separating the ATF areas from the sub-1 and sub-2 areas (subcode areas), each eight blocks long. These subareas contain all the information on time code, tape contents, etc.

Then at both track ends a margin block is inserted, 11 blocks long, and is used to cover tolerances in the tape mechanism and head position.

A single track comprises 196 blocks of data, of which the major part is made up of 128 blocks of PCM data. Other important parts are the subcode blocks (sub-1 and sub-2, containing system data, similar to the CD subcode data), automatic track-finding (ATF) signals (to allow high-speed search) and the IBGs around the ATF signals (which means that the PCM and subcode information can be overwritten independently without interference to surrounding areas). Parts are recorded successively along the track.

The PCM area format is shown in Table 15.3. PCM and subcode parts comprise similar data blocks, shown in Figure 15.6. Each block is 288 bits long.

Each block comprises eight synchronization bits, the identification word ($W1$, 8 bits), the block address word ($W2$, 8 bits), 8-bit parity word and 256-bit (32×8-bit symbol) data. The ID code $W1$ contains control signals related to the main data. Table 15.4 shows the bit assignment of the ID codes. $W2$ contains the block address. The most significant bit (MSB) of the $W2$ word defines whether the data block is of PCM or subcode form. Where the MSB is zero, the block consists of PCM audio data, and the remainder of word $W2$, i.e., 7 bits, gives the block address within

Table 15.3 PCM area format

	Sub-data area 1			Main data area (PCM)											Sub-data area 2	
	8 blocks			128 blocks (8 × 16 blocks)											8 blocks	

Sync	Main ID														Main ID parity	Main data
Sync	8-bit							8-bit							8-bit	32 symbols

		W1							*W2*							M parity	Main data
		B7	B6	B5	B4	B3	B2	B1	B0	B7	B6	B5	B4	B3	B2	B1	B0	M parity	Main data
Sync	Format ID		ID1		Frame address				0			Block address (xxxx000)					M parity	Main data
Sync									0			Block address (xxxx001)					M parity	Main data
Sync	ID2		ID3		Frame address				0			Block address (xxxx010)					M parity	Main data
Sync									0			Block address (xxxx011)					M parity	Main data
Sync	ID4		ID5		Frame address				0			Block address (xxxx100)					M parity	Main data
Sync									0			Block address (xxxx101)					M parity	Main data
Sync	ID6		ID7		Frame address				0			Block address (xxx110)					M parity	Main data
Sync									0			Block address (xxxx111)					M parity	Main data

8 block = 1 unit

MSB   LSB   MSB   LSB

Figure 15.6 PCM and subcode data blocks.

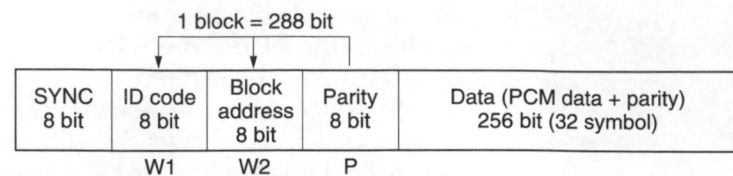

1 block = 288 bit

SYNC 8 bit	ID code 8 bit	Block address 8 bit	Parity 8 bit	Data (PCM data + parity) 256 bit (32 symbol)
	W1	W2	P	

Parity : $P = W1 \oplus W2$ ($\oplus$: MOD 2)

Block address : Corresponds to the PCM data block. The MSB bit indicates an ID bit (subcode or PCM data)

MSB   LSB

0 | | | | | | : PCM block (block address = 7 bits)

Table 15.4 Bit assignment of ID codes

	Usage	Bit assignment	
ID1	Emphasis	B5 0 0	B4 0: Off 1: 50/15 µs
ID2	Sampling frequency	B7 0 0 1	B6 0: 48 kHz 1: 44.1 kHz 0: 32 kHz
ID3	Number of channels	B5 0 0	B4 0: 2 channels 1: 4 channels
ID4	Quantization	B7 0 0	B6 0: 16 bits linear 1: 12 bits non-linear
ID5	Track pitch	B4 0 0	B4 0: Normal track mode 1: Wide track mode
ID6	Digital copy	B7 0 1 1	B6 0: Permitted 0: Prohibited 1: Permitted only for the first generation
ID7	Pack	B5	B4: Pack contents

the track. The 7 bits therefore identify the absolute block address (as 2^7 is 128).

On the other hand, when the MSB of word *W2* is 1, the block is of subcode form and data bits in the word are as shown in Figure 15.7, where a further 3 bits are used to extend the *W1* word subcode identity code, and the four least significant bits give the block address.

The *P*-word, block parity, is used to check the validity of the *W1* and *W2* words and is calculated as follows:

Figure 15.7 Subcode data blocks.

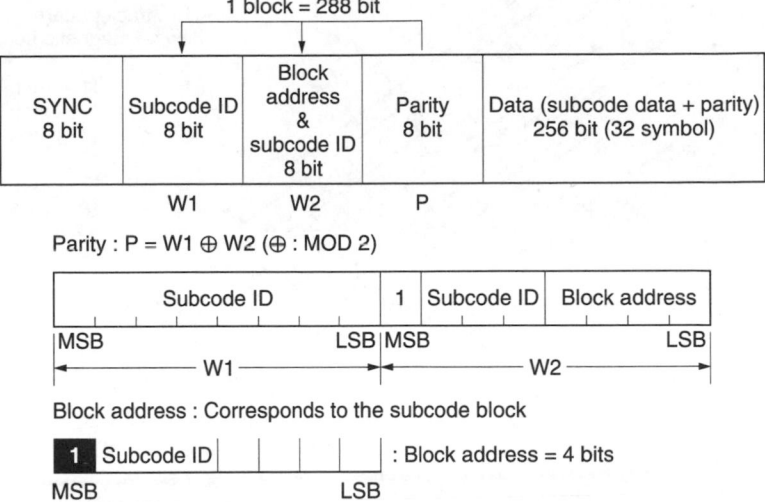

$$P = W1 \oplus W2$$

where $\oplus$ signifies modulo-2 addition, as explained in Appendix 1.

Automatic track following

In the R-DAT system, no control track is provided. In order to obtain correct tracking during playback, a unique ATF signal is recorded along with the digital data.

The ATF track pattern is illustrated in Figure 15.8. One data frame is completed in two tracks and one ATF pattern completed in two frames (four tracks). Each frame has an A and a B track. A tracks are recorded by the head with +20° azimuth and B tracks are recorded by the head with −20° azimuth.

The ATF signal pattern is repeated over subsequent groups of four tracks. The frequencies of the ATF signals are listed in Figure 15.8. The key to the operation lies in the fact that different frames hold different combinations and lengths. Furthermore, the ATF operation is based upon the use of the crosstalk signals, picked up by the wide head, which is 1.5 times the track width, and the azimuth recording. This method is called the area divided ATF.

As shown in Figure 15.8, the ATF uses a pilot signal f_1; sync signal 1, f_2; sync signal 2, f_3; and erase signal f_4. When the head passes along the track in the direction of the arrow (V-head) and detects an f_2 or f_3 signal, the six adjacent pilot signals f_1 on both sides are

199

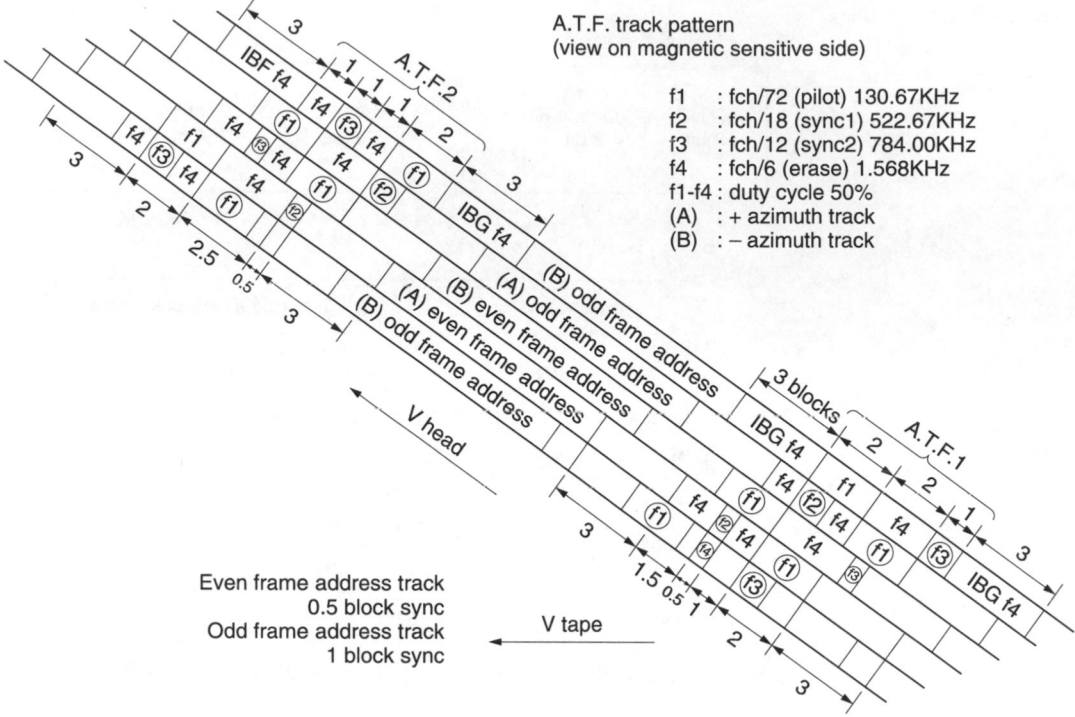

Figure 15.8 ATF signal frequencies.

immediately compared, which results in a correction of the tracking when necessary.

The f_2 and f_3 signals thus act as sync signals to start the ATF servo operation.

The f_1 signal, a low-frequency signal, i.e., 130.67 kHz, is used as low-frequency signals are not affected by the azimuth setting, so crosstalk can be picked up and detected from both sides. The pilot signal f_1 is positioned so as not to overlap through the head scans across three successive tracks.

Error correction

As with any digital recording format, the error-detection and error-correction scheme is very important. It must detect and correct the digital audio data, as well as subcodes, ID codes and other auxiliary data.

Types of errors that must be corrected are burst errors – dropouts caused by dust, scratches and head clogging – and random errors

– caused by crosstalk from an adjacent track, traces of an imperfectly erased or overwritten signal, or mechanical instability.

Error-correction strategy

In common with other digital audio systems, R-DAT uses a significant amount of error-correction coding to allow error-free replay of recorded information. The error-correction code used is a double-encoded Reed–Solomon code.

These two Reed–Solomon codes produce C1 (32, 28) and C2 (32, 26) parity symbols, which are calculated on G_F (2^8) by the polynomial:

$$g(x) = x^8 + x^4 + x^3 + x^2 + 1$$

C1 is interleaved on two neighbouring blocks, while C2 is interleaved on one entire track of PCM data every four blocks (see Figure 15.9 for the interleaving format).

Figure 15.9 ECC interleaving format.

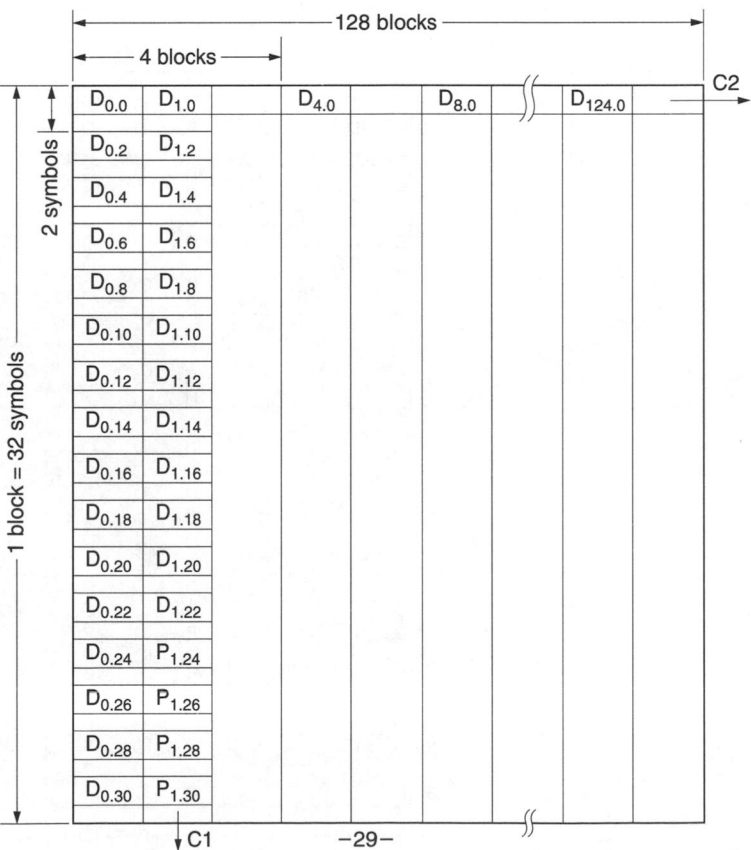

201

In order to perform C1 ↔ C2 decoding/encoding, one track worth of data must be stored in memory.

One track contains 128 blocks consisting of 4096 (32 × 128) symbols. Of these, 1184 symbols (512 symbols C1 parity and 672 symbols C2 parity) are used for error correction, leaving 2912 data symbols (24 × 104).

In fact, C1 encoding adds four symbols of parity to the 28 data symbols C1 (32, 28), while C2 encoding adds six symbols of parity to every 26 PCM data symbols C2 (32, 26).

The main data allocation is shown in Figure 15.10.

This double-Reed–Solomon code gives the format a powerful correction capability for random errors.

PCM data interleave

In order to cope with burst errors, i.e., head clogging, tape dropouts, etc. PCM data are interleaved over two tracks called

Figure 15.10 Data allocation.

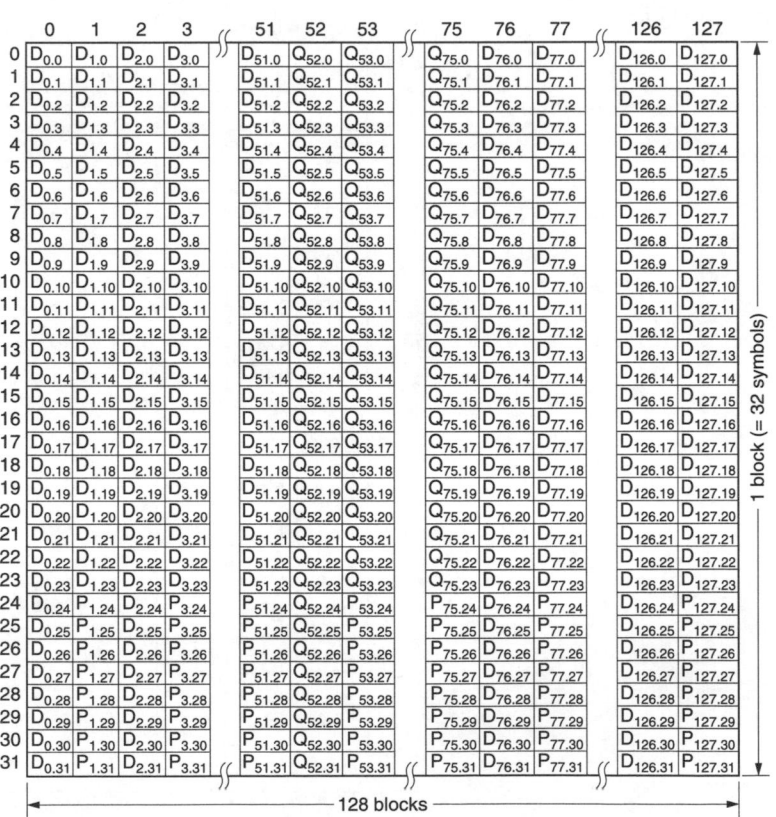

one frame, effectively turning burst errors into random errors which are correctable using the Reed–Solomon technique already described.

To interleave the PCM data, the contents of two tracks have first to be processed in a memory. The memory size required for one PCM interleave block is: (128×32) symbols $\times 8$ bits $\times 2$ tracks = 65.536 bits, which means a 128-bit memory is required.

The symbols are interleaved, based on the following method, according to the respective number of the audio data symbol. The interleaving format depends on whether a 16-bit or 12-bit quantization is used. The interleave format discussed here is for 16-bit quantization, the most important format.

One 16-bit audio data word indicated as A_i or B_i is converted to two audio data symbols each consisting of 8 bits. The audio data symbol converted from the upper 8 bits of A_i or B_i is expressed as A_{iu} or B_{iu}. The audio data symbol converted from the lower 8 bits of A_i or B_i is expressed as A_{il} or B_{il}. Note: A stands for the left channel, B for the right channel.

If the audio data symbol is equal to A_{iu} or A_{il}, let $a = 0$.
If the audio data symbol is equal to B_{iu} or B_{il}, let $a = 1$.
If the audio data symbol is equal to A_{iu} or B_{iu}, let $u = 0$.
If the audio data symbol is equal to A_{il} or B_{iu}, let $u = 1$.

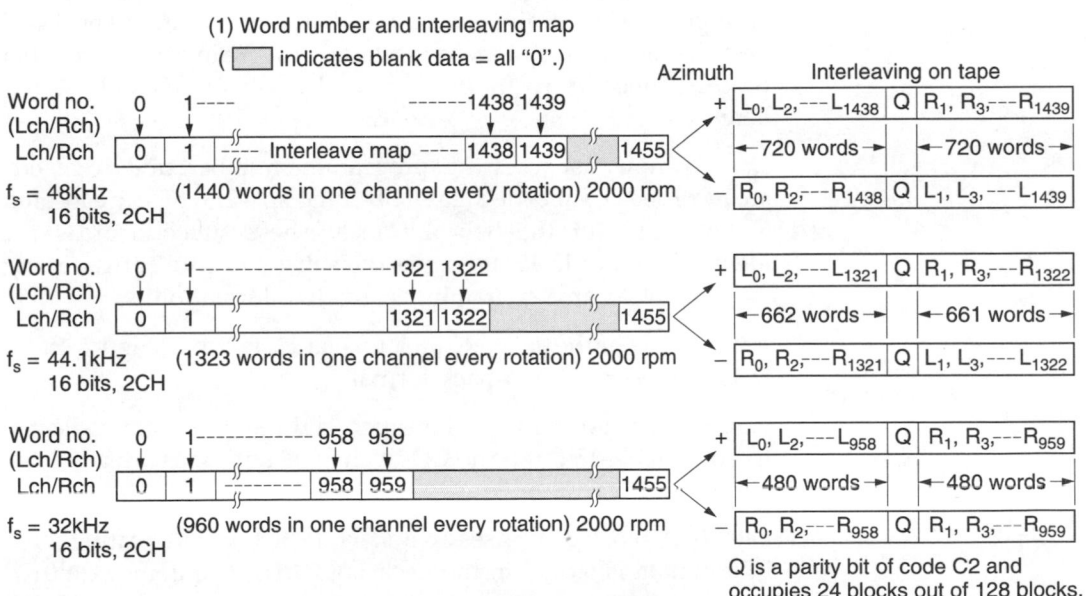

Figure 15.11 PCM data interleave format.

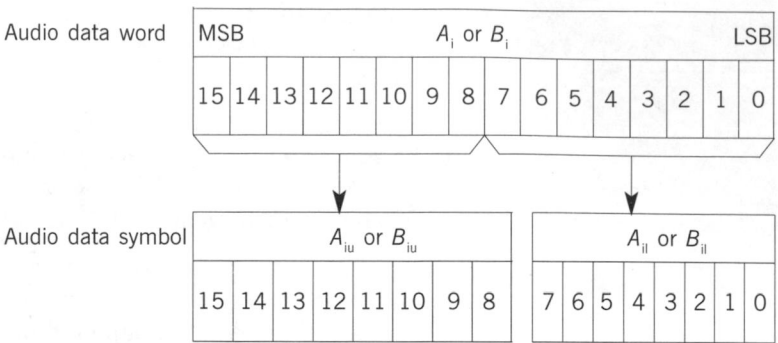

Tables 15.5a and b represents an example of the data assignment for both tracks (+ azimuth and − azimuth) respectively, for 16-bit sampled data words.

Subcode

The data subcode capacity is about four times that of a CD and various applications will be available in the future. A subcode format which is essentially the same as the CD subcode format is currently specified for pre-recorded tapes.

The most important control bits, such as the sampling frequency bit and copy inhibit bit, are recorded in the PCM-ID area, so it is impossible to change these bits without rewriting the PCM data. As the PCM data are protected by the main error-correction process, subcodes requiring a high reliability are usefully stored here.

Data to allow fast accessing, programme number, time code, etc. are recorded in subcode areas (sub-1 and sub-2) which are located at both ends of the helical tracks. These subcode areas are identical. Figure 15.12 illustrates the sub-1 and sub-2 areas, along with the PCM area containing subcode information.

An example of the subcode area format is shown in Figure 15.13. Data are recorded in a pack format.

Figure 15.14 shows the pack format, and the pack item codes are listed in Table 15.6. All the CD-Q channel subcodes are available to be used.

Each pack block comprises an item code of 4 bits, indicating what information is stored in the pack data area. The item code 0100 indicates that the related pack data is a table of contents (TOC) pack. This TOC is recorded repeatedly throughout the tape, in

Figure 15.12

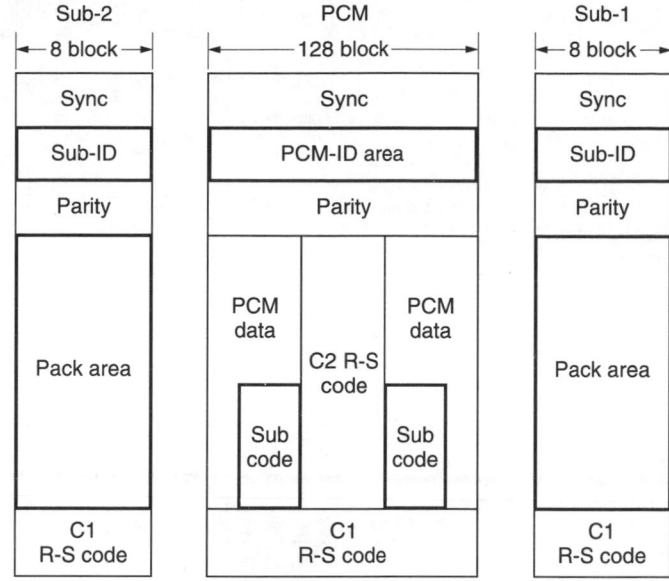

Figure 15.13 Subcode area format.

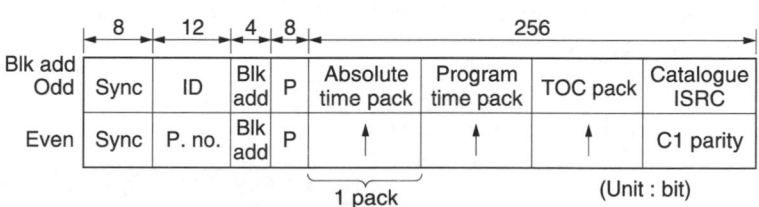

Figure 15.14 Pack format.

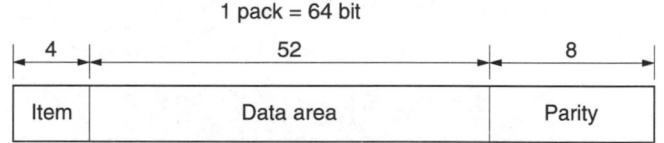

order to allow high-speed access and search (at 200 times normal speed). Every subcode datablock is controlled by an 8-bit C1 parity word allowing appropriate control of data validity.

Subcode data in the subarea can be rewritten or modified independently from the PCM data.

Figure 15.15 shows an example of subcode information for pre-recorded tape. The figure shows the use of different codes and pack data on a tape, such as programme time, absolute time, programme number, etc.

Table 15.5a

Block address

Symbol number	0	1	2	3	35	50	51	52	75	76	77	78	79	111	126	127
0	A 0u	A 832u	A 2u	A 834u	A 866u	A 50u	A 822u	Q 52.0	Q 75.0	B 1u	B 833u	B 3u	B 835u	B 867u	B 51u	B 883u
1	A 52u	A 884u	A 54u	A 886u	A 918u	A 102u	A 934u	Q 52.1	Q 75.1	B 53u	B 885u	B 55u	B 887u	B 919u	B 103u	B 935u
2	A 01	A 8321	A 21	A 8341	A 8661	A 581	A 8821	Q 52.2	Q 75.2	B 11	B 8331	B 31	B 8351	B 8671	B 511	B 8831
3	A 521	A 8841	A 541	A 8861	A 9181	A 1021	A 9341	Q 52.3	Q 75.3	B 531	B 8851	B 551	B 8871	B 9191	B 1031	B 9351
4	A 104u	A 936u	A 106u	A 938u	A 970u	A 154u	A 986u	Q 52.4	Q 75.4	B 105u	B 937u	B 107u	B 939u	B 971u	B 155u	B 987u
5	A 156u	A 988u	A 158u	A 990u	A1022u	A 206u	A1038u	Q 52.5	Q 75.5	B 157u	B 989u	B 159u	B 991u	B1023u	B 207u	B1039u
6	A 1040	A 9361	A 1061	A 9381	A 9701	A 1541	A 9861	Q 52.6	Q 75.6	B 1051	B 9371	B 1071	B 9391	B 9711	B 1551	B 9871
7	A 1561	A 9881	A 1581	A 9901	A10221	A 2061	A10381	Q 52.7	Q 75.7	B 1571	B 9891	B 1591	B 9911	B10231	B 2071	B10391
8	A 208u	A 1040u	A 210u	A1042u	A1074u	A 258u	A1090u	Q 52.8	Q 75.8	B 209u	B1041u	B 211u	B1043u	B1075u	B 259u	B1091u
9	A 260u	A 1092u	A 262u	A1094u	A1126u	A 310u	A1142u	Q 52.9	Q 75.9	B 261u	B1093u	B 263u	B1095u	B1127u	B 311u	B1143u
10	A 2081	A 10401	A 2101	A10421	A10741	A 2581	A10901	Q 52.10	Q 75.10	B 2091	B10411	B 2111	B10431	B10751	B 2591	B10911
11	A 2601	A 10921	A 2621	A10941	A11261	A 3101	A11421	Q 52.11	Q 75.11	B 2611	B10931	B 2631	B10951	B11271	B 3111	B11431
12	A 312u	A 1144u	A 314u	A1146u	A1178u	A 362u	A1194u	Q 52.12	Q 75.12	B 313u	B1145u	B 315u	B1147u	B1179u	B 363u	B1195u
13	A 364u	A 1196u	A 366u	A1198u	A1230u	A 414u	A1246u	Q 52.13	Q 75.13	B 365u	B1197u	B 367u	B1199u	B1231u	B 415u	B1247u
14	A 3121	A 11441	A 3141	A11461	A11781	A 3621	A11941	Q 52.14	Q 75.14	B 3131	B11451	B 3151	B11471	B11791	B 3631	B11951
15	A 3641	A 11961	A 3661	A11981	A12301	A 4141	A12461	Q 52.15	Q 75.15	B 3651	B11971	B 3671	B11991	B12311	B 4151	B12471
16	A 416u	A 1248u	A 418u	A1202u	A1282u	A 466u	A1298u	Q 52.16	Q 75.16	B 417u	B1249u	B 419u	B1251u	B1283u	B 467u	B1299u
17	A 468u	A 1300u	A 478u	A1302u	A1334u	A 518u	A1350u	Q 52.17	Q 75.17	B 469u	B1301u	B 471u	B1303u	B1335u	B 519u	B1351u
18	A 4161	A 12481	A 4181	A12501	A12821	A 4661	A12981	Q 52.18	Q 75.18	B 4171	B12491	B 4191	B12511	B12831	B 4671	B12991
19	A 4681	A 13001	A 4701	A13021	A13341	A 5181	A13501	Q 52.19	Q 75.19	B 4691	B13011	B 4711	B13031	B13351	B 5191	B13511
20	A 520u	A 1352u	A 522u	A1354u	A1386u	A 570u	A1402u	Q 52.20	Q 75.20	B 521u	B1353u	B 523u	B1355u	B1387u	B 571u	B1403u
21	A 572u	A 1404u	A 574u	A1406u	A1438u	A 622u		Q 52.21	Q 75.21	B 573u	B1405u	B 575u	B1407u	B1439u	B 623u	
22	A 5201	A 13521	A 5221	A13541	A13861	A 5701	A14021	Q 52.22	Q 75.22	B 5211	B13531	B 5231	B13551	B13871	B 5711	B14031
23	A 5721	A 14041	A 5741	A14061	A14061	A 6221		Q 52.23	Q 75.23	B 5731	B14051	B 5751	B14071	B14391	B 6231	
24	A 624u	P 1.24	A 626u	P 3.24	P 35.24	A 674u	P 51.24	Q 52.24	Q 75.24	B 625u	P 77.24	B 627u	P 79.24	P111.24	B 675u	P127.24
25	A 676u	P 1.25	A 678u	P 3.25	P 35.25	A 726u	P 51.25	Q 52.25	Q 75.25	B 677u	P 77.25	B 679u	P 79.25	P111.25	B 727u	P127.25
26	A 6241	P 1.26	A 6261	P 3.26	P 35.26	A 6741	P 51.26	Q 52.26	Q 75.26	B 6251	P 77.26	B 6271	P 79.26	P111.26	B 6751	P127.26
27	A 6761	P 1.27	A 6781	P 3.27	P 35.27	A 7261	P 51.27	Q 52.27	Q 75.27	B 6771	P 77.27	B 6791	P 79.27	P111.27	B 7271	P127.27
28	A 728u	P 1.28	A 730u	P 3.28	P 35.28	A 778u	P 51.28	Q 52.28	Q 75.28	B 729u	P 77.28	B 731u	P 79.28	P111.28	B 779u	P127.28
29	A 780u	P 1.29	A 782u	P 3.29	P 35.29	A 830u	P 51.29	Q 52.29	Q 75.29	B 781u	P 77.29	B 783u	P 79.29	P111.29	B 831u	P127.29
30	A 7201	P 1.30	A 7301	P 3.30	P 35.30	A 7781	P 51.30	Q 52.30	Q 75.30	B 7291	P 77.30	B 7311	P 79.30	P111.30	B 7791	P127.30
31	A 7801	P 1.31	A 7821	P 3.31	P 35.31	A 8301	P 51.31	Q 52.31	Q 75.31	B 7811	P 77.31	B 7831	P 79.31	P111.31	B 8311	P127.31

Recording direction →

Symbol number

Table 15.5b

Block address

Symbol number	0	1	2	3	35	50	51	52	75	76	77	78	79	111	126	127
0	B 0u	B 832u	B 2u	B 834u	B 866u	B 50u	B 882u	Q 52.0	Q 75.0	A 1u	A 833u	A 3u	A 835u	A 867u	A 51u	A 883u
1	B 52u	B 884u	B 54u	B 886u	B 918u	B 102u	B 934u	Q 52.1	Q 75.1	A 53u	A 885u	A 55u	A 887u	A 919u	A 103u	A 935u
2	B 0l	B 832l	B 2l	B 834l	B 866l	B 50l	B 882l	Q 52.2	Q 75.2	A 1l	A 833l	A 3l	A 835l	A 867l	A 51l	A 883l
3	B 52l	B 884l	B 54l	B 886l	B 918l	B 102l	B 934l	Q 52.3	Q 75.3	A 53l	A 885l	A 55l	A 887l	A 919l	A 103l	A 935l
4	B 104u	B 936u	B 106u	B 938u	B 970u	B 154u	B 986u	Q 52.4	Q 75.4	A 105u	A 937u	A 107u	A 939u	A 971u	A 155u	A 987u
5	B 156u	B 988u	B 158u	B 990u	B1022u	B 206u	B1038u	Q 52.5	Q 75.5	A 157u	A 989u	A 159u	A 991u	A1023u	A 207u	A1039u
6	B 104l	B 936l	B 106l	B 938l	B 970l	B 154l	B 986l	Q 52.6	Q 75.6	A 105l	A 937l	A 107l	A 939l	A 971l	A 155l	A 987l
7	B 156l	B 988l	B 158l	B 990l	B1022l	B 206l	B1038l	Q 52.7	Q 75.7	A 157l	A 989l	A 159l	A 991l	A1023l	A 207l	A1039l
8	B 208u	B1040u	B 210u	B1042u	B1074u	B 258u	B1090u	Q 52.8	Q 75.8	A 209u	A1041u	A 211u	A1043u	A1075u	A 259u	A1091u
9	B 260u	B1092u	B 262u	B1094u	B1126u	B 310u	B1142u	Q 52.9	Q 75.9	A 261u	A1093u	A 263u	A1095u	A1127u	A 311u	A1143u
10	B 208l	B1040l	B 210l	B1042l	B1074l	B 258l	B1090l	Q 52.10	Q 75.10	A 209l	A1041l	A 211l	A1043l	A1075l	A 259l	A1091l
11	B 260l	B1092l	B 262l	B1094l	B1126l	B 310l	B1142l	Q 52.11	Q 75.11	A 261l	A1093l	A 263l	A1095l	A1127l	A 311l	A1143l
12	B 312u	B1144u	B 314u	B1146u	B1178u	B 362u	B1194u	Q 52.12	Q 75.12	A 313u	A1145u	A 315u	A1147u	A1179u	A 363u	A1195u
13	B 364u	B1196u	B 366u	B1198u	B1230u	B 414u	B1246u	Q 52.13	Q 75.13	A 365u	A1197u	A 367u	A1199u	A1231u	A 415u	A1247u
14	B 312l	B1144l	B 314l	B1146l	B1178l	B 362l	B1194l	Q 52.14	Q 75.14	A 313l	A1145l	A 315l	A1147l	A1179l	A 363l	A1195l
15	B 364l	B1196l	B 366l	B1198l	B1230l	B 414l	B1246l	Q 52.15	Q 75.15	A 365l	A1197l	A 367l	A1199l	A1231l	A 415l	A1247l
16	B 416u	B1248u	B 418u	B1250u	B1282u	B 466u	B1298u	Q 52.16	Q 75.16	A 417u	A1249u	A 419u	A1251u	A1283u	A 467u	A1299u
17	B 468u	B1300u	B 470u	B1302u	B1334u	B 518u	B1350u	Q 52.17	Q 75.17	A 469u	A1301u	A 471u	A1303u	A1335u	A 519u	A1351u
18	B 416l	B1248l	B 418l	B1250l	B1282l	B 466l	B1298l	Q 52.18	Q 75.18	A 417l	A1249l	A 419l	A1251l	A1283l	A 467l	A1299l
19	B 468l	B1300l	B 470l	B1302l	B1334l	B 518l	B1350l	Q 52.19	Q 75.19	A 469l	A1301l	A 471l	A1303l	A1335l	A 519l	A1351l
20	B 520u	B1352u	B 522u	B1354u	B1386u	B 570u	B1402u	Q 52.20	Q 75.20	A 521u	A1353u	A 523u	A1355u	A1387u	A 571u	A1403u
21	B 572u	B1404u	B 574u	B1406u	B1438u	B 622u		Q 52.21	Q 75.21	A 573u	A1405u	A 575u	A1407u	A1439u	A 623u	
22	B 520l	B1352l	B 522l	B1354l	B1386l	B 570l	B1402l	Q 52.22	Q 75.22	A 521l	A1353l	A 523l	A1355l	A1387l	A 571l	A1403l
23	B 572l	B1404l	B 574l	B1406l	B1438l	B 622l		Q 52.23	Q 75.23	A 573l	A1405l	A 575l	A1407l	A1439l	A 623l	
24	B 624u	P 1.24	B 626u	P 3.24	P 35.24	B 674u	P 51.24	Q 52.24	Q 75.24	A 625u	P 77.24	A 627u	P 79.24	P111.24	A 675u	P127.24
25	B 676u	P 1.25	B 678u	P 3.25	P 35.25	B 726u	P 51.25	Q 52.25	Q 75.25	A 677u	P 77.25	A 679u	P 79.25	P111.25	A 727u	P127.25
26	B 624l	P 1.26	B 626l	P 3.26	P 35.26	B 674l	P 51.26	Q 52.26	Q 75.26	A 625l	P 77.26	A 627l	P 79.26	P111.26	A 675l	P127.26
27	B 676l	P 1.27	B 678l	P 3.27	P 35.27	B 726l	P 51.27	Q 52.27	Q 75.27	A 677l	P 77.27	A 679l	P 79.27	P111.27	A 727l	P127.27
28	B 728u	P 1.28	B 730u	P 3.28	P 35.28	B 778u	P 51.28	Q 52.28	Q 75.28	A 729u	P 77.28	A 731u	P 79.28	P111.28	A 779u	P127.28
29	B 780u	P 1.29	B 782u	P 3.29	P 35.29	B 830u	P 51.29	Q 52.29	Q 75.29	A 781u	P 77.29	A 783u	P 79.29	P111.29	A 831u	P127.29
30	B 728l	P 1.30	B 730l	P 3.30	P 35.30	B 778l	P 51.30	Q 52.30	Q 75.30	A 729l	P 77.30	A 731l	P 79.30	P111.30	A 779l	P127.30
31	B 780l	P 1.31	B 782l	P 3.31	P 35.31	B 830l	P 51.31	Q 52.31	Q 75.31	A 781l	P 77.31	A 783l	P 79.31	P111.31	A 831l	P127.31

Recording direction →

Symbol number

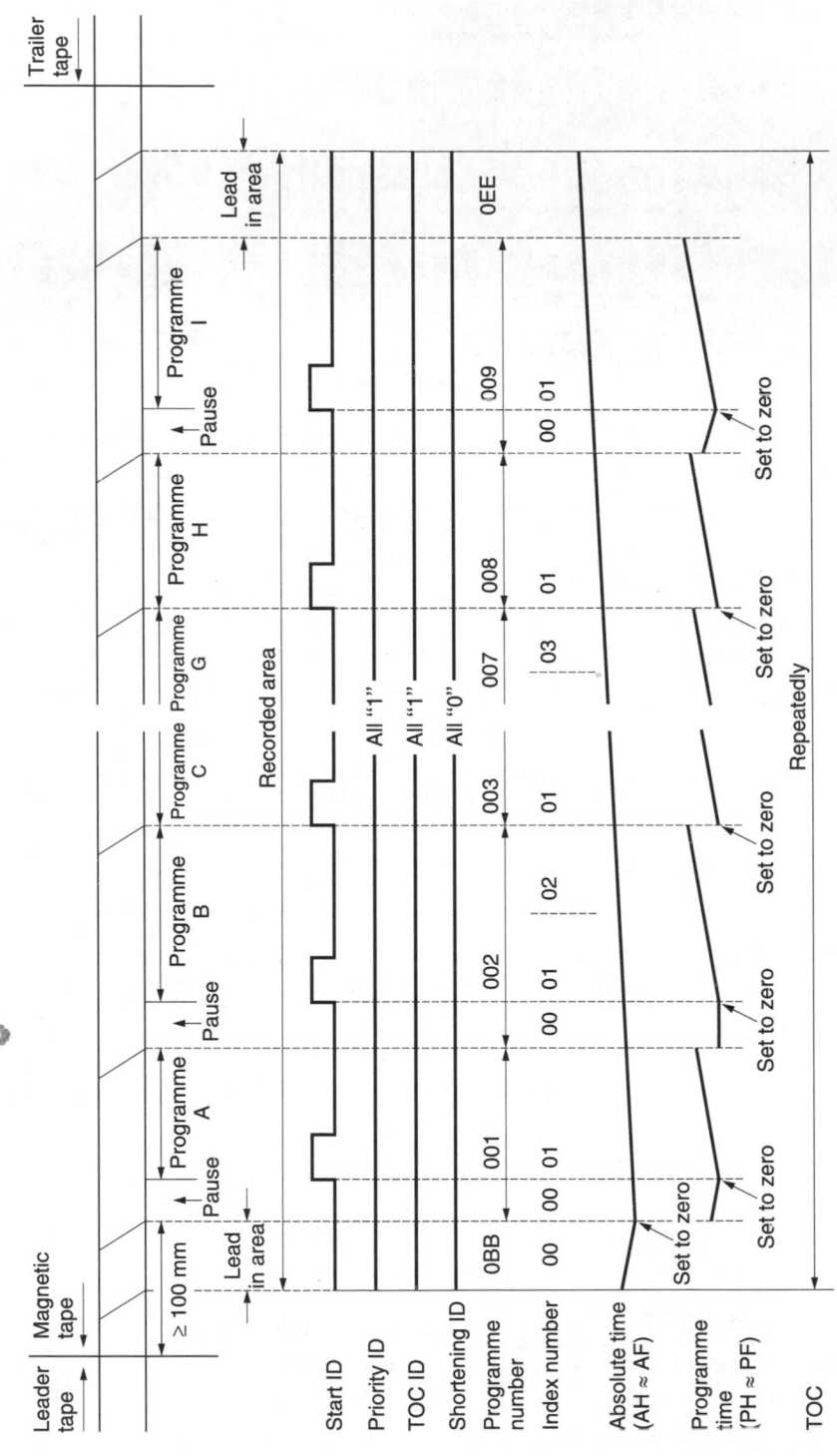

Figure 15.15

Table 15.6

Item	Mode
0000	No information
0001	Programme time
0010	Absolute time
0011	Running time
0100	TOC
0101	Calendar
0110	Catalogue
0111	ISRC
1000 1110	Reserved
1111	For tape maker

Tape duplication

High-speed duplication of R-DAT tapes can be done by using the magnetic contact printing technique. In this method a master tape of the mirror type is produced on a master tape recorder (Figure 15.16, 1).

The magnetic surfaces of the master tape and the copy tape are mounted in contact with each other on a printing machine, as shown in Figure 15.16 (2).

By controlling the pressure of both tapes between pinch drum and bias head, the magnetizing process is performed, applying a magnetic bias to the contact area. Special tape and a special bias head are required (see Figures 15.17 and 15.18).

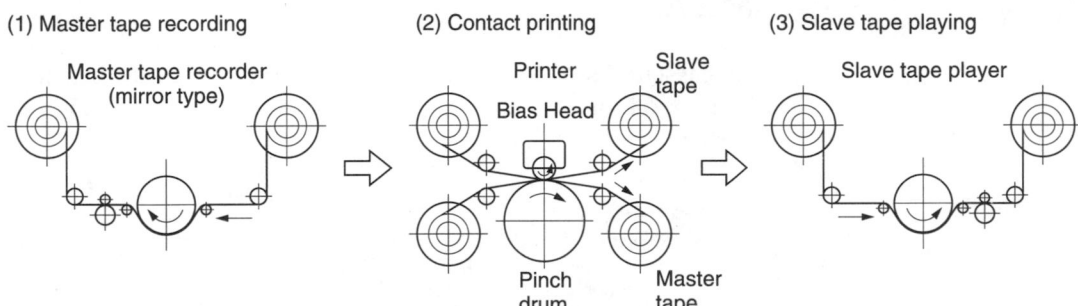

(1) Master tape recording

Master tape recorder (mirror type)

(2) Contact printing

Printer

Slave tape

Bias Head

Pinch drum

Master tape

(3) Slave tape playing

Slave tape player

Figure 15.16

Figure 15.17

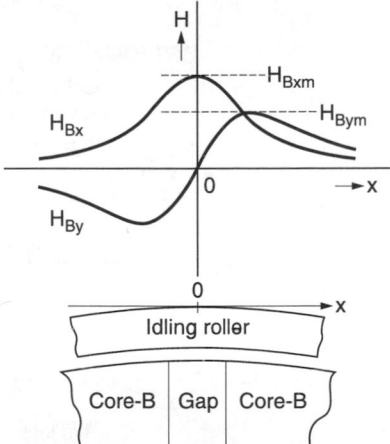

H_{Bxm}

H_{Bym}

H_{Bx}

H_{By}

Idling roller

Core-B | Gap | Core-B

Figure 15.18

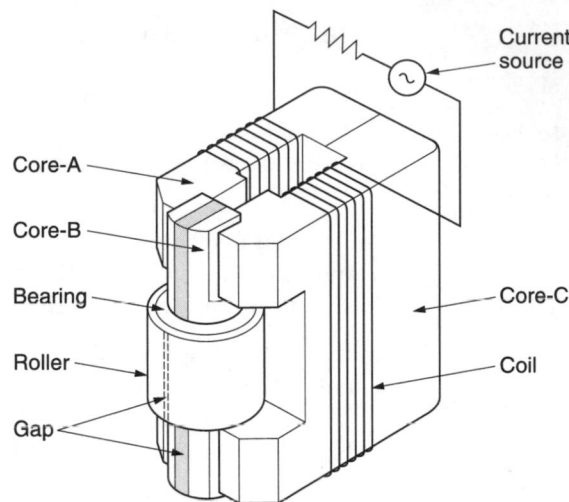

Cassette

The cassette is a completely sealed structure and measures 73 mm × 54 mm × 10.5 mm. It weighs about 20 g (see Figures 15.19a–c).

Figure 15.19a

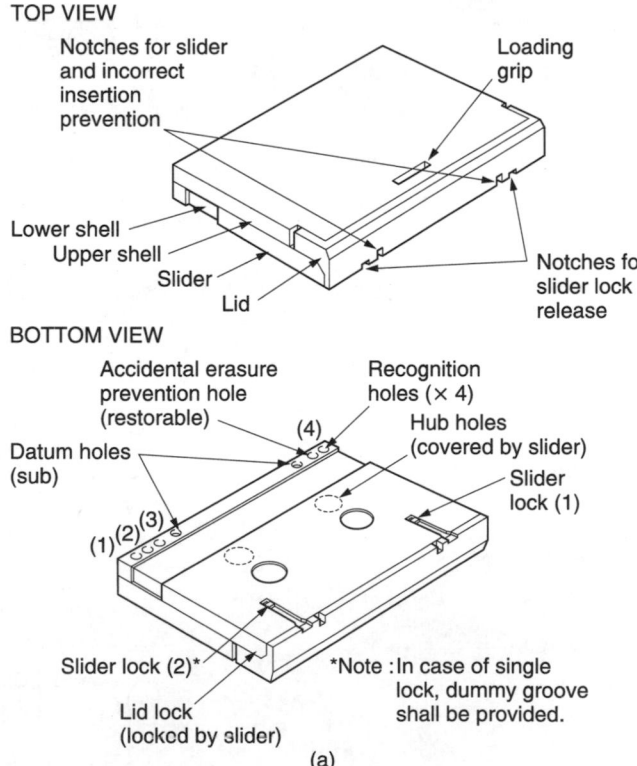

(a)

Figure 15.19b

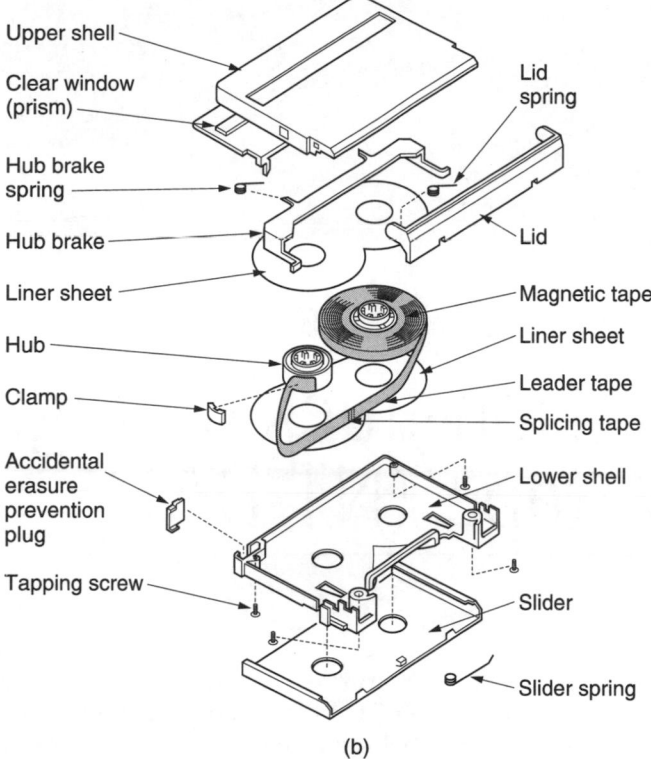

Upper shell

Clear window (prism)

Hub brake spring

Hub brake

Liner sheet

Hub

Clamp

Accidental erasure prevention plug

Tapping screw

Lid spring

Lid

Magnetic tape

Liner sheet

Leader tape

Splicing tape

Lower shell

Slider

Slider spring

(b)

Figure 15.19c

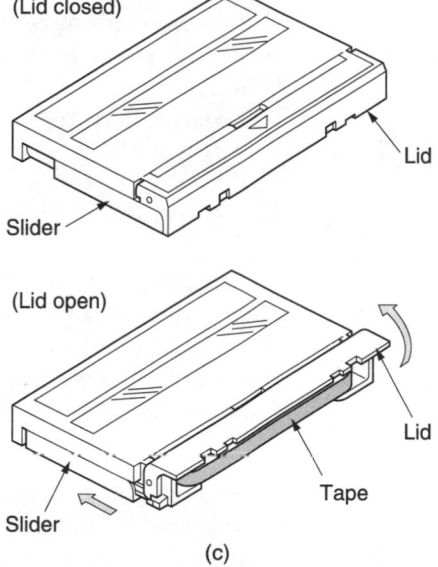

(Lid closed)

Lid

Slider

(Lid open)

Lid

Tape

Slider

(c)

16 Non-tracking digital audio tape (NT-DAT)

A further evolution from the first (tracking) DAT was the NT-DAT. The advantage of this system is the much improved vibration and shock handling capacity. This technology was used in playback-only units such as car DAT players, but also in a miniature version, the so-called stamp-size DAT player. In the latter case, recording with the NT method was also applicable. This technology never became a big commercial success, and is presently virtually at a standstill. The ideas and technology behind this design are still valid and can still be used, and for that reason they are still included in this edition.

NT playback

The original DAT specifications are based on the fact that each rec/pb head has its own azimuth and that each head must run correctly centred over the correct data track, for which purpose the automatic tracking following (ATF) system is included (see Chapter 15).

Whenever vibration occurs, there is the danger that the head loses track and that recovering the correct position immediately is impossible, resulting in loss of read-out or recorded data. Non-tracking DAT does not use ATF tracking; in fact, there is no tracking at all.

The combination of double density scan, high processor capacity and high memory capacity can overcome momentary loss of tracking.

Double density scan

Figure 16.1 shows a comparison of playback waveforms obtained from a 30-mm diameter drum and a 15-mm diameter drum, respectively; however, the drum speed of the latter is 4000 rpm instead of the conventional 2000 rpm. On the other hand, the relative tape to head speed remains the same:

$$V_r = 30 \times \pi \times \frac{2000}{60} = 15 \times \pi \times \frac{4000}{60} = 3.1 \text{m} / \text{s}$$

In the case of the 15-mm drum, the wrap angle is 180°.

Each head will in this case perform read-out, but twice that of the 30-mm drum (the drum speed being doubled), and also not correctly over the centre of the track. The result on the RF level is shown in Figure 16.2.

Figure 16.1 Double density scan.

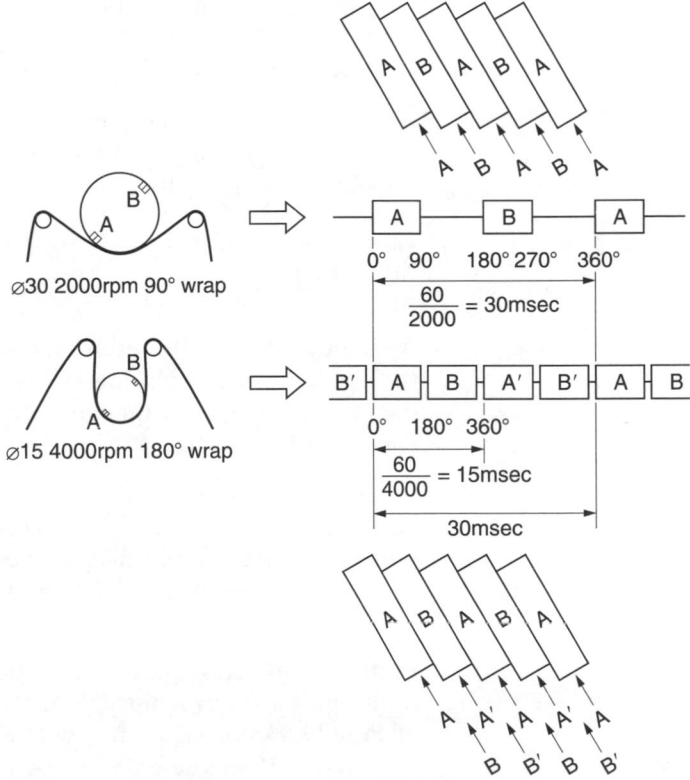

Figure 16.2 Head versus
tape.

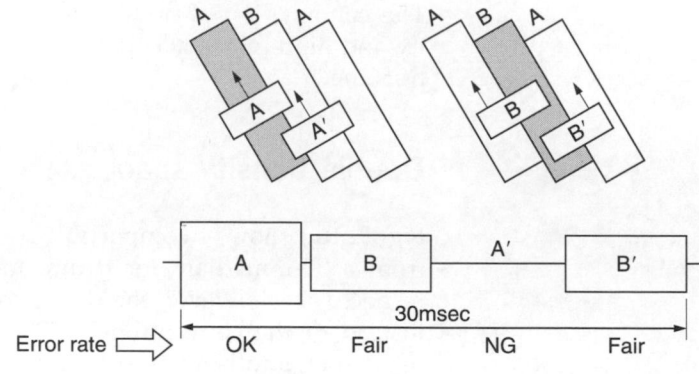

(RF waveform which has been played back)

Depending on the azimuth of the head versus track, it is possible
that some read-out is good, fair or not good at all, but in any
case, all data will be read out at least once with sufficient RF
level.

If the RF level becomes low, it is obvious that the error rate of
the data will increase; this is recovered partly by the error correction and partly by the double read-out.

Another important detail is that the head width in conventional
DAT players is 20 μm, whereas it is 28.5 μm in this type of player;
more RF is therefore obtained.

In long-play mode, the drum speed will be the same and the tape
speed will be halved. This results in a four-times read-out,
thereby decreasing the error rate significantly.

NT read-out

As the data which are recorded on the tape contain address data
(frame address and block address), it is possible to handle these
data correctly even if they are retrieved in a non-logical or discontinuous sequence.

As shown in Figure 16.3, when the tracking angle of the playback head deviates from the recorded track, the read-out order
will be non-logical and maybe even discontinuous. This is not a
problem because of the frame and block addresses of each data
block.

Each data block that has been read out correctly will be stored
in random access memory (RAM). After several head read-outs,
all data blocks of each track will have been read out at least once
and stored; if so, the data can be released.

Figure 16.3 Head locus.

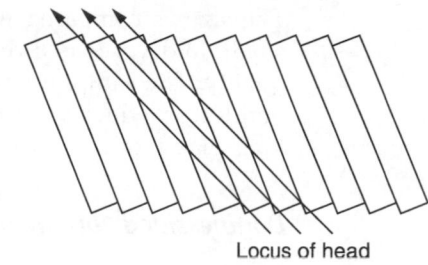

Figure 16.4 Track read-out.

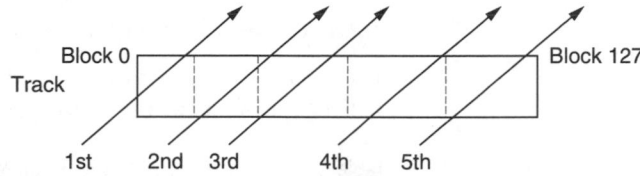

It is obvious that the RAM capacity now becomes important. Supposing an RAM of 1 Mbit is used; this has a capacity of about 12 frames, or 24 tracks, as each frame consists of two tracks.

Each track contains 128 blocks of 32 symbols, and each symbol contains 8 bits, so each track represents $128 \times 32 \times 8 = 32\,768$ bits (or 32 kbit); 24 tracks represent 32 kbit $\times$ 24 = 768 kbit.

The difference between 768 kbit and 1 Mbit is needed for normal processing buffers, such as for error processing and others.

NT stamp-size DAT

NT stamp-size DAT is a miniature DAT format. The dimensions of the cassette are 30 mm $\times$ 5 mm and it weighs 2.8 g.

In this section we will review the essential differences between the conventional DAT and the stamp-size DAT (also referred to as Scoopman).

Figure 16.5 Cassette size comparison.

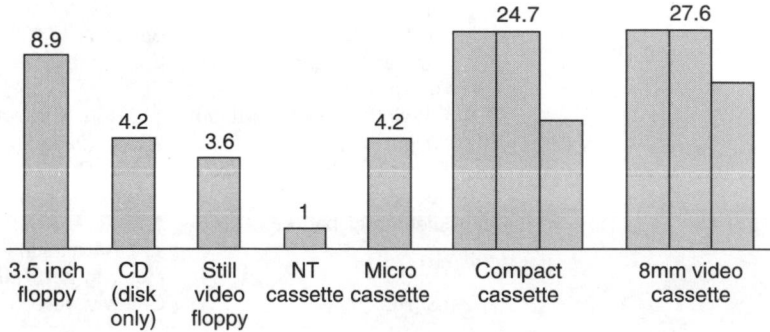

215

The size as compared with other types of cassette is extremely small, and yet it is a digital recordable medium allowing high quality. The sampling rate is 32 kHz, and quantization is 12-bit non-linear, which is comparable to a normal DAT recording in long-play mode.

Double-sided, bi-directional use

Contrary to the conventional DAT type, the stamp-size type is bi-directional; in other words, it is used more like an analog cassette type which has to be reversed.

Tracks are divided into top and bottom halves. In the future, it will be possible to use a bi-directional turnaround mechanism,

Table 16.1 General specifications for NT format

Recording format	Two revolving heads, helical scan, azimuth recording
Tape width	2.5 mm
Tape speed	6.35 mm s^{-1}
Drum diameter	14.8 mm
Drum wrap angle	100 degrees (mechanical wrap angle)
Field frequency	50 Hz
Number of drum rotations	50 Hz (double density scan) = 3000 rpm
Relative head speed	Approx. 23 mm s^{-1} (double density scan)
Track width	Approx. 9.8 μm
Still angle	Approx. 4.4°
Input channels	2 channels stereo
Sampling frequency	32 kHz
Quantization	12 bits non-linear (equivalent to 17 bits)
Error correction	Cross-interleave code
Redundancy	33% (including synchronization word, address)
Modulation	Low deviation modulation (LDM)-2

Table 16.2 Main specifications for NT recorder

Frequency response	10–14 500 Hz (+1 dB, −3 dB)
Dynamic range	Over 80 dB
S/N	Over 80 dB
Total distortion	Below 0.05%
Wow and flutter	Below measurement limits (quartz precision)
Dimensions (width × height × depth)	115 mm × 50 mm × 21 mm
Weight	138 g (including battery and cassette)
Power supply	1.5 VDC
Power consumption	Approx. 270 mW (during battery operation)
Lifetime of battery	Approx. 6 h (for recording and playing; using size AA alkaline battery LR6)
Head	Metal in gap (MIG) head
LSI, IC	6 newly developed, 3 widely used

Table 16.3 General specifications for NT cassette

Dimension	
(width × height × depth)	30 mm × 21.5 mm × 5 mm (with lid)
Weight	2.8 g (120-min tape)
Tape:	
Width	2.5 mm
Thickness	Approx. 5 μm
Length	Approx. 20 m (120-min tape)
Type	Ni–Co metal evaporated tape
Maximal recording time	120 min (round trip)
Maximal recording capacity	Approx. 690 mbytes

which, in combination with a large memory size, will enable a smooth, glitchless autoreverse operation.

The tape format is shown in Fig. 16.6; tape width is 2.5 mm, still angle of each track is approximately 4.4° and track pitch is 9.8 μm.

Non-loading method

All the previously used drum-based systems are based on the idea that the tape has to be taken out of its cassette and wrapped around the drum.

In this case, however, the technology is reversed. The drum assembly will be inserted into the cassette (of which the front cover is tilted away). Figure 16.7 shows the drum versus the cassette.

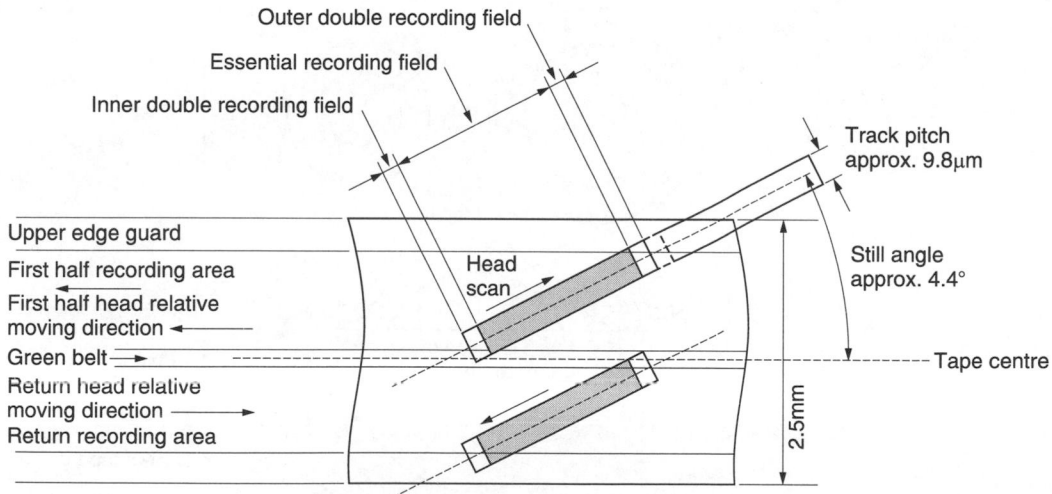

Figure 16.6 Tape format.

Figure 16.7 Drum versus
cassette.

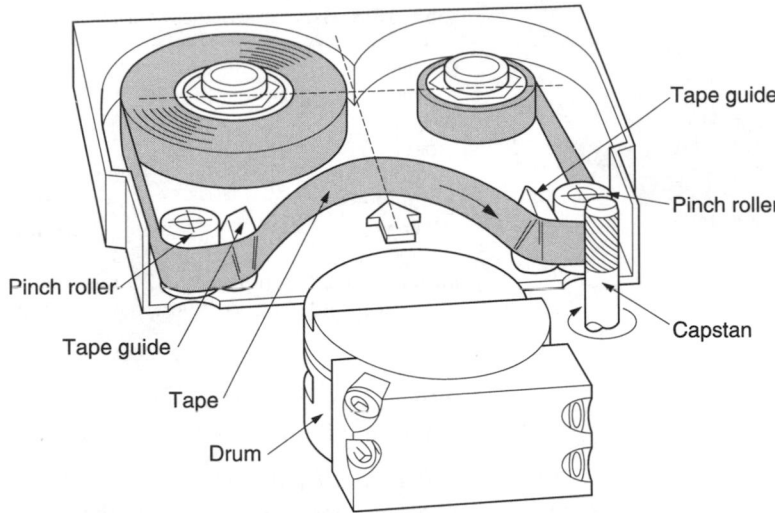

This solution allows a design which is much simpler, smaller
and more cost-effective, as the loading mechanism is extremely
simplified.

One of the major differences is that much of the technology which
used to be built into the recorder/playback set is now built into
each cassette itself. This also has a very positive effect on the
wear of the set. Figure 16.8 shows the non-loading method.

Figure 16.8 Non-loading
method.

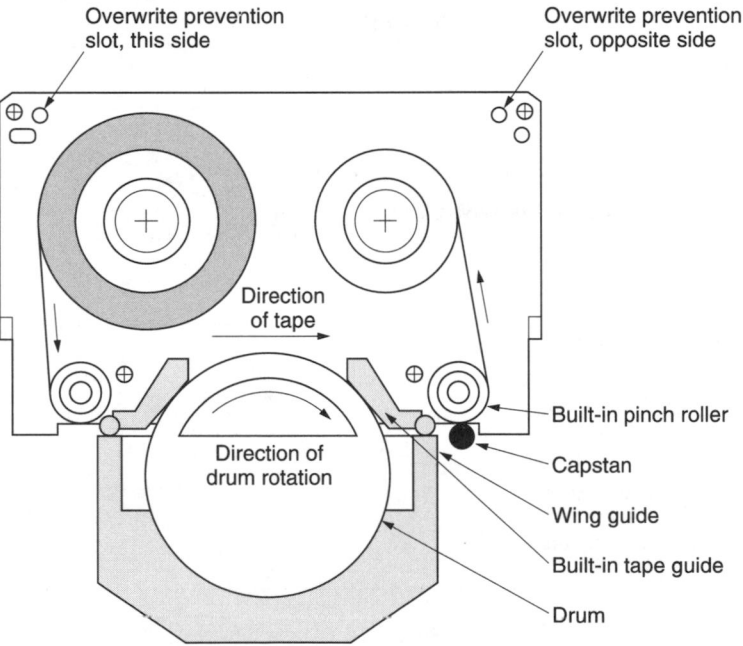

Each cassette contains built-in pinch rollers (also contrary to any conventional design). When inserting the drum into the cassette, the capstan, which is mounted on the same platform as the drum, will automatically be positioned against the pinch roller.

Each cassette also contains tape guides, so that the tape is automatically guided correctly over the inserted drum. Insertion of the drum assembly into the cassette can be performed automatically as well as manually.

17 MiniDisc

MiniDisc is the first real consumer application where the medium is not a tape, but a rewritable disc, with all its benefits.

When Compact Disc came on the market, the conversion from analog audio to digital data was performed in a fairly straight-forward way. In other words, the data are a true representation of the analog audio. This conversion method needs a huge amount of data if we want it to be an accurate representation of the input audio.

For MD, there was a need for a smaller disc to make it even more convenient. However, a smaller disc means automatically that the amount of data that we can put on it is also limited. The only solution then is to compress the audio data. The same has already been performed with computer data for several years, but the main difference, and in fact the big problem, is that audio is an extremely complex type of data; if we were to compress audio in the same way as normal data is compressed in computers (usually these data represent text and drawings), the result would be totally inadequate.

As a result, the compression needed special attention, where psychoacoustic basics are the key factor in the compression (and de-compression) methods.

We should keep in mind that the data which are put on a MiniDisc no longer have a straightforward relation with the analog equivalent. Now the data on the disc have a highly complex algorithm structure, enabling one to recalculate and reconstruct the original input. One major benefit of this method is the fact that we can change and enhance these algorithms without changing the music itself, and without creating compatibility problems.

The following were target items when designing MiniDisc:

- Digital disc format.
- Smaller than Compact Disc, but with the same timewise amount of music.
- High quality, up to Compact Disc level.
- Recordable.
- Enabling quick and random access, similar to Compact Disc.
- High shock-proof capability.
- Lightweight and inexpensive.
- Durable.

The rainbow book

The basic specification book for MiniDisc is called the 'rainbow book'. This name was chosen rather sensibly, as MiniDisc is based upon several other basic specification books, each of them being nicknamed another colour:

- The red book, basic specifications for Compact Disc.
- The yellow book, basic specifications for CD-ROM.
- The orange book, basic specifications for MO drives and WO drives.

All of these books are co-developments of Sony and Philips; they have been the basis for CD, CD-ROM, CD-MO/WO and also for CD-I (its specification book is called the green book; this is also a Sony/Philips co-production). They outline the basic parameters and specifications to be used for all the above-mentioned technologies. The aim of course is to have a standardized format, both for hardware and software. Such standards have enabled the growth of global markets where all manufacturers make hardware and software which is truly compatible.

Figure 17.1 shows the basic relationship between the specification books. The yellow book evolved from the red book; from both specifications the rainbow book 'borrowed' the following items:

- the basic data structure;
- the CIRC system (Cross-Interleave Reed–Solomon Correction),

221

but extended and improved (hence the name ACIRC or Advanced CIRC);

- the basic disc geometry, even if the size of the MD is physically smaller.

From the orange book, mainly optical parameters were used, and other parts used in magneto-optical discs were a source of inspiration, like the ATIP system – a system used to ensure that the laser, while recording and reading out, will correctly follow the data track – from MO, which is closely related to the ADIP in MD.

New, important features:

- the use of a cartridge, similar to a floppy disc;
- ATRAC, the psychoacoustic compression system;
- the use of a shock-resistant memory.

Of course, all the above-mentioned points will be explained. The basic specifications are the following:

- Play/recording time is exactly the same as on a CD, but note the much smaller size of the MiniDisc versus the CD, which has a diameter of 12 cm.
- Similar to a CD, read-out and recording on MD will be performed starting from the inner side, close to the centre.
- The track pitch – i.e. the distance beween tracks – is the same as for CD: 1.6 μm or 0.0000016 m.
- Scanning velocity is also similar to CD, varying between 1.2 and 1.4 m s^{-1}; this is the rotation velocity of the disc. The

Figure 17.1 MD basics.

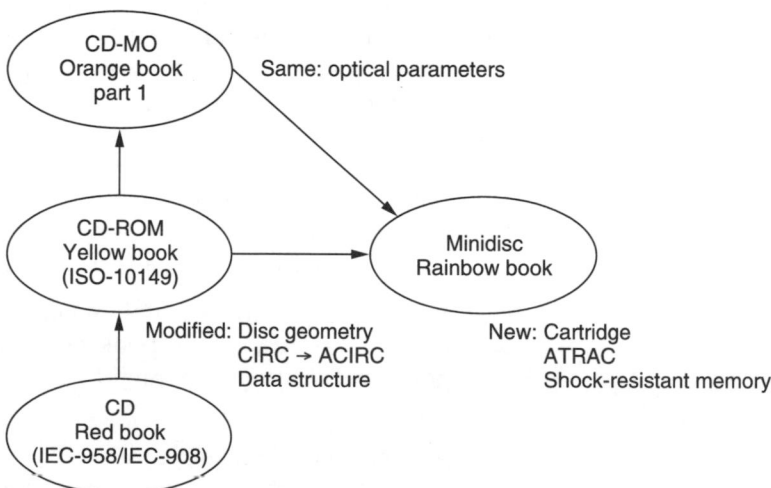

variation is due to the Constant Linear Velocity used. This means that the read-out speed of data is kept constant. However, as these data are writing on a spiral track, starting from the inside to the outside of the disc, when reading (or writing), the diameter of the circle on which the system is working is varying constantly, and therefore the disc speed needs to be adapted continuously.

- The sampling frequency and modulation system are the same as for CD.
- The optical parameters are a mix of optical parameters used in CD, but also in MO; in particular, the higher laser power is typical for systems which are able to make recordings.

Block diagram

The block diagram of a MiniDisc can be compared with the block diagram of a Compact Disc; it is therefore obvious that knowledge of Compact Disc systems makes the introduction to MiniDisc much easier. As CD technology has already been

Table 17.1 MD basics

Main parameters	
Playing and recording time	Max. 74 minutes
Cartridge size	68D × 72W × 5H mm
Disc parameters	
Diameter of the disc	64 mm
Thickness of the disc	1.2 mm
Diameter of centre hole	11 mm
Starting diameter of lead-in area	29 mm max.
Outer diameter of lead-in area	32 mm
Diameter of program end	61 mm max.
Track pitch	1.6 μm
Scanning velocity	1.2–1.4 m s^{-1}
Audio performance	
Number of channels	Stereo and mono
Wow and flutter	Quartz crystal precision
Signal format	
Sampling frequency	44.1 kHz
Coding	ATRAC (Adaptive Transform Coding)
Modulation	EFM (Eight-to-Fourteen Modulation)
Error correction system	ACIRC (Advanced CIRC)
Optical parameters	
Laser wavelength	780 nm typical value
Lens NA (Numerical Aperture)	0.45 typical value
Recording power	2.5–5 mW (main beam)
Recording strategy	Magnetic field modulation

explained in previous chapters, we can mostly concentrate on the differences at this point.

The differences between CD and MD are mainly due to:

- The recording capability of MiniDisc, which needs an A/D converter, an EFM/ACIRC encoder, a head drive and a magnetic recording head.
- The use of compression in the ATRAC encoder/decoder.
- The shock-resistant memory controller, although a similar system, is nowadays also included in some CD Discman types, but in such cases only in playback.
- The address decoder is typical for the read-out of the recordable MiniDisc type.

Most of the other blocks can be related back to the Compact Disc format.

In order to have a quick understanding of the operation of MiniDisc, we will now trace the most important operation paths, and give a brief explanation; in the next chapters most of these items will be repeated in more detail.

REC/PB path

- Recording of any disc starts of course with audio input; this can be an analog input or a digital input (typically S/PDIF format).
- The analog input will be A/D-converted and fed to the ATRAC encoder. The digital input is directly fed to the ATRAC encoder.
- In the ATRAC encoder, the digital audio data will be compressed according to several psychoacoustic rules. The compression ratio is about 1:5.
- The compressed data are sent to the EFM/ACIRC encoder through the shock-proof memory; shocks during recording which might disable proper recording will be absorbed here.
- The EFM/ACIRC encoder will handle the digital audio data similarly to the Compact Disc encoding; error encoding and interleaving are performed, subdata are added and all this is EFM-modulated. ACIRC and EFM circuits have to process the audio data such that they are in a correct state to be recorded.
- Recording is then performed through the magnetic recording head while the laser unit is used for the spot heating of the disc.
- Recording is not possible on the pre-mastered disc; only read-out is possible. Unlike recordable discs, read-out at the beginning of pre-mastered discs is not completely the same.

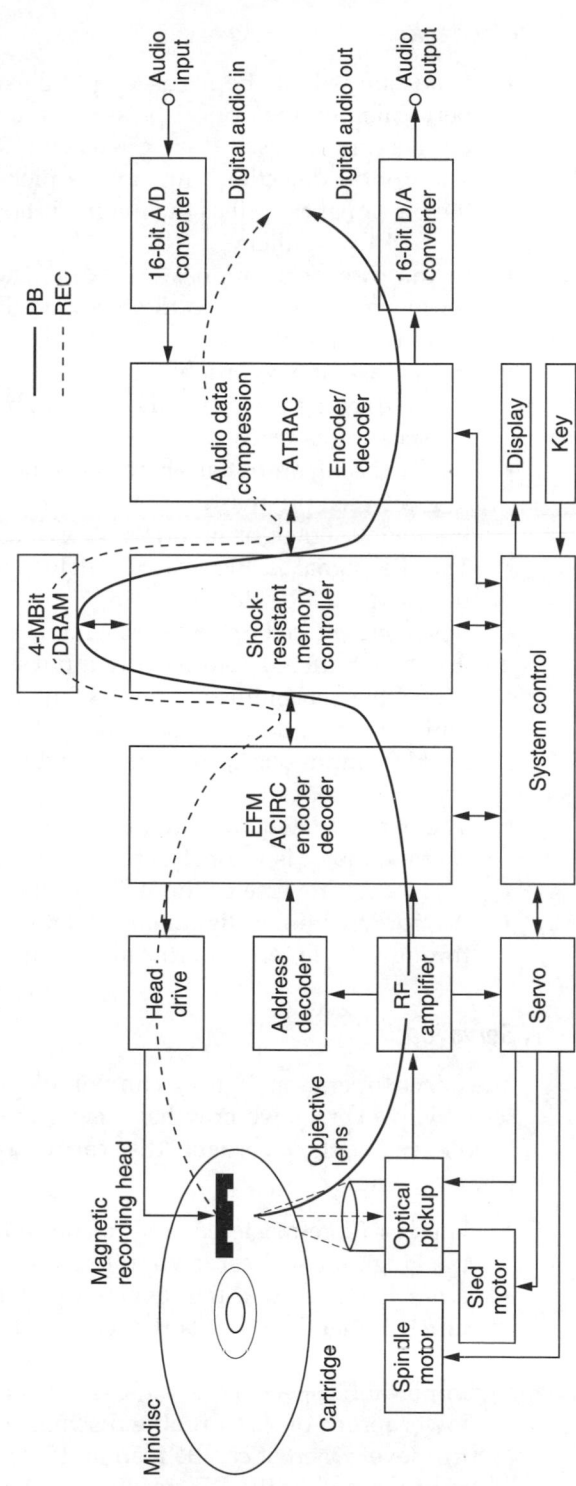

Figure 17.2 REC/PB path.

Playback

- Read-out will in both cases (pre-mastered/recordable) be performed by the optical pick-up unit; the magnetic recording head is not used during read-out.
- The optical detecting unit (in the pick-up) is able to detect the pit signal as well as the magnetic signal; both will be sent to the RF amplifier.
- In the case of a pre-mastered disc, the focus and tracking servo signals, similar to the Compact Disc system, will also be retrieved at this point.
- In the case of a recordable disc, not only the already known focus/tracking servo signals are seen, but also the pre-groove information is read out.
- As for the RF amplifier, there is no more difference in handling between pre-mastered and recordable read-out RF (radio frequency signal, or high-frequency signal).
- This RF signal is EFM-demodulated, and error-decoded in the EFM/ACIRC decoder.
- It is then sent to the shock-resistant memory controller, where it will be buffered before being input in the ATRAC decoder. When the set endures shocks while reading out and track jumps occur due to these shocks, the buffer memory will enable continuous music output while the system recovers from the shocks.
- The demodulated data are then ATRAC-decoded; in fact, the original music is reconstructed in a transform system, which is the exact reverse of the ATRAC encoding format.
- As from this point digital audio data output is possible, and through the D/A converter analog audio is output.

Servo path

The servo operation is also comparable to the Compact Disc servo. However, as we now have more than one type of disc to handle, and as the recordable disc carries a pre-groove, the servo is more complex.

- For both pre-mastered and recordable disc, the focusing, tracking and sled servo circuits are similar.
- Note, however, that the reflectivity of a pre-mastered disc is much higher than that of the recordable disc. Due to this fact, correct Automatic Gain Control (AGC) handling is very important, especially in the case of the recordable disc. The lower return of a recordable disc has to be boosted correctly to a level where it can be read and interpreted correctly. This is performed in the RF amplifier.

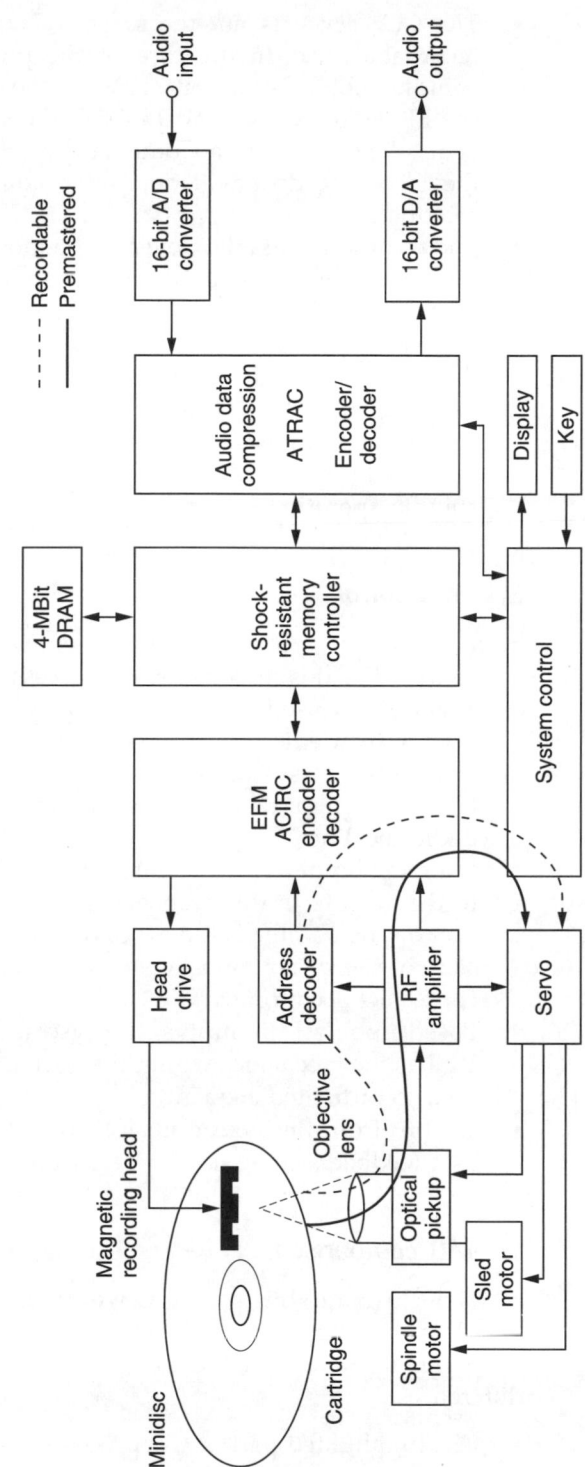

Figure 17.3 Servo path.

- The CLV servo is not the same for a pre-mastered and a recordable disc. In the case of the pre-mastered disc, the spindle motor is driven through the RF signal input in the EFM/ACIRC IC, and through the servo control. This is again the same as in a Compact Disc. Synchronization with the incoming data will determine adaptations to the disc rotation speed.
- The recordable disc, however, is the most complex one. Here the pre-groove is also used. The pre-groove signal is detected by the optical pick-up main spot detector, and retrieved in the RF amplifier. A specific address in pre-groove (ADIP) decoder will decode this pre-groove signal; decoded address data will then be output from the RF IC as well as clock signals also retrieved from the pre-groove. This output will be used by the EFM/ACIRC and servo side to control the spindle operation.

System control

- It is obvious that such complex actions need correct system control. For this reason, a microprocessor (also referred to as a syscon) is used; this can be a general-purpose processor driven by dedicated software, or a totally purpose-built processor. The syscon reads and writes from and to the main ICs through a dataline system, but also through many dedicated lines.
- User key input as well as internal detection switches are fed to the syscon. In this way, and also helped by the information on the datalines, the processor can determine the actual state of the system and the actual wishes of the user, and of course act accordingly.
- Based upon all the inputs, the syscon can drive each IC as well as the mechanics within the proper timing, and in a correct, co-ordinated sequence.
- Output from the syscon is also given to the display section as a feedback to the user.

CD–MD comparison

As a conclusion to this part, it is worthwile comparing the main data handling between a CD and MiniDisc. Note that this is a simplified approach for the sole purpose of pointing out major differences.

- The highlighted part of CD processing is only performed on the disc manufacturer's side.

- This comparison clearly shows that even though a MiniDisc player is very similar to a Compact Disc player, the complexity is much higher and the amount of possible operations and processings to be performed is significantly higher.
- Note also that the above table starts from and ends with analog audio input. It is of course possible to start from and end with digital audio, basically in the S/PDIF format (Sony/Philips Digital Interface format).

Table 17.2 CD–MD comparison

Compact Disc	MiniDisc
Analog audio input	Analog audio input
A/D conversion	A/D conversion
	ATRAC encoding
	Data buffering
CIRC encoding	ACIRC encoding
EFM encoding	EFM encoding
Disc recording	Disc recording
Disc playback	Disc playback
EFM decoding	EFM decoding
CIRC decoding	ACIRC decoding
	Data buffering
	ATRAC decoding
D/A conversion	D/A conversion
Analog output	Analog output

Physical format

Disc types

Here, immediately one of the special features of MiniDisc stands out: there is not one type of MiniDisc, but in fact there are two types.

- The first type is the non-recordable MiniDisc; this is a pre-mastered disc using CD technology.
- The second type is the user-recordable disc, using CD-MO technology.

When explaining the physical format of the MiniDisc, we should remember that, although the dimensions are always the same, we do have these two different disc types.

The outlook of a MiniDisc is similar to a floppy disc; the pre-mastered/recordable medium is a 2-inch disc housed in a cartridge, closed by a shutter.

Figure 17.4 Photograph of discs, one recordable (left) and one pre-mastered (right).

The main dimensions are the following :

Cartridge size 72×68 mm $\times 5$ mm
Weight   30 g (including disc)
Centre hole in cartridge 18 mm diameter
Disc diameter   64 mm
Clamping area   16.4 mm
Disc thickness   1.2 mm

There is a distinct difference in physical appearance between the pre-mastered and recordable disc types.

- In the first case, only read-out is needed; therefore, only one side of the disc needs to be opened. In this case, the shutter is only on one side, leaving room on the other side for a label (showing graphics, information, etc.).
- In the case of a recordable type, there is a need for opening both the top and the bottom sides; therefore, the shutter opens on both sides.

As is the case with most disc and cassette media, some holes and/or switches are used for information on the disc itself. On the MiniDisc cartridge, there are two important information detection holes, one to distinguish between high/low reflectivity and another for write protect/record enable.

The pre-mastered disc is of the high-reflectivity type, similar to

Figure 17.5 Cartridge exploded view.

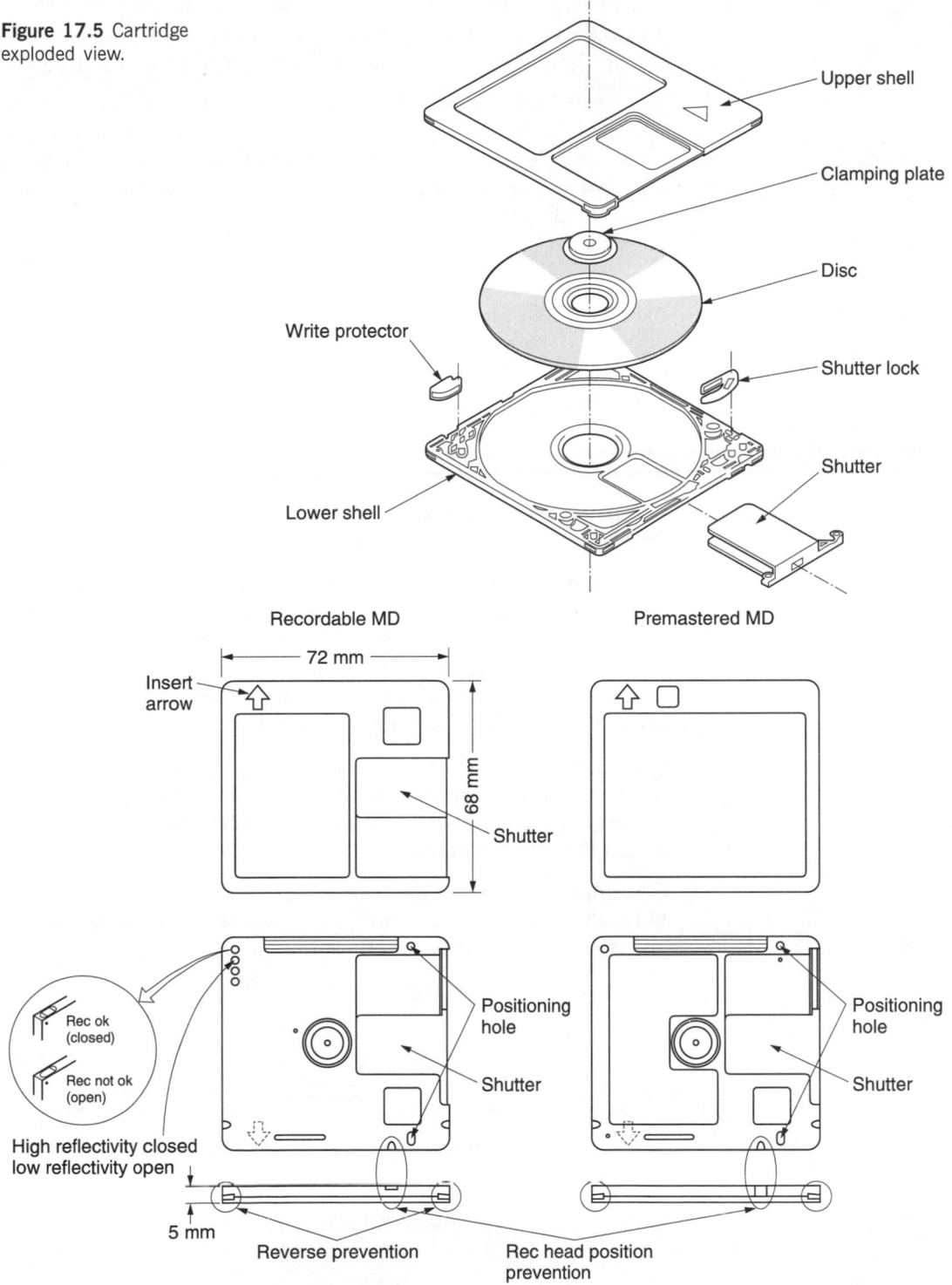

Upper shell

Clamping plate

Disc

Write protector

Shutter lock

Shutter

Lower shell

Recordable MD

Premastered MD

Insert arrow

72 mm

68 mm

Shutter

Rec ok (closed)

Rec not ok (open)

High reflectivity closed low reflectivity open

5 mm

Positioning hole

Shutter

Positioning hole

Shutter

Reverse prevention

Rec head position prevention

Figure 17.6 Cartridge layout.

a Compact Disc, and it cannot be recorded. It therefore has no low-reflectivity hole and a fixed (open) write protect hole.

The recordable disc is of the low-reflectivity type and can be recorded. It therefore has a low-reflection hole and next to this there is a record inhibit switch. The position of some more holes have been defined, but they are not used now; they are included for possible future use.

Note: the reflectivity factor of a disc is used to quantify the amount of light that it will reflect; this depends of course on the material used. There is no specific unit for this item. In the case of a CD, 70% of the incident light has to be reflected; for MO drives (low reflectivity), it ranges between 15% and 30%; for MiniDisc, MO drive reflectivity is between 15% and 25%.

Clamping

When remembering the cross-section of a Compact Disc, we see that the clamping area in the case of a MiniDisc is somewhat different.

Clamping means the actual holding of the disc while playing in the player; it is obvious that a disc that needs to be rotated must be held correctly without wobbling or eccentricity. For this purpose, discs are clamped. In a Compact Disc recorder, a disc is squeezed between the disc motor (for rotation) and a top clamping plate, sometimes containing a magnet.

For MiniDisc, clamping is performed through a magnetic clamping plate mounted in the disc itself and one single counterplate on top of the disc motor. In this way, correct centring is ensured, which is very important.

Clamping as it is done in MiniDisc enables one to catch and stabilize the disc from the bottom side only, which is more efficient and avoids the need for another hole in the cartridge.

Figure 17.7 MD clamping plate.

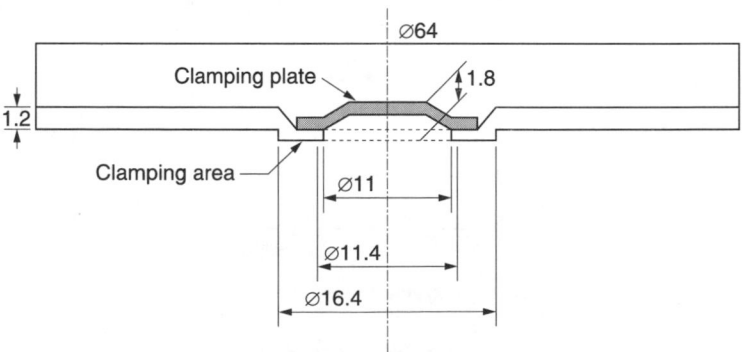

232

Cross-sections

The pre-mastered disc is totally comparable with the Compact Disc. When looking at the cross-section, reflection of the laser beam is also ensured by the pits/bumps.

The laser return level of such a disc is considered to be of the high-reflectivity level. The recordable disc has a different cross-section.

On a substrate, an MO layer (TbFeCo) is caught between dielectric layers. These are protection layers. On top of this there is still a reflective layer, but the reflectivity in this case is much lower compared with the reflectivity from a Compact Disc. For this reason, the recordable disc is also referred to as low-reflectivity type and the signals retrieved from such a disc are handled likewise. On top of the reflective layer there is a protective layer. Over this last protective layer there is a lubricant, silicone grease.

Note that this lubricant can be deleted; it has been included to enhance the smoothness of contact between magnetic head and disc.

Address in pre-groove (ADIP)

Another important physical feature on the recordable disc is the wobble or pre-groove, containing address information; hence the name Address In Pre-groove. This is totally different from the Compact Disc, but it is similar to the CD-MO format, where there is also a wobbling pre-groove containing time information (ATIP = absolute time in pre-groove).

Note that this pre-groove is **only used on the recordable disc**, as the need for additional addressing (as explained hereafter) does not occur in the pre-mastered disc.

Figure 17.8 CD cross-sections.

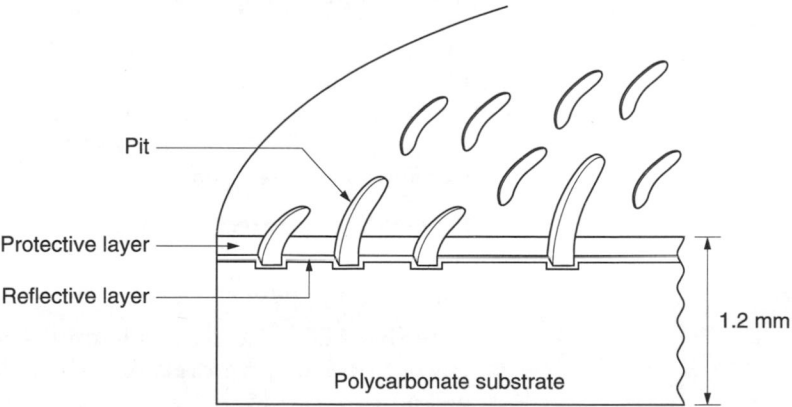

Figure 17.9 MD cross-sections.

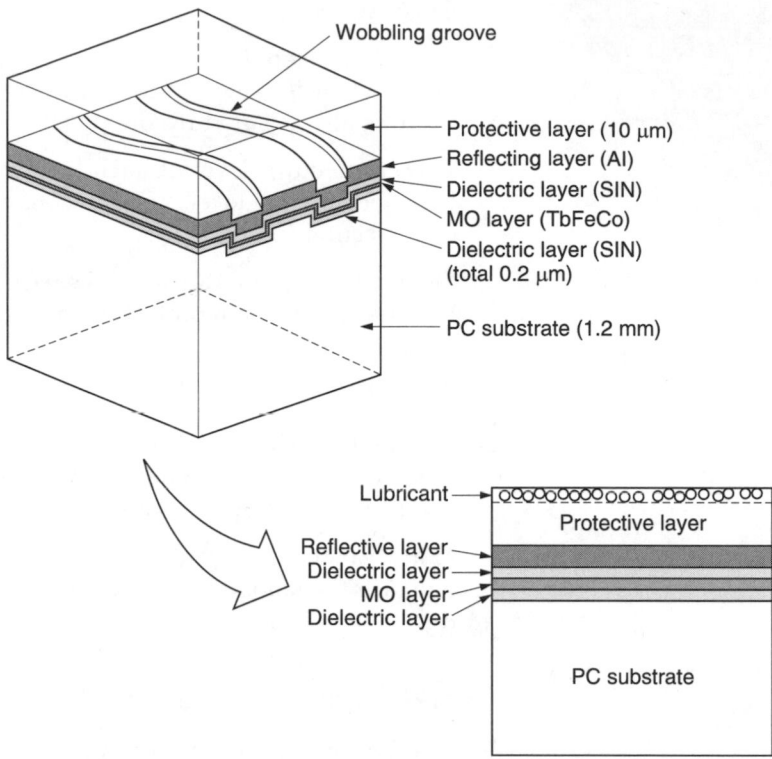

A sine-wave wobble is physically included into the disc; this wobbling groove coincides with the data track on its spiral way. It is a U-shaped depression, physically stamped on the disc. This groove is the physical representation of an electronic signal, comparable to the 'old' analog phonogram recording.

The size of the laser beam when hitting the disc is somewhat larger than the pre-groove, but it will be modulated by the wobble of the pre-groove.

This wobble signal is a 22.05 kHz frequency signal which is FM-modulated with address information. This address information is exactly the same on each disc. It enables the system to control the CLV motor (the disc motor) and to know the exact location of its laser unit.

It would be incorrect to say that on a recordable disc there are pre-recorded data. A better expression would be that the pre-groove modulation enables the detection of address information.

Why this ADIP, and why only on the recordable disc? All reading out of data, regardless of the medium, is highly dependent on the correct addressing. In other words, we can read out as

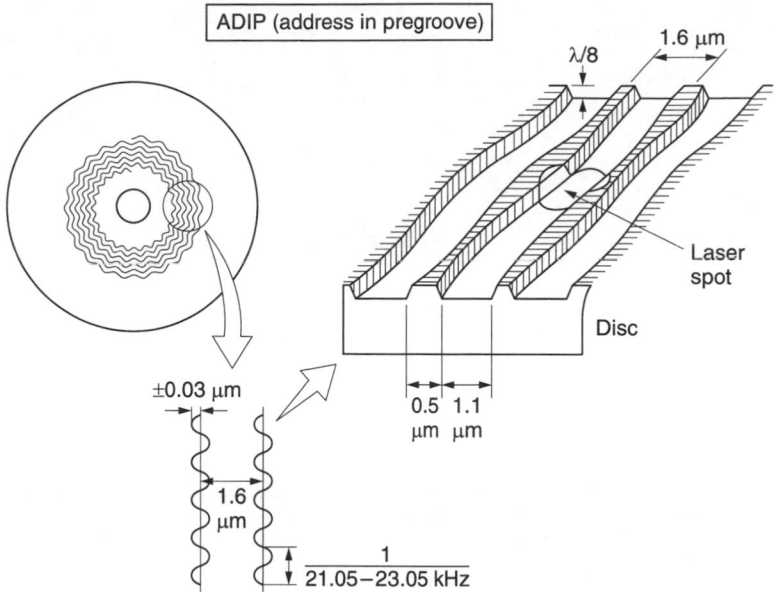

Figure 17.10 Pre-groove.

much data as we want, but if we are not able to find out what the logical place and order of these data are it will be useless. Any pre-recorded medium contains address data, which are of course included during encoding and manufacturing. Therefore, from the first read-out, any system is able to detect the addresses and correctly follow the data track.

However, the situation on a recordable disc is different. Initially, such a disc will not contain any information, so during recording, how is the recorder supposed to know whether it is following the correct track? Or when a recording is under way, and the system would endure a vibration, how would it be able to recover if there is nothing at all on the disc?

For these reasons, there has to be a fixed addressing on the disc, physically included (non-erasable) before any recording takes place. Another benefit is the fact that, as recordable discs are of low reflectivity, and as the addressing in the pre-groove will coincide with the addressing in the data stream, this double address read-out possibility may overcome errors.

ADIP encoding is of course performed prior to making the disc. The ADIP data (i.e., the address data) are bi-phase-modulated with a 6300 Hz clock. This clock is derived directly from a 44.1 kHz source; a 22.05 kHz carrier is also derived from this source.

The bi-phase signal is then FM-modulated onto the 22.05 kHz carrier; the resultant signal is used for the pre-groove. Bi-phase

Figure 17.11 ADIP modulation.

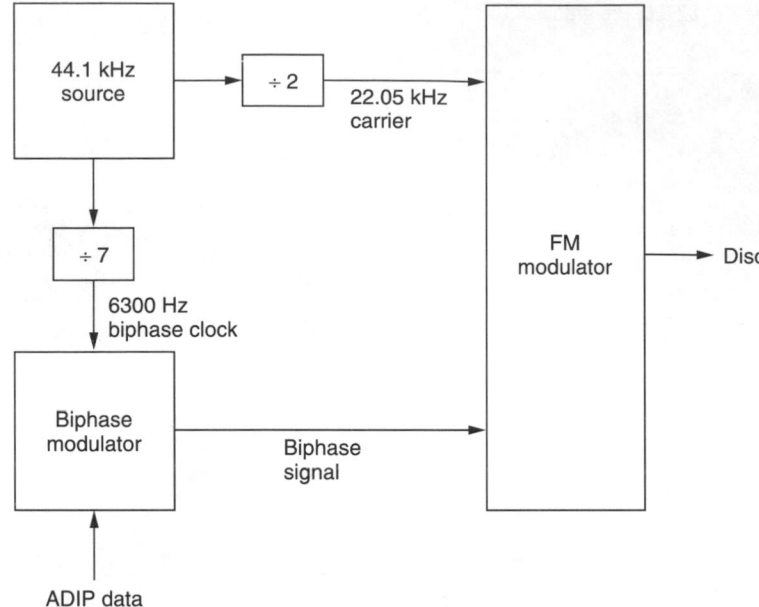

modulation means that the carrier will only be modulated by two possible modulation inputs, a digital '1' and a digital '0'. The modulated bit rate is 3150 bit s^{-1}. The frequency shift due to the modulation input is 1 kHz. In other words, depending on the input modulation data (logic '1' and '0'), the 22.05 kHz carrier frequency will shift 1 kHz up or down.

The MiniDisc will use the carrier frequency as well as the address data for CLV operation and track addressing. The contents of ADIP data will be explained later.

Physical track layout

The pre-mastered disc can again be compared with a Compact Disc. The information area consists of a lead-in area (table of contents), a programme area and a lead-out area. All these areas are pre-recorded; none of them can be changed.

The recordable disc can be divided into a non-recordable area and a recordable area. The non-recordable area is similar to the lead-in area on the pre-mastered disc. It contains information concerning locations on the disc (for example, the start and end address of User Table Of Contents, UTOC), but also information such as laser power level, disc type, etc. The lead-in area of a recordable disc is a pit signal, similar to a pre-mastered disc, but in this case the reflectivity is much lower compared to a CD type.

The MiniDisc set will adapt to this lower reflectivity. The recordable area starts with the UTOC. The UTOC is an area similar to a TOC, but in the case of a recordable disc, the allocation of addresses is not constant, it depends on the user. The way UTOC is handled is very similar to the way a computer allocates addresses on a floppy disc.

One important note must be made here: similar to the use of a computer floppy disc, writing, erasing, dividing, editing, separating and all other possible data manipulations can only be considered complete when the UTOC area is updated. If, for example, we try to record data, this will of course be recorded in the data area, but when the recording is ended, the start and end addresses need to be written in UTOC. If, for one reason or

Figure 17.12 Track layout.

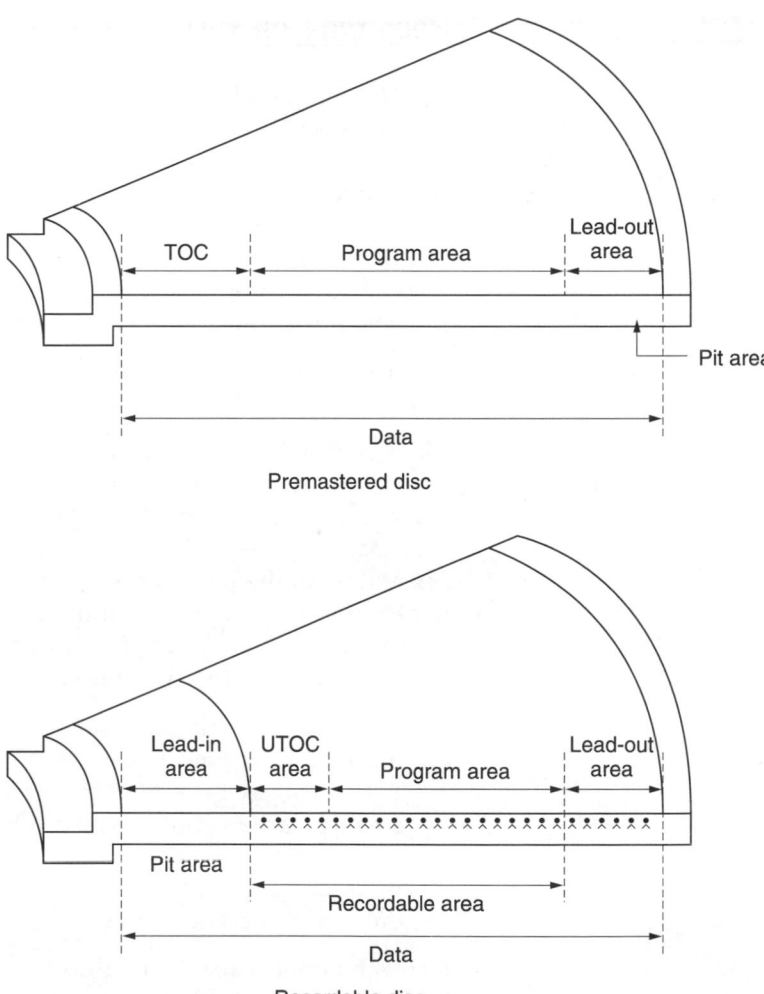

Premastered disc

Recordable disc

another, this is not performed (the writing in UTOC), the system will not be able to consider the recorded data as valid.

Another example: when we erase a music track (data track) we do not erase the audio data at all; only the start/end address in the UTOC area will be erased. The area taken by the music data then becomes available and will be overwritten during a next recording session. The same applies for all other possible recording operations. The UTOC area also includes other information such as disc info, copy protect codes, recording time and date, start and end addresses of tracks, etc. After the UTOC area, the recordable user area takes most of the space on the disc and at the end of course a lead-out area, which in this case is recordable and also contains the pre-groove.

Recording on MD

- The pre-mastered disc is manufactured similarly to the Compact Disc; it is pre-recorded, and the user can only perform read-out.
- The recordable disc only contains address information in the ADIP pre-groove; on this disc it will be possible to record user audio.
- On these recordable disc types it is possible to re-record a nearly unlimited amount of times.

Recording on a MiniDisc is based upon well-known physical laws.

Many metals and alloys, once they reach a certain temperature, called the Curie point, can be magnetically influenced by external magnetic fields. In other words, by heating and applying external magnetic fields we can change the magnetic orientation of the particles of metals. The substance that we will use for this purpose is a terbium–ferrite–cobalt alloy. Many other alloys can of course be used, but this one is particularly apt as it has some features that were needed for MiniDisc:

- A fairly low Curie point (about 185°C), enabling quick heating with little power.
- A low coercivity of about 80 Oersted (6.4 kA m^{-1}), enabling stable polarity reversal with a relatively weak field for which also no high power is needed. (The higher the coercivity, the more power is needed to impose an external magnetic field onto the material.)

Here, the critical reader might object that magnetic recordings have been performed for decades already, and that in most of

Figure 17.13 Track layout.

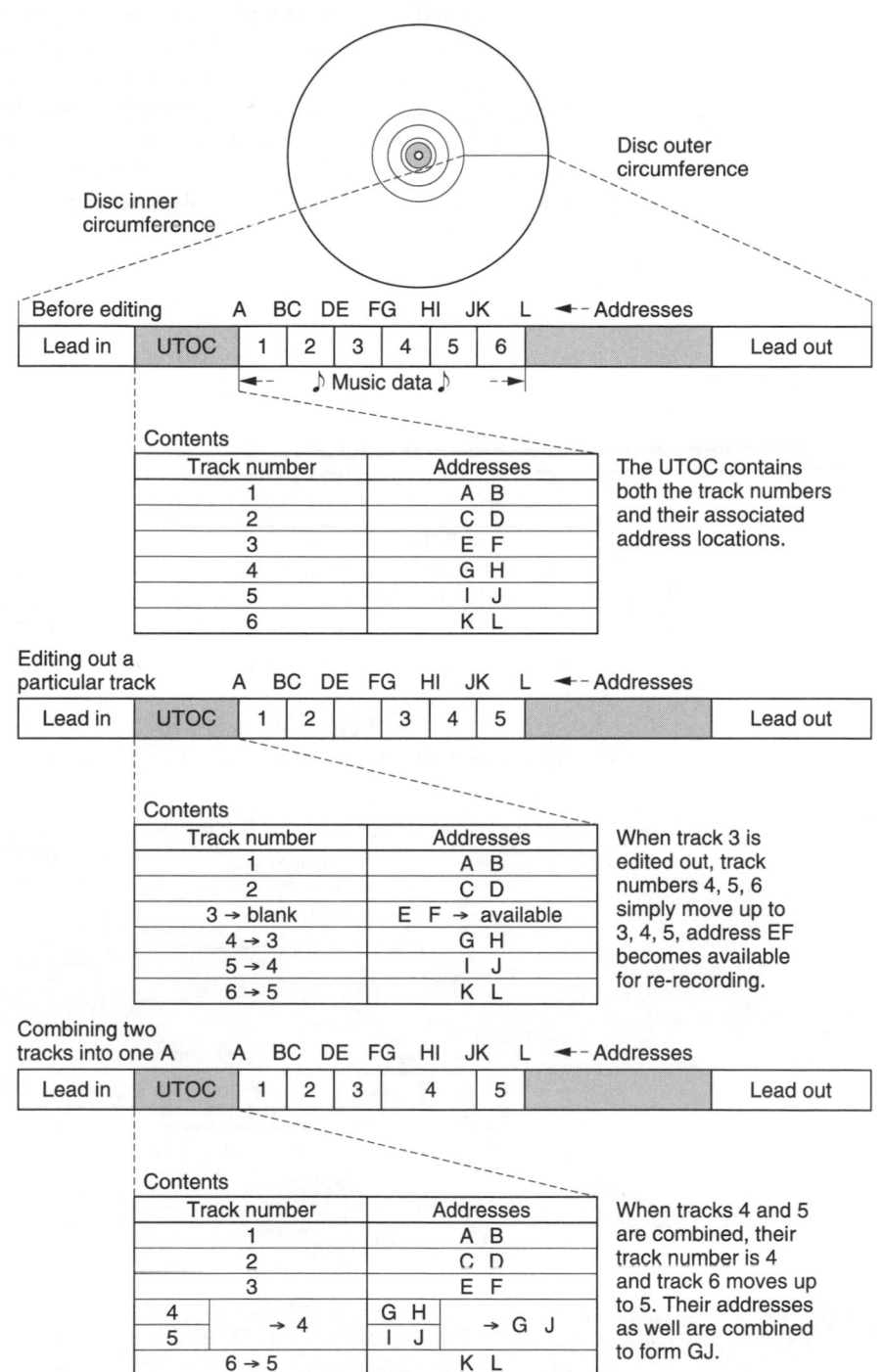

Disc outer circumference

Disc inner circumference

Before editing

Lead in | UTOC | 1 | 2 | 3 | 4 | 5 | 6 | | Lead out

A   BC DE FG   HI   JK   L ◄‒‒ Addresses

♪ Music data ♪

Contents

Track number	Addresses
1	A B
2	C D
3	E F
4	G H
5	I J
6	K L

The UTOC contains both the track numbers and their associated address locations.

Editing out a particular track

Lead in | UTOC | 1 | 2 | | 3 | 4 | 5 | | Lead out

A   BC DE FG   HI   JK   L ◄‒‒ Addresses

Contents

Track number	Addresses
1	A B
2	C D
3 → blank	E F → available
4 → 3	G H
5 → 4	I J
6 → 5	K L

When track 3 is edited out, track numbers 4, 5, 6 simply move up to 3, 4, 5, address EF becomes available for re-recording.

Combining two tracks into one A

Lead in | UTOC | 1 | 2 | 3 | 4 | 5 | | Lead out

A   BC DE FG   HI   JK   L ◄‒‒ Addresses

Contents

Track number		Addresses	
1		A B	
2		C D	
3		E F	
4	→ 4	G H	→ G J
5		I J	
6 → 5		K L	

When tracks 4 and 5 are combined, their track number is 4 and track 6 moves up to 5. Their addresses as well are combined to form GJ.

those cases only a magnetic field was necessary, not the combination of heat and a magnetic field. This is a correct observation, but this simple way of magnetic recording has one tremendous drawback: it is not safe! Any magnet or magnetic field which might – by accident or not – be brought into the area of such a simple magnetic recording can erase it. The amount of valuable information that has been lost over so many years in this way is impossible to quantify. The combination of a specific temperature which is definitely outside the range of normal environmental temperatures and a magnetic field will result in a recording that can only be erased or overwritten when required, which is of course much safer. Another fact is that recordings made in this way are extremely stable; they will not change significantly over a long period of time.

The recording uses Magnetic Field Modulation. Recording is performed at one time, the laser will on one side to heat the magnetic layer up to the correct temperature, and on the other side, the magnetic head will impose the correct polarity. At this point, it should be remembered that the information that is written on the disc is digital information; in other words, there is only a need for two states: either logic 1 or 0. Translated to the magnetic information, this means north or south magnetization. Recording and re-recording are performed by overwrite; the previous data – if there was any – are not erased first. A special magnetic head was developed which enables extremely quick

Figure 17.14 Disc recording.

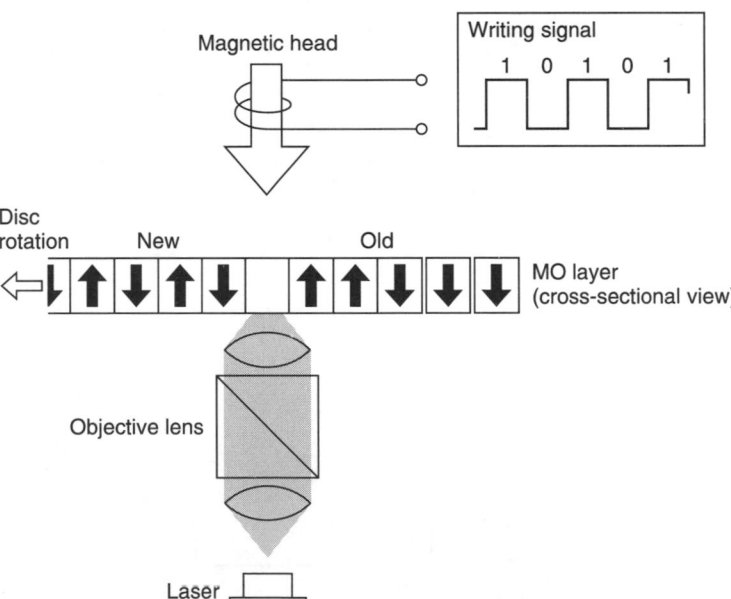

flux reversals of approx. 100 ns. This is important to ensure precise recording, but it also implies that recording is really performed in one stretch. The layers surrounding the magnetic layer are adapted for high-temperature handling.

Read-out of the disc

Pre-mastered disc

The pre-mastered disc is obviously read out similarly to a compact disc. The laser light level reflected from the compact disc or pre-mastered disc depends on the pits stamped in the disc.

The laser objective shown here is of course very simplistic; this will be explained later. Note that the double detector blocks (PD1 and PD2) are needed for both pre-mastered and recordable disc.

Read-out of recordable disc

On the recordable part of the recordable disc, another physical law is involved: the laser light sent to the disc will hit the disc surface, pass through the magnetic layer and hit the reflective layer. In this way, the laser light will also be reflected similarly to the Compact Disc or pre-mastered MiniDisc. However, as there

Figure 17.15 Pit read-out.

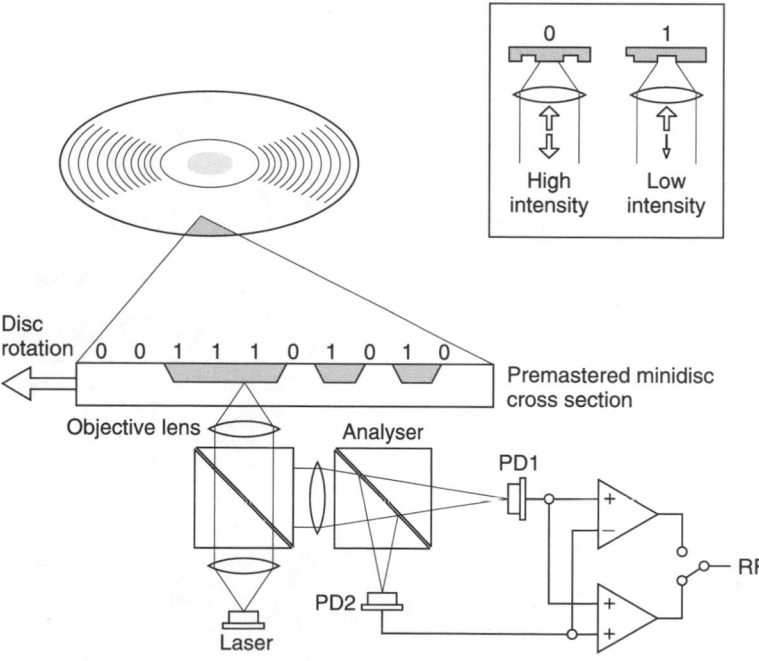

are no pits, the reflected laser light level is constant. When passing through the magnetic layer, the so-called Kerr effect takes place: the polarization of a light beam is changed when it passes through a magnetic medium.

As the magnetic layer was modulated magnetically during recording, it is obvious that the modulated contents will now pass on to the laser light.

The magnetic recording included only two possible states (north/south). The read-out polarization will also be predominantly either one of two main polarizations.

Note the following:

- During read-out, the magnetic head is not used; this part will only be used during recording.
- The recordable disc is a low-reflectivity disc, but the reflected level will be fairly contstant, contrary to the pre-mastered disc.
- The same laser is used for recording as well as for read-out; for recording, the laser power will be significantly higher than when reading out.

Figure 17.16 Recordable MiniDisc read-out.

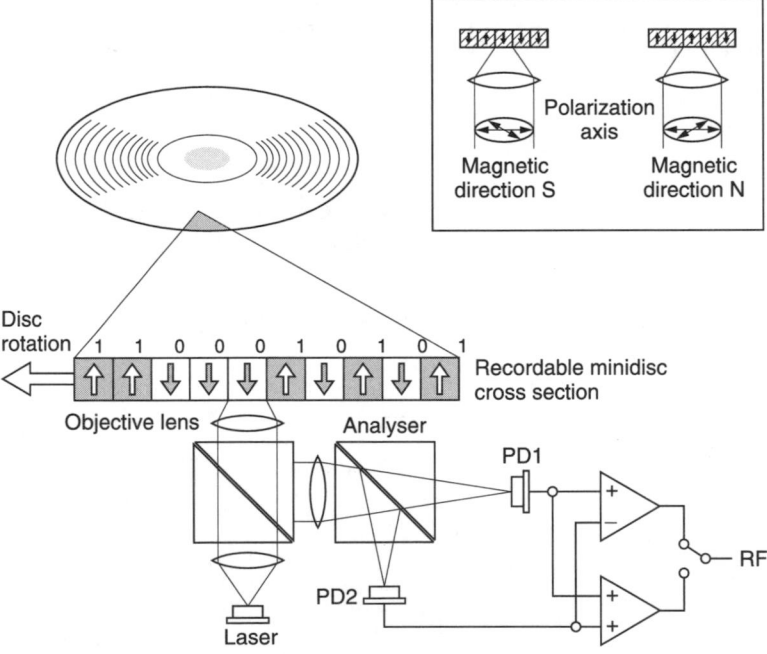

The MiniDisc optical block unit

The optical block unit used in MiniDisc resembles the one used in Compact Disc and was explained in previous chapters.

The main similarities are the concept used, basic design of the laser unit, the laser type used and wavelength. Even if it has already been explained that read-out of the recordable disc is based upon a different basis than the pre-mastered disc, this does not really affect the way the optical block operates during read-out; the differences appear mostly in view of recording.

Figure 17.17 MO read-out.

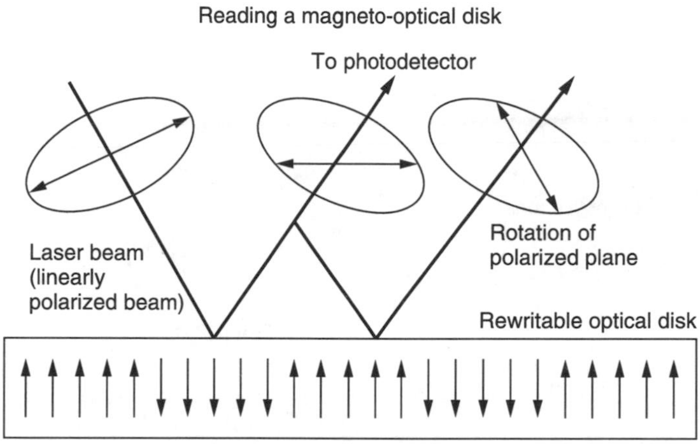

Reading a magneto-optical disk

To photodetector

Laser beam (linearly polarized beam)

Rotation of polarized plane

Rewritable optical disk

In view of recording, the laser power used is much higher: up to 7 mW. However , for playback, the laser power used is the same or similar as for a CD read-out: 0.5 mW.

One optical block assembly is used to read out both signals (from pre-mastered as well as from recordable disc). It is obvious that, in order to obtain this dual read-out possibility, as well as the higher power handling, new technologies need to be used. These will now be explained.

A note must be made concerning the laser power: when discussing laser power we must distinguish between total laser power and the power of the main beam. When using total laser power, we refer to the original three-beam system, as mostly used in Compact Disc, and we add the power of main beam and side beams. This is logical, as all beams are emitted from one laser diode. In the case of the MiniDisc, the total power is about 7 mW. When using the main beam power only, we have of course a lower power (in MiniDisc, about 5 mW), as we only take a part of the total power into account. This distinction should be remembered, as there may exist some confusion when comparing publications.

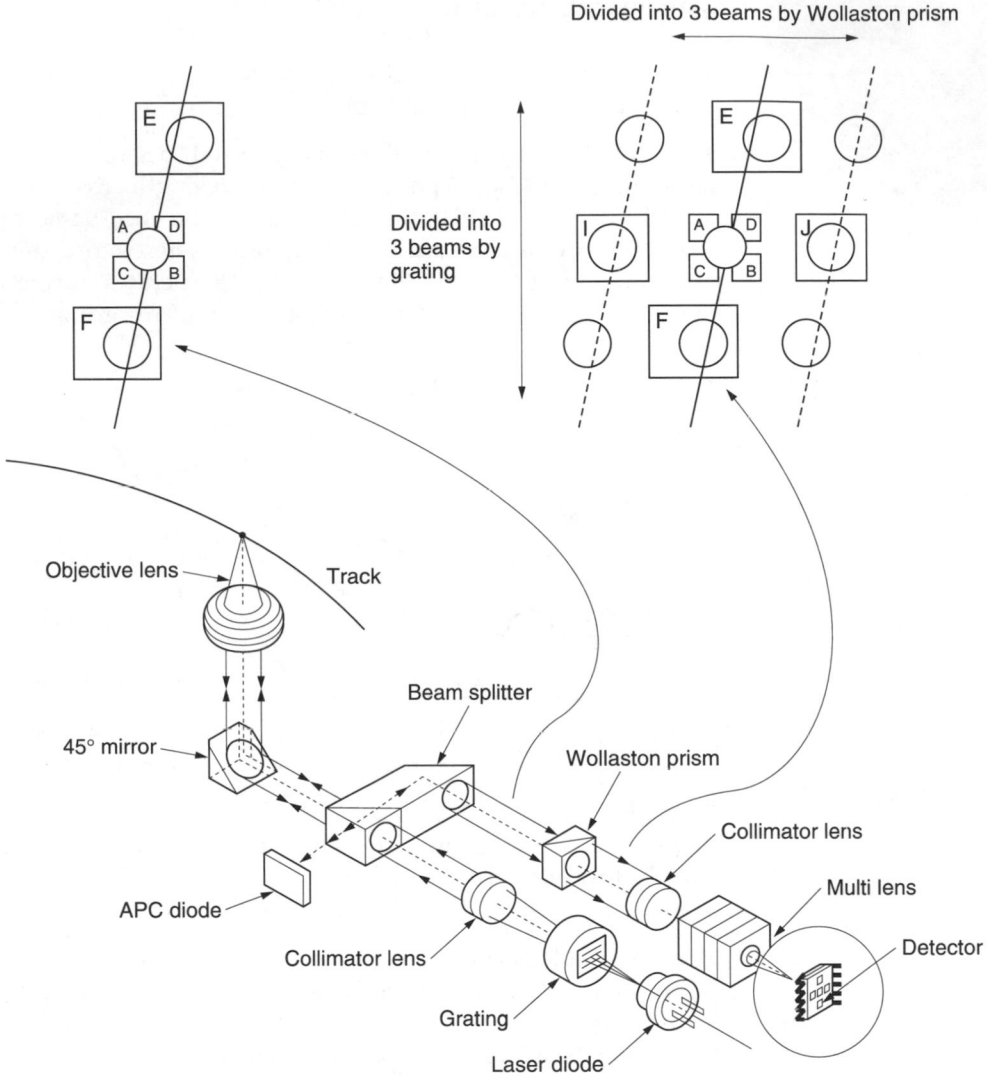

Figure 17.18 MD optical block.

When following the laser beam from laser unit to detector unit, we see initially the same path as in the compact disc OPU.

- The beam is emitted from the laser and passes through a diffraction grating, where the side beams (E and F beams) are created.
- As from this point we have the three-beam laser light. The collimator lens is used to create a parallel beam.
- The beams pass through a beam splitter, but at this time the beam splitter is of no real use.

- A 45° mirror will point the laser correctly to the disc.
- An objective lens is used for focusing purposes.

As explained in the disc section, on the disc there are two possibilities, either the pit signal or the magnetic signal (Kerr effect). However, in both cases we will have a reflection of the laser light which will carry the information we need. Also, for the recordable disc, the return will also carry a component from the pregroove (also referred to as 'wobble').

- The return laser light goes through the objective lens and the 45° mirror.
- In the beam splitter it is sent to the photodetector side.
- After the beam splitter, the return laser light follows a different path from the emitted (from laser diode) laser light.
- The beams are sent through a Wollaston prism; this special type of prism is used to extract from the main laser beam the polarization (north/south or X/Y) components as set by the magnetic layer on the disc.
- After the Wollaston prism, the beams are sent through a multi lens, which is a combination of a concave and a cylindrical lens, resulting in a correct spot configuration on the detector.
- The detector unit is similar to the compact disc detector unit, but two more detectors are included, the I and J detectors. These two detectors will catch the side beams, extracted through the Wollaston prism and containing the data.

Note also the automatic power control (APC) photodiode, which is located at the left side of the beam splitter. This module is necessary to monitor the power emitted by the laser, in order to control it correctly, as such lasers are easy to break down if they emit too high a power level. In the CD optical block, the APC diode was located at the back of the laser emitter. In the laser unit which is now used in MiniDisc, a back APC photodiode would not give a correct feedback signal; the front APC diode gives the correct feedback signal (i.e., the return is directly related to the laser power).

The Wollaston principle

Before getting to the operation of the Wollaston principle, remember which kind of signals we can expect here.

We have two types of disc, pre-mastered and recordable.

- The pre-mastered disc gives a main beam related to the pit structure, and the E and F side beams related to the tracking.

- The recordable disc gives in the lead-in area a pit signal as well as E and F signal, but with a decreased level.
- After the lead-in area, the main beam polarization has been influenced by the magnetic layer on the disc (Kerr effect).
- The main beam has also been modulated by the (wobble) pre-groove.
- Besides the main beam, the E and F side beams are still present.

The Wollaston prism is a combination of two rock crystals bonded together at a precise angle of 45°.

Any incident beam will result in four outputs:

- Two main outputs, each containing a main polarization part (north/south or X/Y in the drawings, according to an orthogonal structure); these two beams are so close to each other that they can be considered and treated as an unchanged main beam, as they will practically recombine.
- The two other beams are highly important. The Wollaston prism separates the north- and south-oriented components of the incident beam and emits these components as side beams at a 45° angle.
- These side beams are called the I and J beams.

Note:

- Each beam originally has north as well as south orientation. The Kerr effect will make one of the two more dominant.
- The E and F beams will each create side beams due to the Wollaston effect, but these side beams are not used as they are insignificant.

The I an J outputs are both derived from the main input (P vector in previous figures); the right part of Figure 17.20 shows the polarization angles caused by the Kerr effect on the disc. If a laser light was polarized (either north or south), the P vector will be shifted slightly to one of both sides (I or J side). The angle is fairly small (below 0.5°), but this is enough to detect a difference between I and J outputs. θK shows the Kerr angle in both directions.

Figure 17.19 Wollaston prism.

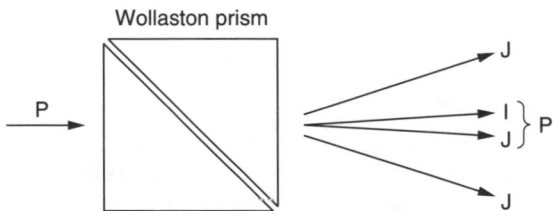

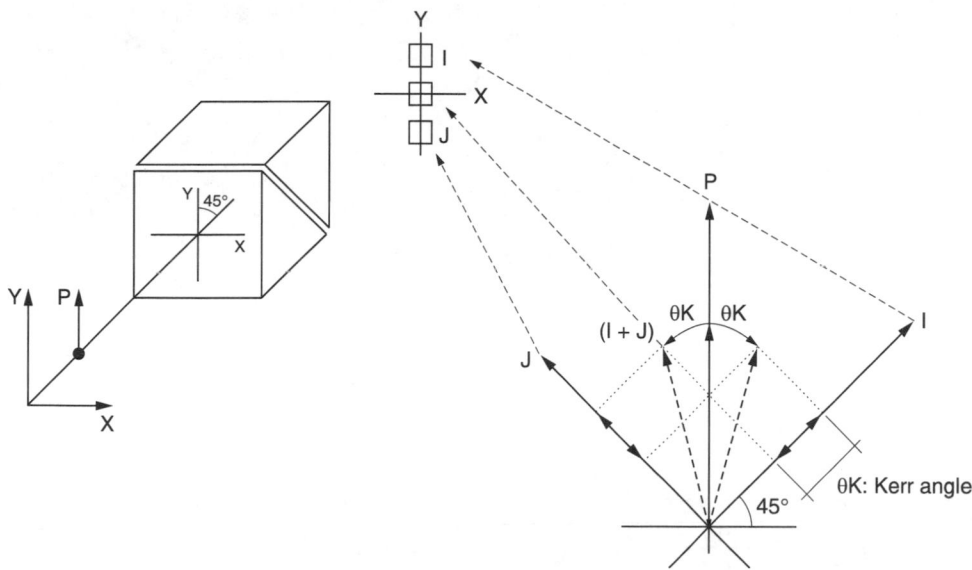

Figure 17.20 Wollaston prism operation.

The operation should be clear now; the magnetic surface on the disc polarizes the laser light. This polarization is seen by an angle shift of the original P vector to either the I or J side, due to the Wollaston prism. In this way, one of the two side beams will be larger, which can be detected. The north/south differences on the disc can therefore be detected by the differences in I and J.

The I and J beams now become the most important beams. For the recordable disc the use is obvious, as the beam reflected from the disc contains a north/south polarization related to the data. The extraction of north/south components by the Wollaston prism will enable the read-out data correctly.

If, for example, the laser beam has been momentarily north-polarized by the Kerr effect on the disc, the north-polarized component as extracted by the Wollaston prism will be significantly larger than the south component. In that case, the I beam will be detected with a higher level than the J beam. If, on the other hand, the south polarization is dominant, the J beam will become bigger. The subtraction I − J will therefore enable the MiniDisc system to interpret the magneto-optical signal and convert it to a correct data stream.

However, the pre-mastered disc system also uses the I and J side beams, contrary to the Compact Disc system. It is now very important to remember that the Wollaston prism is such that the I and J components come at a 45° angle: no matter what the

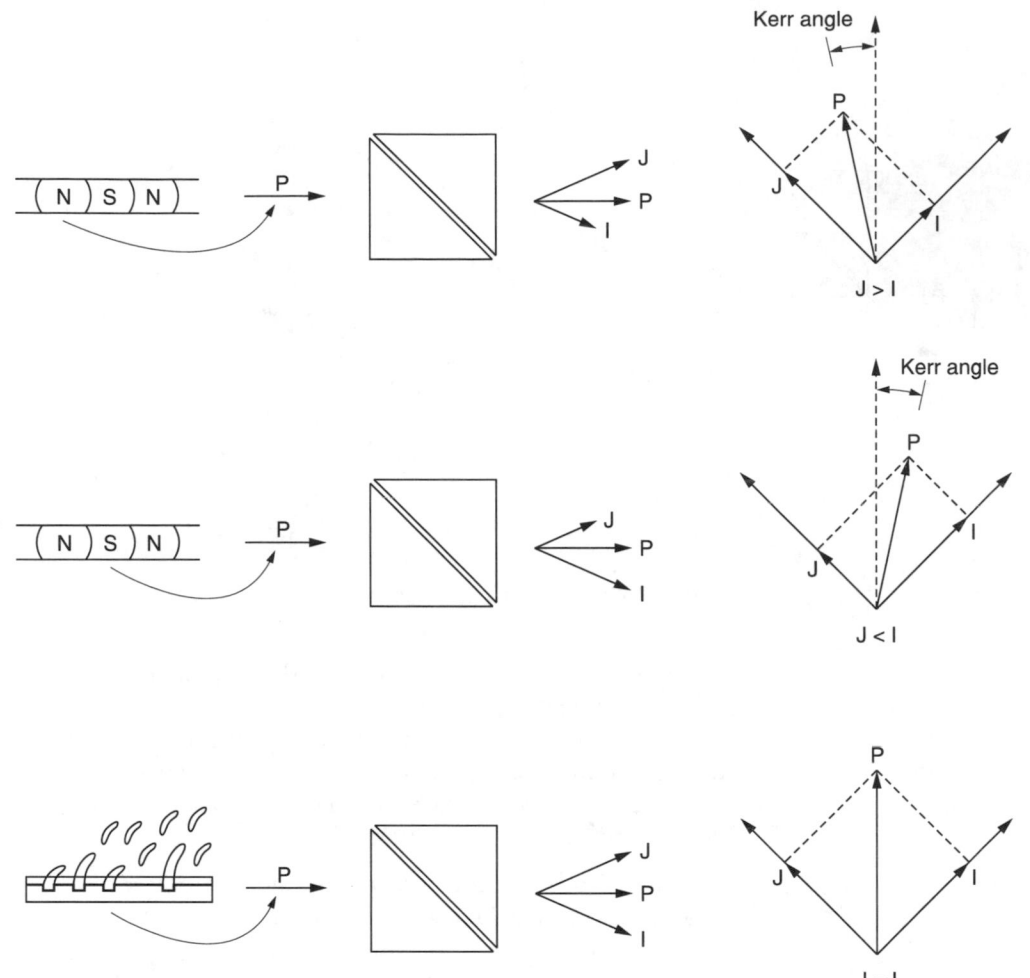

Figure 17.21 Wollaston prism operation.

polarization of the incoming beam is, the sum of the I and J side beams will always give an indication of the level which is effectively wanted for the pit signal. With a laser signal which has not been influenced by the Kerr effect, we can expect the north/south component to be equal and thus the summing of I and J will give a perfect reconstruction of the original signal.

The detector block

Again, the detector unit is similar to the one used in a Compact Disc, but the I and J detectors have been added.

Figure 17.22 CD detector.

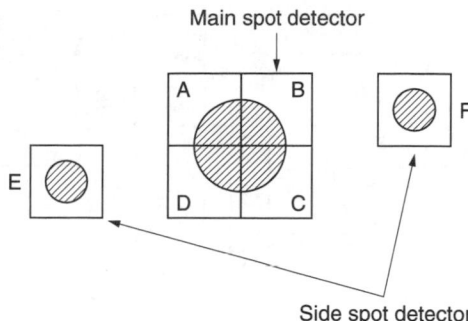

In a Compact Disc optical block, the detector layout is as shown above.

- The main spot will be detected by a block of four segments, named detectors A, B, C and D.
- This quadruple set-up enables one to detect correct focusing of the main beam. For the main (audio data) signal, the signals from each segment are added.
- For focusing the combination of (A + C) − (B + D) is made. The fact that this calculation is zero, smaller or larger than zero reveals the relative focusing of the beam (correct, focusing lens too close or lens too far). If the beam is correctly focused, it will appear on the ABCD detectors as a circle, with an equal amount of light on either side of the detectors. If focusing is not correct, the return beam, due to the special lenses on the return path, will not be a circle, but an ellipse. The calculation algorithm will then be higher or lower than zero.
- The detectors E and F will be used for tracking. These beams are directed partly on and partly off the data track, each on one side. In this case, the calculation of E − F signals shows whether the laser is correctly centred (E − F = 0) or at the left or right from the track (E − F0 or E − F).

As mentioned before, the E and F beams also create side beams in the Wollaston prism, but these will not be detected as there are no detectors at the place where they will arrive. We now have ABCD, E, F, I and J detectors. The use of these detectors is not totally the same as in a Compact Disc:

- I and J will be used to detect the magneto-optical (I − J) as well as the pit (I + J) signal.
- E and F will still be used for tracking.
- ABCD will be used for several purposes.
- Automatic Gain Control (A + B + C + D) will give the MiniDisc player an indication of the relative strength of the return, in order to set the gains of following amplifiers.

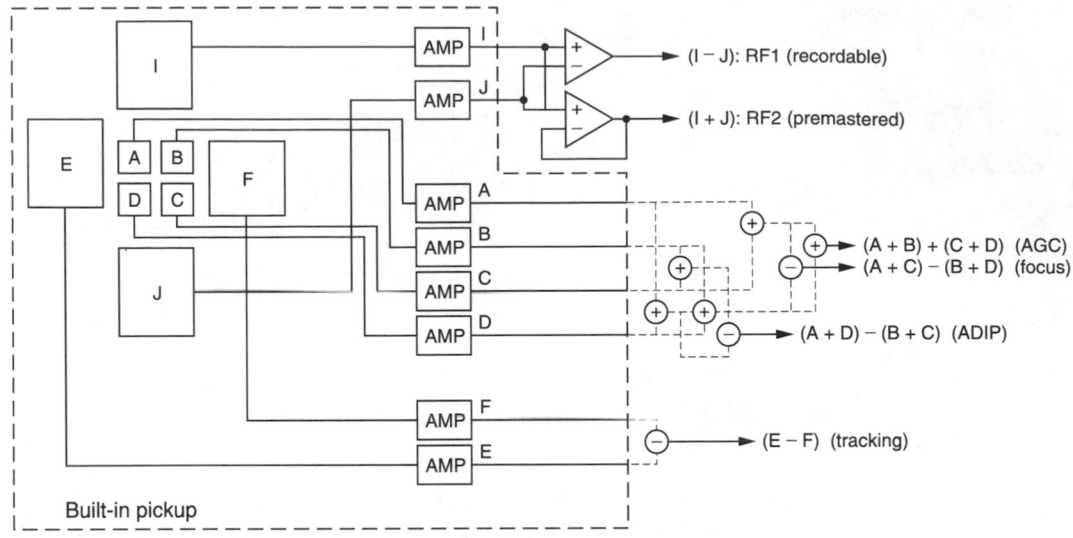

Figure 17.23 MD detector.

- Focus $(A + C) - (B + D)$ is performed similarly to the Compact Disc.

Pre-groove (wobble) $(A + D) - (C + B)$ detection can also be explained similarly to the focus operation. When checking the relative position of pre-groove and laser beam, and comparing this with the mathematical algorithm as shown, it should become clear that the calculation is effectively checking the difference between the left and right edges of the returned beam, and as these edges have been modulated slightly by the wobbling

Figure 17.24 Main spot on MD pre-groove.

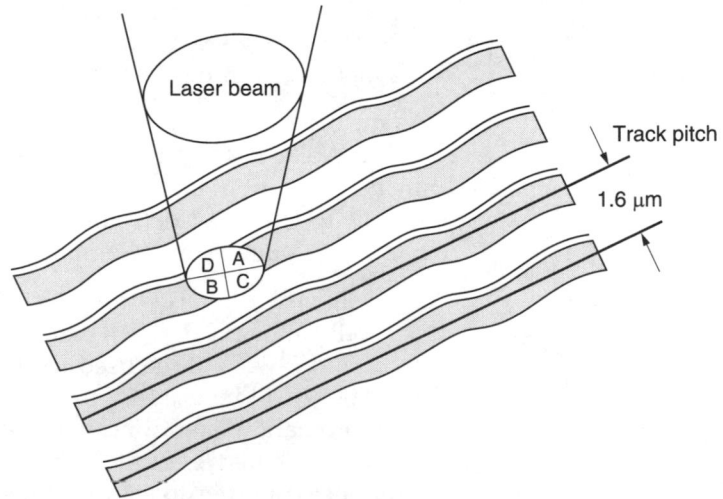

pattern of the pre-groove, the calculation will effectively enable detection of the ADIP sine wave.

Each signal is detected (i.e. light input is converted to electric current) and amplified. Next, according to all the required outputs, the detected signals are added or subtracted and fed to the MiniDisc player for further processing.

Psychoacoustics

The MiniDisc specifications show that, on a small disc, it is possible to store the same amount of time (74 minutes) of audio data as on a CD. It is obvious that some data compression is needed to attain this goal. Such compression has to be performed while maintaining the full audio quality. This is performed in the ATRAC compression system, but before explaining the ATRAC compression system, some notes on psychoacoustics have to be made, as the compression is based upon psychoacoustics.

Psychoacoustics is related to the way we perceive sound, or in other words, the relation between a physical stimulus and the subjective sensation that this stimulus will provoke. Here, immediately the complexity becomes clear: it is related to each individual, and it also depends highly on a multitude of surrounding factors. Most measurements in this field have been performed on series of persons, this over several years. The results obtained at these test sessions can be considered good averages.

It is beyond the scope of this book to cover all of the human hearing process, but some items should be remembered:

- sound propagates through air as pressure waves;
- these pressure waves hit our ear drum;
- the combination of outer ear and ear canal, up to the ear drum, partly acts as a horn, and also partly as a resonator (open pipe);
- after the ear drum the pressure will be transposed to the inner ear, which consists of fluid-filled structures;
- in these structures, there will be a transition from sound as pressure to sound as electrochemical signals, and at the same time the sound is already analysed into its constituent frequencies;
- the electrochemical signals are transported to the auditory systems in our brains, where they will be processed; in this way our brains perceive sound;
- processing will of course be based upon cross-relating input from our two ears.

Figure 17.25 Human
hearing process.

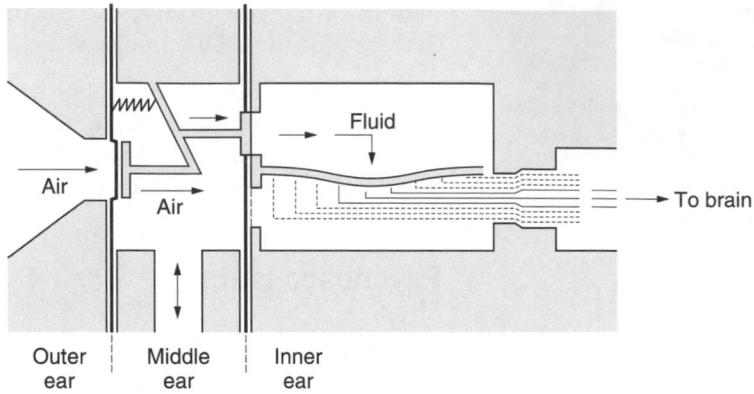

The outside hearing system, outer ear and ear canal including
the drum, are built similarly to, and react as, an analog sound
system. In this respect, it is obvious why a significant part of our
hearing system was named the 'drum'.

However, from the moment that sound is analysed and transposed to elecrochemical signals, and also the processing that
follows, this system is no longer reacting as a known analog
sound system. In this respect, it should not really come as a
surprise that the way digital systems such as ATRAC operate will
appear similar to the way our hearing system processes sound.

The hearing range of a human being starts at about 20 Hz and
can go up to just below 20 000 Hz (= 20 kHz). Here again, it
should be stated that this is a purely theoretical approach, as large
variations are possible depending on:

- the individual;
- ear to ear (left/right);
- age;
- health;
- fitness;
- environmental aspects.

Psychoacoustics related to MiniDisc

Some major facts in psychoacoustics have been used to enable
designing the MiniDisc compression systems.

Thresholds

The sensitivity of our hearing system is not equal or linear over
the total frequency range. A sound at a given frequency must

have a minimum level in order to be perceived. This minimum level is called the threshold of hearing.

Figure 17.26 shows the audio frequency spectrum and the threshold level for each frequency. Any sound above the minimum level will be perceived, and any sound below the threshold level will not be perceived.

This phenomenon can be used to our benefit. Suppose we are able to analyse sound into its constituent frequencies with their respective levels. When comparing this with the threshold curve it is possible to delete all frequencies whose levels fall below the threshold curve. In most of the cases these frequencies are noise frequencies anyway.

In this way, it will be possible to lower the amount of sound information that has to be recorded, without touching the audio quality itself.

Masking

Masking means that one frequency, with a higher level, can make another frequency with a lower level inaudible. This is easy to explain: suppose two persons are having a conversation, and a

Figure 17.26 Frequency spectrum and psychoacoustic effect.

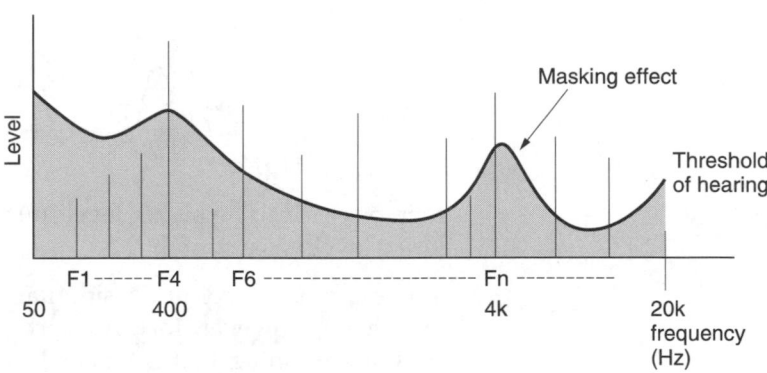

Figure 17.27 Frequency components and levels extracted by ATRAC.

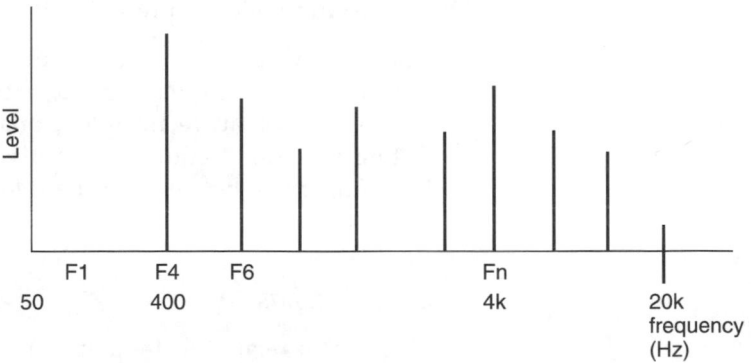

Figure 17.28 Critical bands.

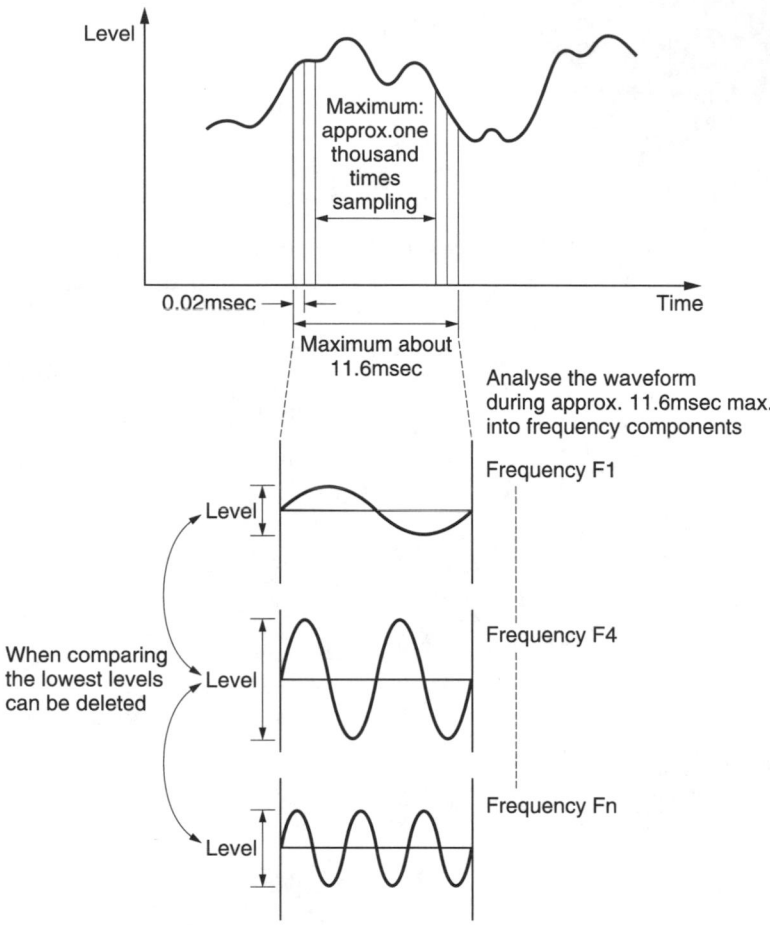

train passes by. The higher level sound of the train will mask the lower level voice sound.

This example is typical for simultaneous masking. Strange as it may seem, however, forward and backward masking are also possible, meaning that a higher level sound can mask a lower level sound which is preceding or following.

Sound as we perceive it is usually the composition of a multitude of single frequencies, each with its own specific level. Suppose now that we are able to analyse sound such that we can distinguish each frequency at a given time with its level. Here, the comparison between all these levels will enable us to delete those frequencies which are masked.

Critical bands

This is also related to frequency resolution, which is the ability of our hearing system to distinguish and separate two simultaneously present signals.

Critical bands are parts of the audio spectrum where signals appear not to be separated. The sensitivity of our ears in the frequency domain within a critical band is about equal, but different compared to another critical band.

Over the whole hearing range, about 24 critical bands have been distinguished. The width of each band is not constant; these widths range from 100 Hz at lower frequencies (Hz) up to several kilohertz at higher frequencies, or in other words, our hearing is more sensitive in certain areas than in others. The critical bands approach can also be used if we can break down sound into its constituent frequencies; we are able to adapt the sensitivity of processing and thereby reduce the amount of data.

Conclusion

- Supposing we are able to analyse sound in a correct way, it is possible to apply psychoacoustic facts to it, and if this is done correctly we can use this to reduce the amount of audio data.
- This data reduction can be performed without noticeable changes to the original sound.

ATRAC

ATRAC is an acronym for Adaptive TRansform Acoustic Coding. When reading this carefully, the most important actions of the ATRAC system become apparent.

- Audio input is encoded according to a transform method which is adaptive. This means that the transform method can adapt to the input signal; this is in accordance with acoustic phenomena.
- The main target of the ATRAC encoding system is to decrease the information density and in that way increase the recordable time on a small disc; all this, however, without degrading the sound quality.
- The decoder, on the other hand, will restore the original audio data based upon the compressed data input. The word compression is often used for this kind of action, although the meaning in this application is not totally the same as in a computer environment.

In a normal CD system, the basic bit stream is 16 bits, two (L/R) channels and 44.1 kHz sampling rate:

$$16 \times 2 \times 44.100 = 1.4 \text{ Mbit s}^{-1}$$

After further processing (EFM, CIRC, etc.), this becomes about 4 Mbit s^{-1}. With this bit rate, we can put 74 minutes of music on the 12-cm disc format. If we want to put the same amount of music time on a disc of nearly half that size, it is obvious that the bit stream has to be reduced drastically. There are several ways to do this, but not many of these give an acceptable result. In most cases (example: reducing the amount of sampled bits or the sampling rate) the result will be a degraded sound.

The ATRAC system is highly sophisticated, as it is based upon complex mathematical and psychoacoustic methods. Several complex manipulations are performed similarly and in connection with one another. It is therefore very difficulty to give a simple and clear overview of this system.

Figure 17.29 CD–MD size comparison.

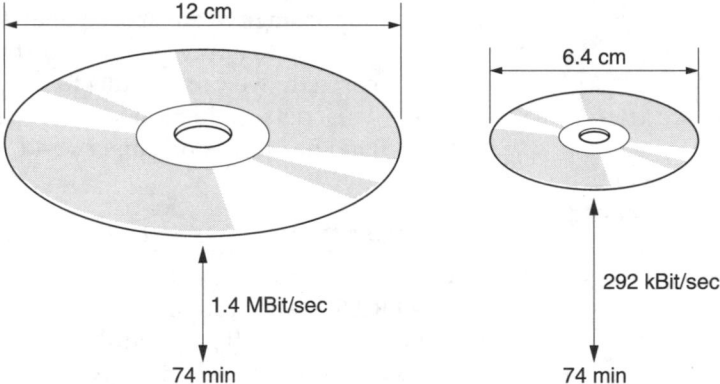

Basics on Fourier analysis

Some basic notes on Fourier analysis are necessary for a good understanding of digital sound processing:

- Sound can be analysed and described as a mathematical function (sine and cosine waves).
- One single frequency, seen mathematically, is one single sine function, with an amplitude (A), angle (ω) and phase (φ):

  $f(t) = A \sin (\omega t + \varphi)$

  where t is a function of time.

- From this basic mathematical formula, one can also derive that sound is a function of frequency and of time. We can describe sound as a function of frequency and also as a functon of time; accordingly, we term this as frequency domain and time domain. A 'transform' is a mathematical description of the transition from one domain to another.

- Starting from the mathematical formula of one basic frequency, we can define harmonics, which are derived from this basic frequency, but whose angle is a multiple of the basic angle:

$f = A \sin(\omega t + \varphi)$ basic frequency
$f = A \sin(2\omega t + \varphi)$ first harmonic

- 'Sound' is a composition of basic sine and/or cosine waves, each basic wave also having series of harmonics. Each sound can be analysed and written into its mathematical components by using Fourier of other (usually derived) methods. Differences in sound can be seen as differences in composition of sine/cosine waves and harmonics.
- Normally, Fourier analysis handles infinite series, but as the spectrum that we handle is limited (up to 20 kHz), we will use finite series.

Figure 17.30 is a very simple example of a waveform (a) and its constituent elements (b + c + d), consisting of one basic wave and two harmonics. A Fourier analysis performed on the waveform should – if performed correctly – reveal these constituent frequencies with their respective amplitudes.

Figure 17.30 Fourier analysis.

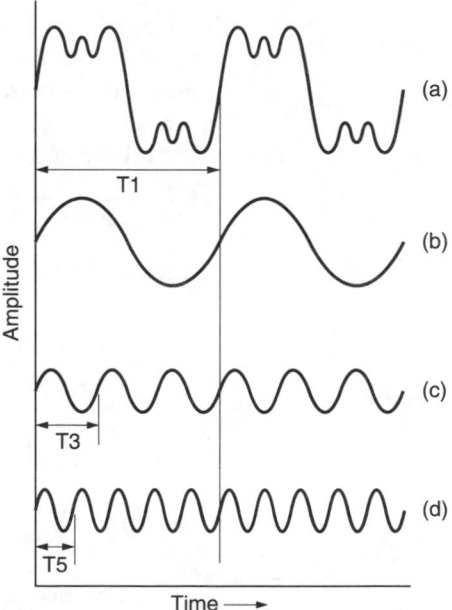

ATRAC input versus output

- The input to the ATRAC encoder is still the 16-bit, two-channel, 44.1 kHz sample rate bit stream, exactly the same as in a CD system.
- The output, however, will not be the 1.4 Mbit s^{-1} data stream, but it will be a 292 kbit s^{-1} data stream. This is about five times less than the CD system.
- For this reason the MiniDisc compression rate is about 1:5.

ATRAC operation

We already mentioned the basic psychoacoustic phenomena that are taken into account and some basics on Fourier analysis; now both will meet.

At first, the incoming (digital) sound is analysed for frequency bandpassing. The full audio spectrum will be split into three bands: high, middle and low. This simplifies the processing and more detailed and correct handling of each specific part. This splitting is performed in so-called Quadrature Mirror Filters (QMF filters).

- The high-frequency (above 11 kHz) band is separated first.
- Next the middle- and low-frequency bands (0–5.5 kHz/5.5–11 kHz) are separated.
- While the middle/low-band analysis takes place, the high-frequency band is delayed to keep correct timing.

The three bands will now pass the Modified Discrete Cosine Transform (MDCT) blocks separately. This is a highly complex mathematical transform method, comparable to a Fourier analysis. The input sound which is still in the time domain will be analysed, and the constituent frequencies and their levels over a

Figure 17.31 ATRAC block diagram.

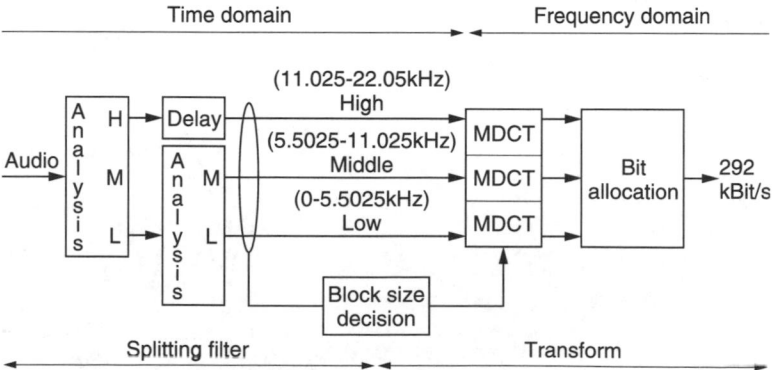

certain time slot will be extracted. At this time, a transition is made from time domain to frequency domain.

Before MDCT processing is performed, a decision needs to be taken about the block size. For this purpose, the output of each band will be monitored and evaluated. This block size relates to the time/frequency blocks that will be allocated in the MDCT processor. Simply put, it relates to the amount of processing power, processing time and resolution that will be used. Depending on the type of input sound, a decision can be taken to allocate more or less time to it, and then analyse it on a higher or lower level. This is also a way to decrease the amount of data bits and enhance the overall efficiency.

A simple example can clarify this: when we encode music for a Compact Disc, it does not make any difference whether there is sound or not; even silent passages or passages with no sound at all will be sampled, processed and stored. Therefore, we can expect a high amount of samples (each taking 16 bits) which only mean: no sound. It is obvious that this is a waste of time and data, as well as storage space. Suppose we were able to detect these 'useless' samples and modify the amount of bits allocated to such meaningless inputs (for example, use only a couple of bits instead of 16 for each sample which has no real content), we would gain much processing time and storage space. The block decision taken in ATRAC is similar to the one of the example, but it goes a lot further.

Analysis and transform is performed on blocks of data, i.e., a certain number of samples are taken and processed as one block. When we vary the block size, we consequently also vary the number of samples, which means the time and space allocated to the processing of that particular block.

The maximum time slot for analysis as one block is 11.6 ms. There are eight possible other timings, the smallest being 11.6/8 = 1.45 ms. This separation in time blocks is called Non-Uniform Time Splitting.

The idea behind this block decision is the following:

• There are several types of sound. The human hearing system recognizes this and adapts to it.
• Take, for instance, a short-impulse type sound; on the time axis this is a very short burst, therefore we do not need to allocate a big time block to analyse it. Such a short burst needs no detailed analysis, as the sound level changes are very sharp, and only the changes or the change ratio need to be known. This is a typical effect: when the differences in

levels between consecutive sounds are large, our hearing system will concentrate more on the differences, not on the details.

- The second example is the opposite: take a slowly varying sinus-type sound as input; on the time axis this takes a long time, and we will need to allocate a long time slot to analyse it. At the same time it is then possible to analyse this sound in a very detailed way. This is necessary as our hearing system will do the same with such sounds. In these cases we will become very sensitive for details.

Apart from the explained non-uniform time splitting, we also use Non-Uniform Frequency Splitting.

This is based upon the critical bands concept, as explained in the section on psychoacoustics. There is, however, a difference: Non-Uniform Frequency Splitting as used in ATRAC uses more bands; 52 bands are specified.

- For the lower frequency band, 20 critical bands are used.
- The middle- and high-frequency bands each use 16 critical bands.

It should be obvious that by using a higher amount of critical bands the precision can be very high. Also, in the area where a human ear is the most sensitive (up to approx. 4 kHz) the amount of bands will be the highest.

The decision on time splitting is of course very important. It is basically performed upon comparison of a number of adjacent blocks, where the peak values are measured and compared. Based upon this comparison, the decision is taken about the splitting mode to be used.

What is the practical use and benefit of this time/frequency splitting process?

- We can use a processing method which is similar to the processing as performed in a human hearing system.
- For each frequency band, we can define critical bands, adapted to our sensitivity in that area, and we can adapt the processing accordingly.

Figure 17.32 Pulse transient (a) and sine-wave constant (b).

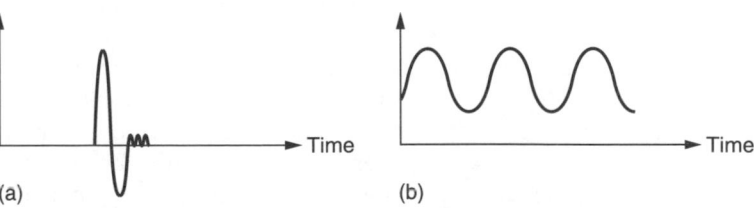

Figure 17.33 Non-uniform splitting.

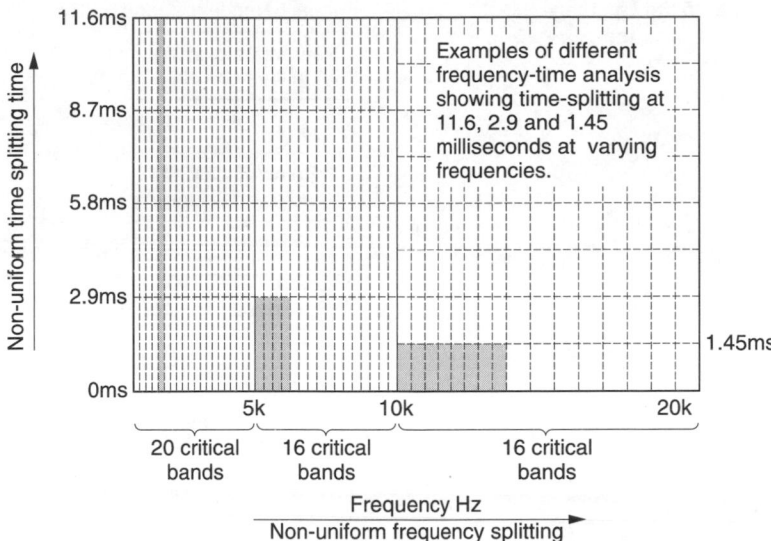

Non-uniform time splitting time

Examples of different frequency-time analysis showing time-splitting at 11.6, 2.9 and 1.45 milliseconds at varying frequencies.

20 critical bands   16 critical bands   16 critical bands

Frequency Hz

Non-uniform frequency splitting

- For each frequency band, and adapted to the input, we can define the amount of time that will be used for processing.
- Within each frequency band, and according to the time decision, a number of so-called spectra will be calculated. This is in fact the processing of input sound and it should be obvious that the more spectra calculated, the higher the resolution will be.
- Time/frequency splitting can in this way really adapt processing to the input sound and at the same time keep the output bit rate constant at 292 kbit s^{-1}.

It all comes down to making a correct choice between allocated time and processing resolution. For those parts of the sound where precision is needed, we can allocate more processing power (typically lower frequencies); for those parts of the sound where speed is of more importance (typically higher frequencies), we can allocate smaller time slots with smaller resolution, and so on.

The output of the MDCT blocks is frequency domain information, and as mentioned before, it can be seen as a frequency/level analysis of the input over a certain time block.

On this output the mentioned psychoacoustic phenomena can now be used to reduce the bit rate.

- threshold checking can be performed;
- masking checking can be performed.

The last stage is then the bit allocation (see Figure 17.35, page 263); the significant information is not transmitted as audio data,

Figure 17.34 Splitting decision algorithm.

For each frequency band

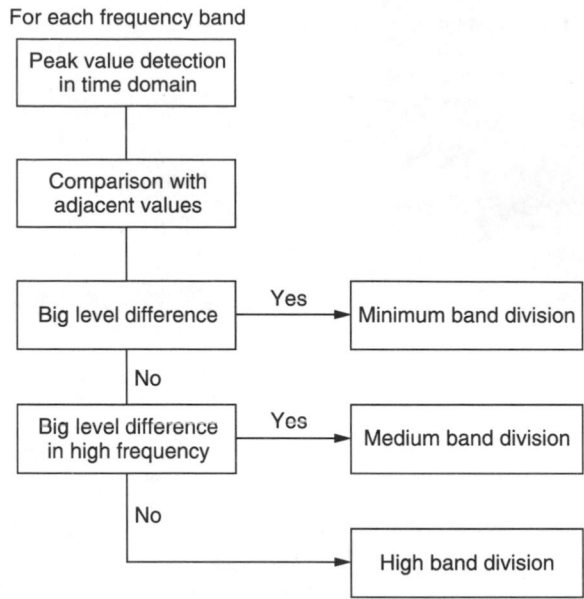

comparable to a CD format. Each frequency/time data block is now known, along with its level. This is not represented by a fixed amount of bits, but by a variable amount of bits, from 0 to 15, representing the dynamic range of that part. The scale factor gives the relative level of the signal; the rest of the bits can be deleted as quantization noise.

The output of the ATRAC encoding system can in no way be compared with the data as used in a CD system. In the MiniDisc system, the data recorded on the disc describe word lengths, scale factors and spectrum data (data in each block), comparable with the floating-block data calculations in computers.

The encoded output, 292 kbit s^{-1}, will be further processed (EFM, ACIRC, etc.) and recorded on the disc.

ATRAC decoding

During read-out, the reverse processing is performed; after EFM and ACIRC decoding, the data are again sent to ATRAC for decoding.

The input bit stream will be demultiplexed and checked for errors; based upon the same psychoacoustic principles as used for encoding, decoding will reconstruct and synthesize the original data. These data (in each frequency band) will be sent to an inverted MDCT block, after which they are synthesized with the other frequency bands to obtain the final correct (still digital) audio output.

Figure 17.35 Bit allocation.

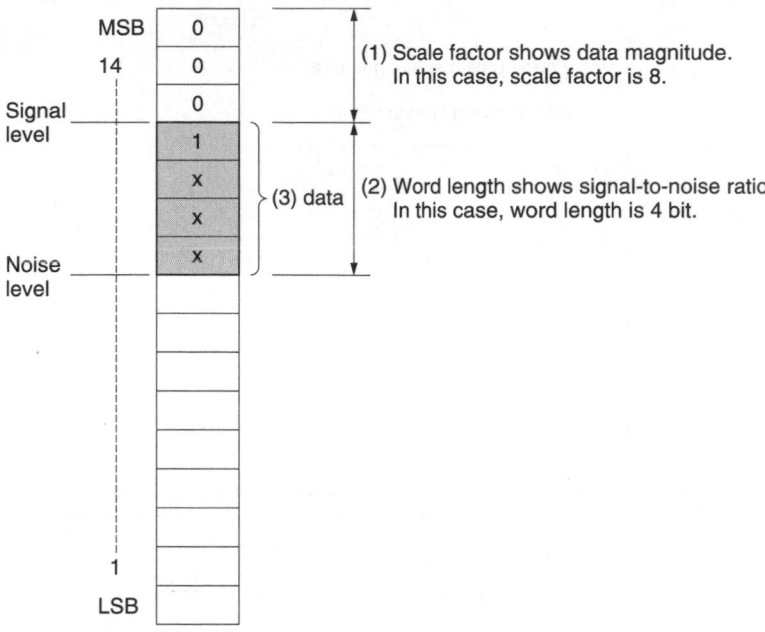

(1) Scale factor shows data magnitude. In this case, scale factor is 8.

(2) Word length shows signal-to-noise ratio. In this case, word length is 4 bit.

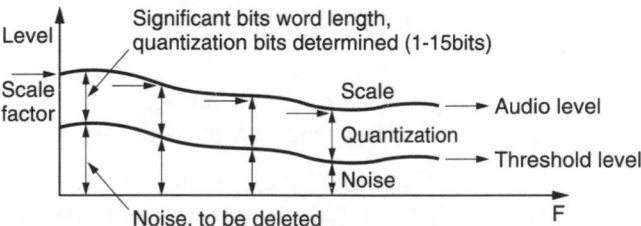

In this way, the original signal will be reconstructed, minus the deleted parts, but these parts were not necessary in the first place, so the difference between original sound and ATRAC reproduced sound is minimal.

Another benefit of transferring the audio in this way is the possibility to change some of the parameters (in order to improve it even more) in the future without creating compatibility problems. The system design is indeed such that algorithms can be changed, without creating compatibility problems, as long as basic formats are used.

ATRAC versions

The first MiniDisc sets used ATRAC version 1; since then, we have witnessed versions 2, 3, 4, 4.5 and ATRAC-R.

Each new version was obviously an improvement on the

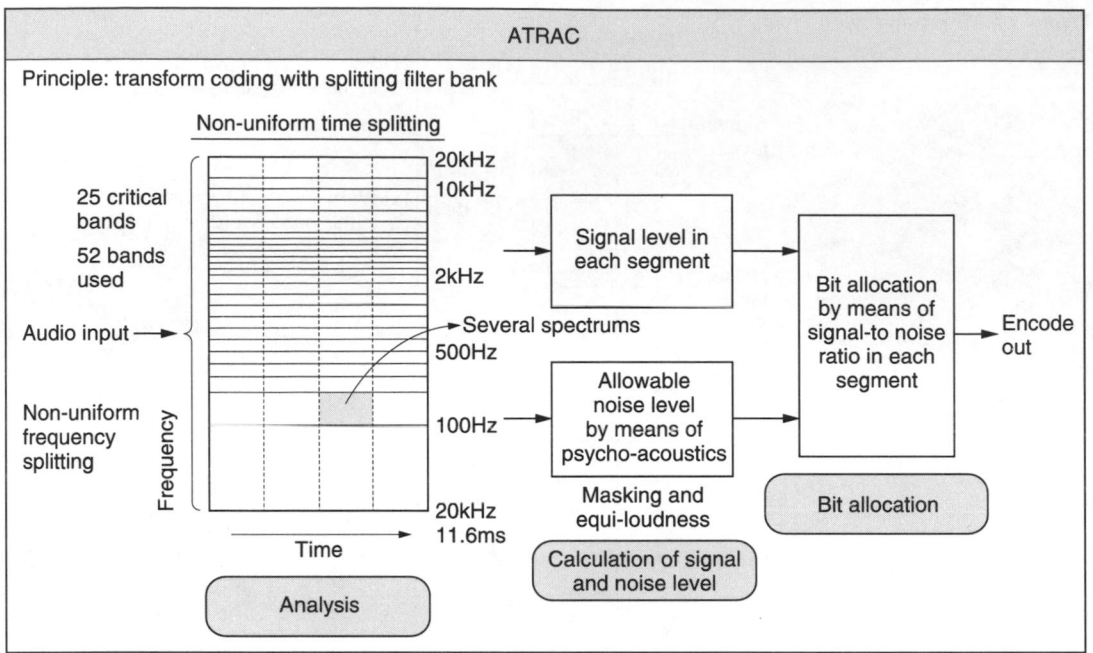

Figure 17.36 ATRAC overview.

previous one, step by step improving the sonic qualities of the encoding and compression algorithms.

Where ATRAC version 1 was justifiably perceived as not being within full Hi-fi standards (this version was only used in the first sets), version 2 already showed noticeable improvement, and as for version 3, the sound of a MiniDisc was getting very close to CD standards. Current ATRAC versions are of such remarkable quality that it needs extremely good ears or specific test equipment to be able to distinguish between CD and MiniDisc.

The ways to improve ATRAC from version to version can only be described in rough terms, as they concern proprietary technology of which the design is ongoing. Algorithms such as ATRAC (or MPEG as another example) allow different manufacturers to develop their own versions, which can be very different from one another. The question arises if it leads to incompatibilities when different manufacturers would use different encoding algorithms on disc which can be recorded and played on sets from different brands. The answer is no, as long as the basic code language is used, and some basic data specifications are upheld. A comparison can be made with the PC world: using the same basic software code and some basic

Figure 17.37 MD decoding (a) and ATRAC decoding (b).

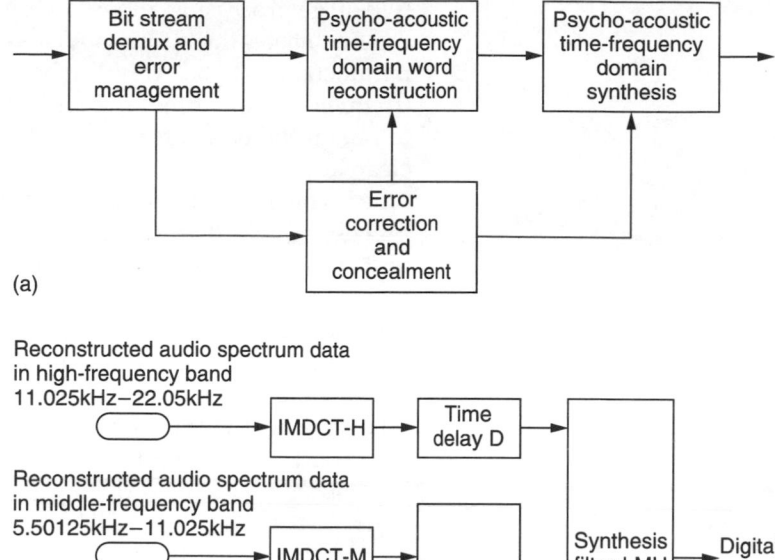

(a)

(b)

standards, various software companies can produce programs which run on the same PC without creating too many problems.

Most improvements between ATRAC versions are related to the following facts:

- Numbers and sizes of critical bands were reviewed and refined.
- Time and frequency splitting parameters were reviewed and made flexible.
- Scale factors and the number of bits to describe scale factors increased.
- Psychoacoustic parameters were refined and adapted more according to user needs.
- The speed and processing power of the processors used for ATRAC encoding and decoding were – similar to the PC world – raised drastically, which allowed faster and more complex calculations.
- The knowhow and technological progress as mentioned in previous points currently allows us to reprocess the audio input a second time in order to make the most of it. This

two-stage processing allows a very accurate encoding: during the first stage, the audio is ATRAC processed to set its main parameters. At that time, it becomes clear at which point (frequency, scale, time) the encoder has to spend most effort to obtain the best result, but at that moment it also becomes clear at which point not that much effort is needed, or in other words, where some capacity is available. The second processing will then use this information gained during the first stage to reprocess the audio data for a better result. This two-stage process can easily be compared with a sorting exercise: suppose you needed to sort a number of items in a short time according to a combination of factors (size, colour, smell, etc.), it is obvious that you will get a better result if you were able within the same time to do a first rough sorting, and then a second one starting from the result and the experience you gained during the first one.

ATRAC3/MDLP

Apart from the previously mentioned ATRAC versions 1, 2, 3, etc. a separate ATRAC3 (not to be confused with ATRAC version 3) algorithm has also been developed and used to address the need for long-play recording and playback in MiniDisc, this being abbreviated as MDLP (MiniDisc Long-Play).

The recording modes shown in Table 17.3 are possible.

ATRAC3 mode encoded discs can be played back in any MiniDisc player produced after the start of ATRAC3 (ca 1999); it should be obvious that backward compatibility was hard to achieve, so it was not done. The 160-min mono mode was foreseen from the start, although not all first sets were programmed for mono playback or recording ability.

It is possible to mix music tracks which have been encoded with different ATRAC modes onto one disc, as information on the recording mode is indeed information which is recorded for each track individually.

Table 17.3 Recording modes

Rec. mode	Channel	Max. rec. time with 80-min disc (min)	ATRAC mode	Data transmission rate (kbps)
Mono	Mono	160	ATRAC	146
Stereo	Stereo	80	ATRAC	292
LP2	Stereo	160	ATRAC3	132
LP4	Stereo	320	ATRAC3	66

The LP2 and LP4 modes start from the same algorithms as all other full rate ATRAC modes. A higher compression is obtained through stronger data reduction; this can be obtained in various ways: less bits for scale factors or timing, less critical bands, different thresholds, etc. It is obvious, however, that this will affect audio quality. It should therefore be known that the LP2 and LP4 modes are a tradeoff between more quality and quantity. Of course, development never stops, and even within these very limited transmission rate modes there is a continuous search for improvement.

Data format

Basic format

The data format of a MiniDisc is similar to the mode 2 of a CD-ROM, yellow book specification.

In this CD-ROM, mode 2 data are divided into sectors. The same sector format is used in MiniDisc, but another concept was included; the CD-ROM sector format is now extended to cluster format. A simple comparison of the basic MD data format would be that of a book:

Book	MiniDisc
Chapter	Cluster
Paragraph	Sector
Sentence	Sound group
Word	Data

One cluster comprises 36 sectors, in which 32 sectors are used for compressed data and four sectors are used for link and subdata. These four sectors (link/subdata) are different in premastered and recordable discs.

- In pre-mastered discs, these four sectors are filled with subdata, which can be graphic data (for display), information on the disc, lyrics, etc. In other words, it is a feature area.
- In recordable discs, three sectors need to be used as link sector and one for subdata; link data area will be explained later.

We can now further analyse the data format: each sector contains in total 2352 bytes, of which 2332 bytes are data bytes (comparable to the sector format of CD-ROM). The remaining 20 bytes are used for addressing and control.

A further division is the sound group; a sound group is the smallest division, containing 424 bytes (212 left channel, 212 right

Figure 17.38 Cluster layout.

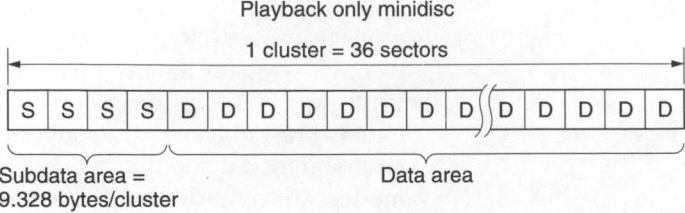

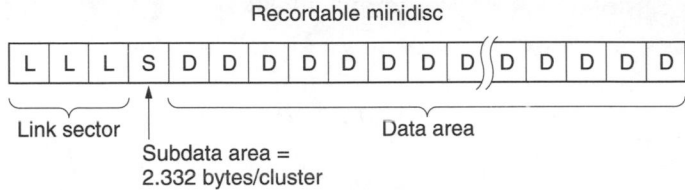

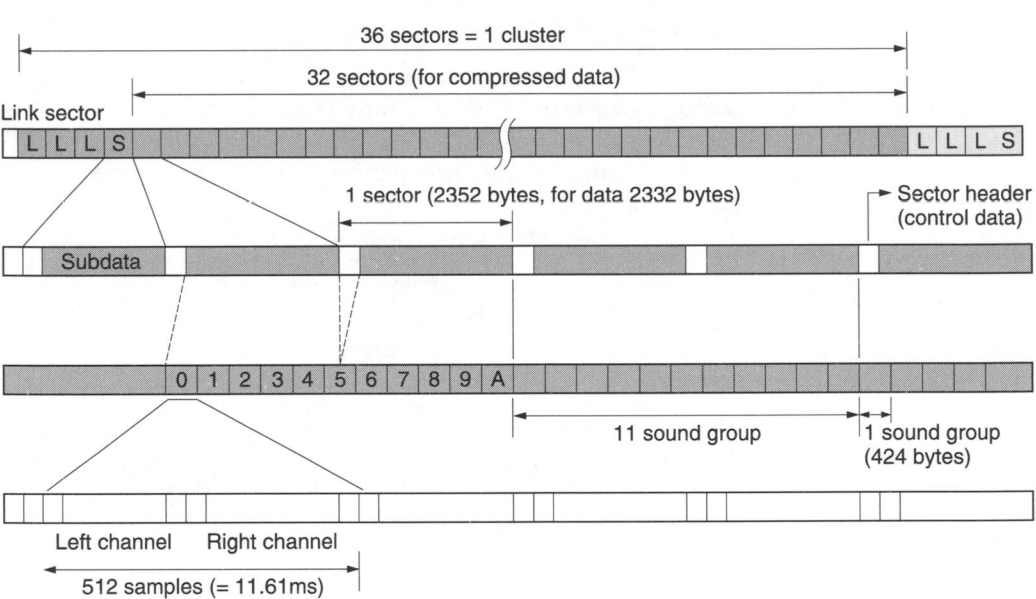

Figure 17.39 Detailed cluster layout.

channel). A sound group gives ATRAC-compressed audio data equal to 512 uncompressed samples at 44.1 kHz sampling rate.

In two sectors, we find 11 sound groups. The first sector contains five full sound groups and the left channel half of the sixth sound group. The next sector starts with the right channel of the sixth sound group and continues with five more sound groups. In this way, we can calculate the amount of bytes per sector to be:

$$424 \times 5 + 424/2 = 2332 \text{ bytes}$$

At this point we can confirm the compression rate used in MiniDisc:

- Each ATRAC-compressed sound group contains 424 bytes.
- Each of these sound goups represent 512 samples of 16 bits and two channels after compression: $512 \times 16 \times 2 = 16\,384$ bits = 2048 bytes.
- **Conclusion: 2048 bytes were compressed into 424 bytes, which shows a compression ratio of about 1:5.**

The link concept

In MiniDisc, sectors have to be 'linked' to each other; this means that a transition area between sectors has to be included to avoid erroneous overwriting of the beginning of one sector by the end of the previous sector.

The reason for linking sectors comes from the **recordability**. In CD and CD-ROM (both formats being non re-recordable), CIRC encoding and EFM modulation are used; in MiniDisc, the same CIRC encoding is used as well as EFM modulation, but CIRC is extended (ACIRC = Advanced CIRC).

CIRC was taken as the starting point, but advanced interleave was included, to protect even more against burst errors.

As Figure 17.40 shows, the difference between CIRC and ACIRC is that more delay and more interleave was included to improve the error correction capability. One result from this additional delay, however, is the fact that one sector which is input to the encoder still takes 98 frames, but the interleave takes 108 frames. As a result, there must be a buffer zone between consecutive clusters to allow proper recording of this longer interleave sequence. One cluster is the minimum area used when recording on a MiniDisc. As the interleave of the last sector is not complete, and the interleave of the first following sector has to start, the linking sectors are needed to separate one cluster from the next one.

Imagine a cluster layout without linking area: we record one cluster, but as the interleave is not complete at the last sector of the cluster, these last interleaved and delayed data words will be recorded into the first sector of the next cluster and of course destroy the information in that sector. Obviously, we must avoid this situation.

Three link sectors are sufficient for this purpose. The first link sector and a part of the second link sector are used for the

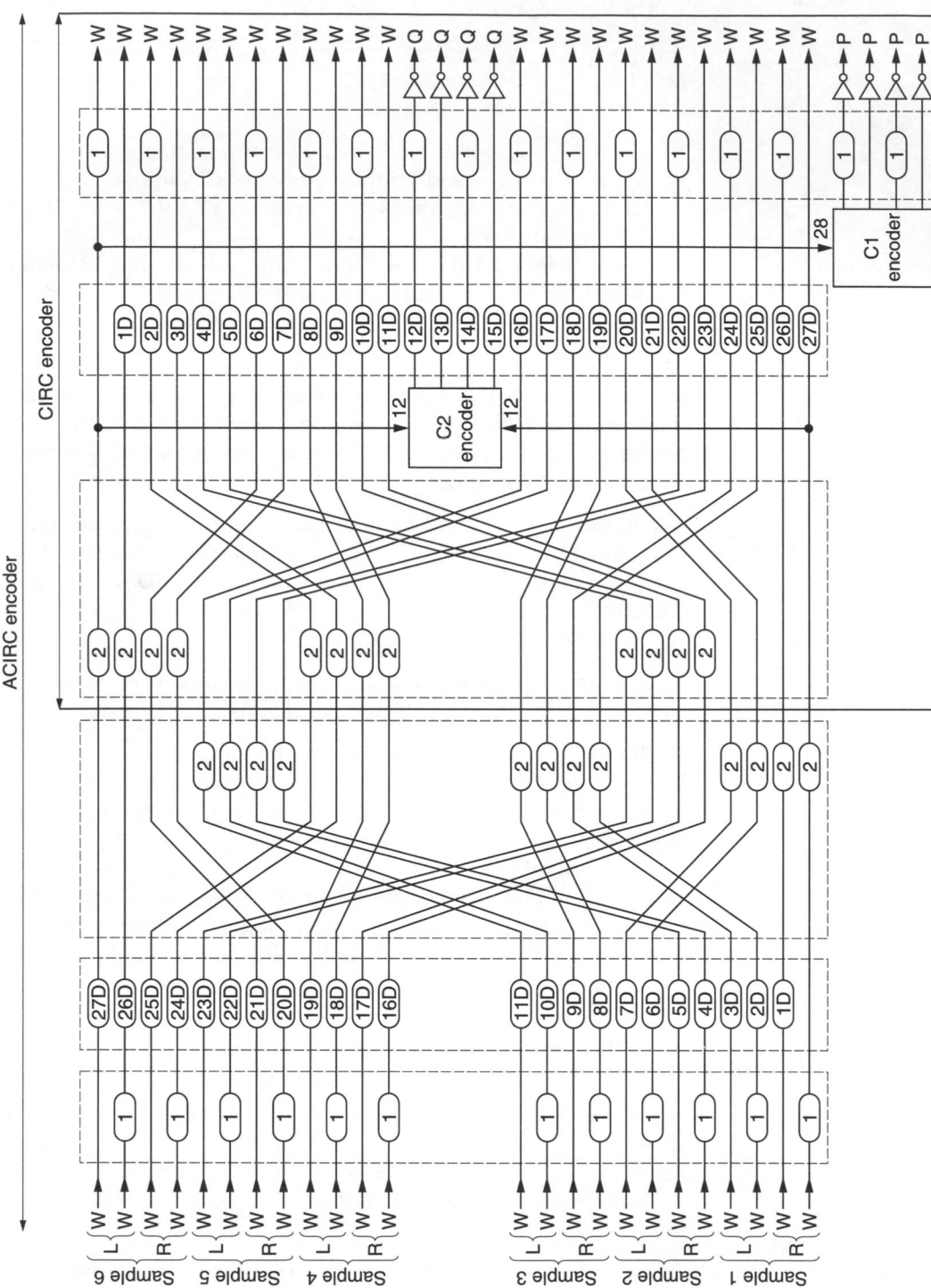

Figure 17.40 ACIRC–CIRC block diagram.

remaining interleave of the last cluster; another part of the second link sector and the third link sector are used for interleave of the next sector.

This linking format is only necessary in recordable MiniDisc. In the pre-mastered format, data are written in one continuous stroke; it is not re-recordable, and therefore there is no need for linking sectors. As already mentioned, in this case the free space is used for subdata.

Address structure

First of all, why is addressing needed? It is obvious that any data, be it audio data or computer data or whatever, which is recorded sequentially will need some addressing to make correct decoding possible afterwards. Just imagine a street with identical houses; no postman will be able to distribute the mail correctly if he has no way of checking addresses, names and numbers. During the encoding of the disc formats, an address structure is inserted in the audio data. In CD, for example, address data are included in the main data just after CIRC encoding. After each CIRC-encoded frame, a control word is included. In this control-word structure, address (and timing) information is included.

- As mentioned before, the cluster format being used in MiniDisc, the basic addressing structure is therefore based upon this cluster format.
- Another very important point is the pre-groove on the recordable disc, which uses this same address format.

Earlier in this chapter, it was explained that due to this pre-groove the laser return shows a 22.05 kHz carrier frequency modulated pattern with ADIP data. These ADIP data are the cluster and sector address. We now have to differentiate again between the pre-mastered and the recordable disc.

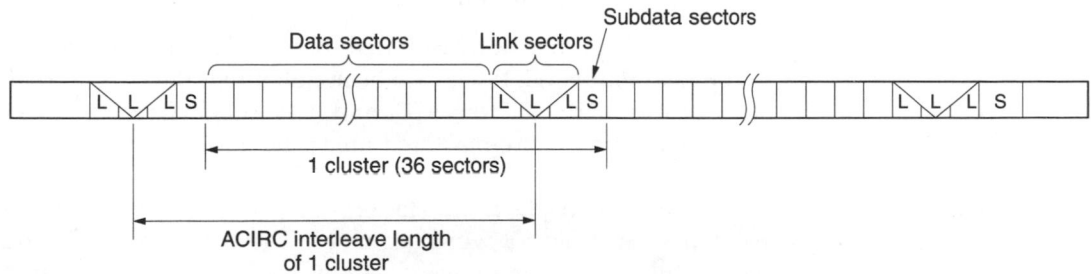

Figure 17.41 Cluster linking.

- The pre-mastered disc will carry the cluster/sector address in its subcodes which are in the main data stream.
- Likewise, the recordable disc will carry the cluster/sector address in the main data (including the pre-mastered lead-in area), but also in the pre-groove.
- In the case of the recordable disc, it is obvious that the location of the address data in the pre-groove, which is physically included in the disc and therefore not changeable, has to coincide with the (re-recordable) address in the recorded data.
- As mentioned previously, each sector contains 20 bytes for address and control data.
- The cluster address is a 2-byte format.
- The sector address is a 1-byte format.

During lead-in, in the pre-mastered disc as well as the recordable disc, the cluster address increases and has to end in the following address: FFFF;1F (cluster;sector, expressed in hexadecimal). The lead-in area of the disc is the inner area where read-out starts, and where still no audio data are recorded.

After lead-in, the cluster/sector addressing starts at 0000;FC and will increase sequentially and uninterrupted to the end of the disc.

As we already know that each cluster contains 36 sectors, of which the first 32 sectors are the data sectors, the sector address will start at 00 and go up to 1F. The link sectors are then addressed starting from FC, up to FF. (All addresses are hexadecimal.)

In the case of ADIP, the full sequence is the following: each ADIP address is a 42-bit block, containing four sync bits, 16 cluster address bits (2 bytes), eight sector address bits (1 byte) and a 14-bit cyclic reduncancy check. These 42 bits are bi-phase-modulated with a 6300 Hz bi-phase clock, and then FM-modulated onto the 22.05 Hz carrier. The bit rate is 3150 bit s^{-1}, the sector rate is in that case 3150/42 = 75 Hz, which is the same as for CD-ROM and CD.

Address in pre-mastered disc

The block diagram already revealed that the data in MiniDisc are also EFM-modulated. It has also been noted that the general data structure is derived from CD and CD-ROM. It is therefore obvious that in all pre-mastered areas the address data, cluster and sector data will be inserted in the CD-like format. As already mentioned, each CD frame (after CIRC encoding) is supplemented with a control word sequence. These control words are used (still in CD) for time indication, track number indication and so on. For MiniDisc, we do not need the same amount of bits as used in CD, but in order to stay within the same format, the number of bits remains unchanged. The bits which are not needed in MiniDisc will therefore be zero bits.

Address in recordable area

The addresses in the recordable area are included in the data stream. Each sector has a fixed format with a fixed area for synchronization, addresses and data bytes. Refer also to Table 17.4.

Data structure:

Each addressable block (each sector) contains 2352 bytes.

- The first 12 bytes are a unique sync pattern.
- The next 4 bytes are header bytes, containing three address bytes (two clusters, one sector) and a mode byte.
- The mode byte describes the nature of the data fields.
- Between the header bytes and the actual data bytes, four all-zero bytes are inserted.

Sound groups in the data structure

In each sector there are five full and one half sound group. Each sound group can be split into a left-channel and a right-channel

Figure 17.42 Subcode in pre-mastered disc.

Subcode Q format (lead-in, pre-mastered and lead-out areas):

0	0	00	00	00	00	00	00	Cluster	Sector	CRC

Control ADR TNO   X   Time   Zero

* Cluster: 2 bytes binary, MS Bit is first out
* Sector: 1 byte binary, MS Bit is first out
* CRC: 16 bit CRC (Cyclic Redundancy Check)

Table 17.4 Data structure

	←1 byte→			
	Sync	Sync	Sync	Sync
	Sync	Sync	Sync	Sync
	Sync	Sync	Sync	Sync
	cluster	cluster	sector	mode
	0000000	0000000	0000000	0000000
↑ Audio Block 2332 bytes ↓	D	A	T	A

sound group. The data structure of a sound group depends on the ATRAC encoding. Please also refer to the ATRAC section (pages 255–267) for more information on this topic.

Each sound group is divided into sound parameter bytes, audio spectrum bytes and again sound parameter bytes. The second sound parameter bytes group is exactly the same as the first one, but reversed in order.

Data structure in TOC and UTOC

The data structure in TOC and UTOC is basically the same as on the rest of the disc. The general structure remains the same, the sync is the same, the cluster, sector address and mode bytes are the same, but in the data field of course there is a different type of data. **However, remember that TOC is pre-recorded and UTOC is recordable.**

*   A pre-mastered disc only contains TOC.
*   A recordable disc contains TOC and UTOC.

In the TOC there is initially an indication of disc type. This information is needed from the start as there is a difference in TOC information between pre-mastered and recordable discs. In the case of pre-mastered discs the TOC will contain the data shown in Table 17.6.

Table 17.5 Sound group structure

1 SECTOR	1 SECTOR
L R L R L R L R L R L R L R L R L R L R L R	

SOUND GROUP

	Sound group byte numbers	Meaning
		Sound Parameter Bytes
	B0	Block Size Mode
	B1	Sub Information Amount
	B2…Bxxx	Word Length
	Bxxx…Bxxx	Scale Factor
L channel	Bxxx…Bxxx	**Audio Spectrum Data Bytes**
		Sound Parameter Bytes
	Bxxx…Bxxx	Scale Factor
	Bxxx…B209	Word Length
	B210	Sub Information Amount
	B211	Block Size Mode
R channel	B212…B423	

In the case of recordable discs, the TOC will contain the data shown in Table 17.7.

The UTOC area also contains a power calibration area and a reserved area, both after the real UTOC data.

Table 17.6

←1 byte→			
Sync	Sync	Sync	Sync
Sync	Sync	Sync	Sync
Sync	Sync	Sync	Sync
cluster	cluster	sector	mode

2336 bytes

Disc type
First/last TNO (track number)
Lead-out start ADS (address)
Used sector indication
Pointers for track numbers
Start/end of tracks
Track mode – copy protect (SCMS)
        – audio/other
        – stereo/mono
        – emphasis off/on (50/15 μs)
Pointers for track name table
Disc name (ASCII code)
Track names (ASCII code)
Pointers for date/time table
Disc REC date/time
Track REC date/time
Pointers for track ISRC table
Bar code (UPC/EAN code)
ISRC code (DIN code)

Table 17.7

1 byte			
Sync	Sync	Sync	Sync
Sync	Sync	Sync	Sync
Sync	Sync	Sync	Sync
cluster	cluster	sector	mode

2336 bytes

Disc type
Laser REC power
Lead-out start ADS
Used sector indication
Power calibration area start ADS
UTOC start ADS
REC user area start ADS

Table 17.8

←1 byte→			
Sync	Sync	Sync	Sync
Sync	Sync	Sync	Sync
Sync	Sync	Sync	Sync
cluster	cluster	sector	mode

2336 bytes

First/last TNO
Used sector indication
Disc-ID (optional)
Pointer for defective area start ADS (optional)
Pointer for empty position on parts table
Pointer for the rest of REC area start ADS
Pointers for start ADS
Start/end ADS
Like positions of start/end ADS
Track   – copy protect (SCMS)
       – audio/other
       – stereo/mono
       – emphasis off/on (50/15 µs)
Pointers for empty position on name table
Pointers for track name table
Disc or track name (ASCII code)
Link positions on previous item
Pointers for date/time table
Disc REC date/time
Power calibration area
Reserved area

Lead-out

Lead-out contains pre-mastered pits on the pre-mastered MD; again, this is the same as on a Compact Disc. On the recordable disc, the lead-out area contains no data, there is just the pre-groove. Detection of the start of lead-out can therefore be performed on the pre-groove ADIP data.

Anti-shock operation

One of the main features of the MiniDisc is its anti-shock capability, which depends on the amount of memory that is included in the set.

- The data rate needed by the MiniDisc ATRAC decoder is about 0.3 Mbit s^{-1} (292 kbit s^{-1}) to produce a correct, continuous audio output.
- Remember that this 0.3 Mbit is the compressed data corresponding to full-scale digital audio as used in a CD player.

- The amount of input data to the audio decoder depends on the linear rotation speed of the disc (the Constant Linear Velocity).
- The disc rotation speed as used in MiniDisc is the same as for a CD player.
- The CLV speed of a Compact Disc is sufficient to read out 1.4 Mbit s^{-1}, which is the minimum amount of data necessary to obtain a correct CD read-out.
- Due to this same CLV speed in MiniDisc, we have an input data rate that is nearly five times higher than the neccessary input rate (1.4 Mbit s^{-1} input, but only 0.3 Mbit s^{-1} needed).

We can benefit largely from this fact by using a buffer memory. CD and CD-ROM players are susceptible to shocks and vibration; there is of course a certain amount of recovery range (otherwise there would not be any portable or car CD player), but when such systems endure heavy shocks or vibration, there is a possibility that they lose correct tracking – i.e., the laser forcibly jumps to the wrong track, and incorrect or no data are read out – and if the system is not quick enough to recover – i.e., the laser has to return to the place from which the last correct data was retrieved and read-out has to start again correctly – this will result in momentary loss of output, usually referred to as 'skipping'.

Figure 17.43 Anti-shock block diagram.

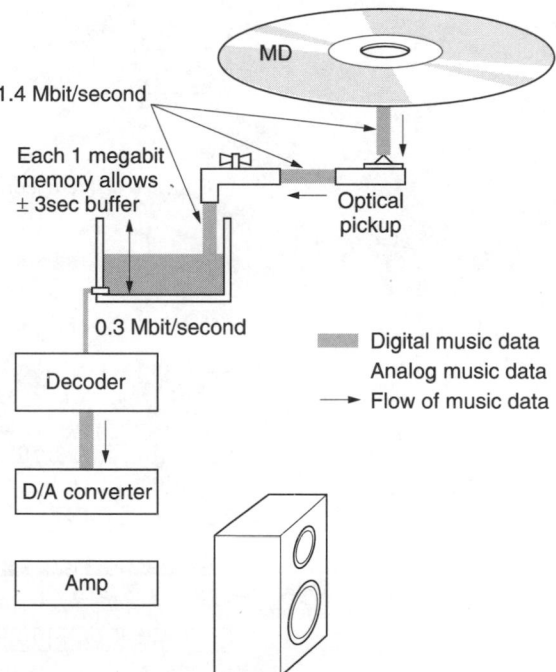

This loss can be recovered through the use of a buffer RAM. The first generation of MD players uses a 4-Mbit RAM, enabling a shock-proof time of about 10 seconds (some parts of the RAM are also used for other operations). Theoretically, it is correct to state that the more RAM will be used, the more shock-proof time will be available, limited of course by the design possibilities and the cost aspect.

The operation is as follows:

- When the system starts to operate, initially the buffer RAM will be empty.
- Data are read out by the laser unit and demodulated in the RF electronics. Next EFM and ACIRC decoding will be

Figure 17.44 Principle of intermittent read-out.

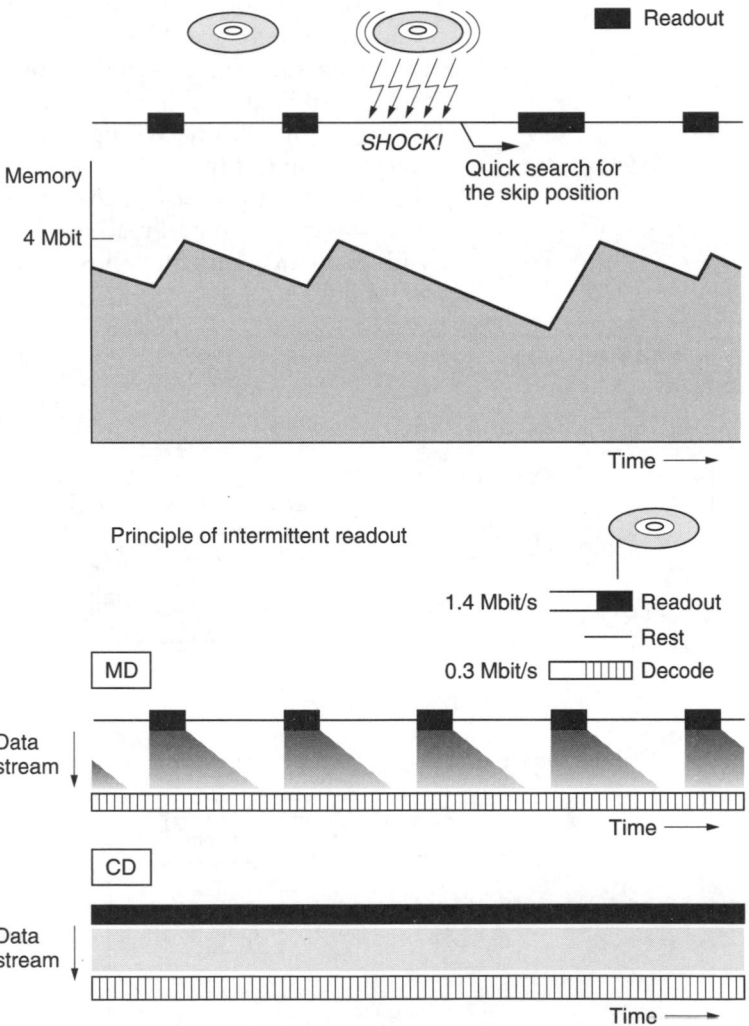

performed. At that time, data read-out is performed at 1.4 Mbit s^{-1}.

- The data should now be ATRAC decoded, but as the needed decoder input is only 0.3 Mbit s^{-1}, instead of being fed directly to the ATRAC decoder, data are fed to the buffer RAM.
- It is obvious that the buffer will be full after a short while, the input being about five times higher than the output.
- For this reason, the microcontroller – the main steering and logic system – must be able to control the flow of input data.
- The microcontroller of the MD player controls tracking servo and buffer RAM in such a way that the amount of data in the buffer RAM is always as high as possible.
- Practically speaking, the MD will go into PAUSE mode whenever the buffer RAM is nearly in overflow status.
- Pause mode means that the laser objective will remain at the same track, that no read-out is performed and that the ATRAC decoder still decodes data from the RAM, and still outputs music. It should not be confounded with Pause mode as seen by the user, when no audio is output.
- When this internal Pause occurs, the RAM level decreases, then again the PAUSE will be released until a high level has been reached and so on.

If a shock causes a track jump, the buffer RAM continues to send data to the ATRAC decoder, as there is still enough data available. In the mean time, the microcontroller knows that it is under a shock condition (can be found through many ways; refer also to Compact Disc), and will start to recover.

Figure 17.45 Shock recovery.

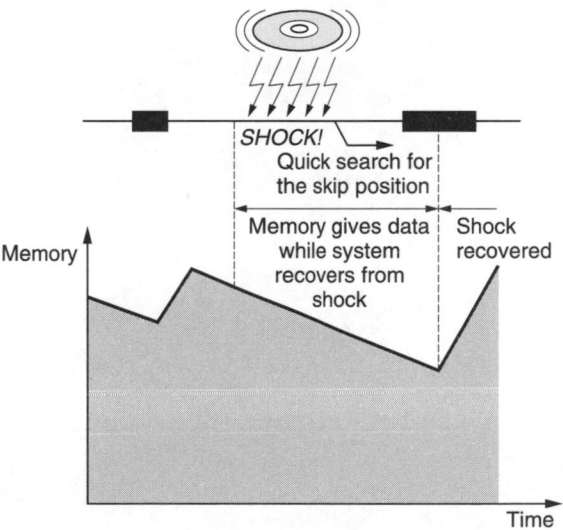

This recovery action is as follows:

- The system has in its RAM the remaining data along with the addresses; based upon the addresses of the last correct data, the system will try to put the laser unit back onto the correct position and restart read-out.
- When this last address is found on the disc (in other words, the system has recovered from the shock), data will be read out as from the next address location and sent to the buffer RAM.
- If this operation is performed within the timing limit of the RAM contents, the audio signal is reproduced continuously, glitchless and noiseless.
- The same buffer RAM is used while recording, to store multiple clusters, and also when a shock occurs during recording the system will try to recover through RAM operation.

It should be noted, however, that when recording the anti-shock operation is more critical.

PART FOUR
Advanced Digital Audio Technologies

18 Super Audio CD (SACD)

Introduction

Since the CD, developed in the early 1980s, has almost reached its upper limitations of sound quality, Sony and Philips joined once again to develop a new system with an extremely high sound quality, surpassing the limit of CD. To review proposals, the three major trade associations of the recording industry formed the International Steering Committee (ISC). These three associations are the Recording Industry Association of America (RIAA), the Recording Industry Association of Japan (RIAJ) and Europe's International Federation of Phonographic Industry (IFPI).

In May 1996, the ISC made a list of recommendations about the new format. Key points to these recommendations are:

- Active Copy Management System (ACMS)
- Copyright protection
- Anti-piracy measures
- Compatibility
- Audio, video and data storage
- Conditional access
- High-quality sound and multi-channel possibilities
- Archive and master transfer without loss of sound quality
- Extended disc functions, including text
- Packaging (may not include caddy or cartridge)
- Durability (more resistant or protected to scratches than CD)

- Single-sided disc is preferred
- 12 cm diameter is preferred

At the same time as SACD was developed, based upon the same recommendations from the same group, the **DVD-Audio** format was developed. These formats are very similar and are competing to gain acceptance within the same public. The underlying design and technology are also similar, and as both technologies will be explained within this book, a number of cross references cannot be avoided.

At the end of 1999, Sony launched its new Super Audio CD (SACD) player, meeting all the requirements of the ISC. The new SACD format allows mainly a much higher dynamic and frequency range compared to the conventional CD. The introduction of Direct Stream Digital (DSD) essential to SACD (see overview of the A/D conversion systems) implies a higher quality of A/D–D/A conversion.

Besides SACD, the Super Audio CD players are capable of playing conventional CDs.

Starting with the SCD1 and SCD777ES, Sony introduced the first generation of SACD players that were capable of reproducing signals of over 100 kHz with a dynamic range of above 120 dB in the audible range. To avoid high-frequency problems when connecting the SACD player to a conventional sound system, a switch to select standard and custom mode is mounted on the set. The standard mode has to be selected when the set is connected to a conventional amplifier. In this position, the high-frequency components above 50 kHz will be attenuated, avoiding possible damage to audio systems not designed to handle such high frequencies.

The custom mode can be selected when the set is connected to the Sony pre-amplifier TA-E1 and power amplifier TA-N1 in combination with the speaker system SS-M9ED. The maximum frequency range will be output.

With the SCD-XB940 (April 2000), Sony introduced the second generation of SACD players.

Technical specifications and dimensions

In Table 18.1, a comparison of the main CD and SACD parameters is given.

Table 18.1 CD and SACD parameters

Item	SACD	CD
Disc diameter (cm)	12	12
Disc thickness (mm)	1.2	1.2
Playback side	One	One
Disc type	3 (single, dual, hybrid)	1
Coding system	1-bit DSD	16-bit PCM
Sampling frequency	2.8224 MHz	44.1 kHz
Data capacity (MB)	4700 (single layer)	780
	8540 (dual layer)	
Min. pit length (μm)	0.40	0.83
Track width (μm)	0.74	1.6
Laser wavelength (nm)	650	780
Lens aperture rate	0.60	0.45
Playback frequency range	DC – over 100 kHz	20 Hz–20 kHz
Dynamic range	More than 120 dB	96 dB
Max. recording time (min)	Approx. 110	74
	(two channels/single layer)	
Max. track number	255	99
Max. index number	255	99
Linear velocity (m s^{-1})	3.49	1.2 or 1.4
Modulation method	EFM+	EFM
Additional functions	Text, graphics, video	CD text

Optical block structure

Compared to the optical bock of the CD player, the two-axis device is much finer. The numerical aperture (NA), on the other hand, is higher. The effect of a higher NA is, besides a narrower focal depth and a smaller focal point, a lower tilt tolerance and a disc thickness that becomes stricter.

Another difference is that instead of one main photo detector with four fields (for RF detection in CD) and two side spots for tracking, two identical photo detectors (PD1 and PD2) with four fields are used for SACD.

Figure 18.1 Numerical aperture.

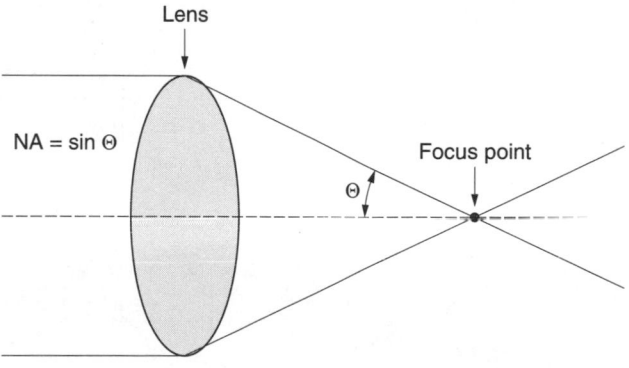

Figure 18.2 Photo coupler.

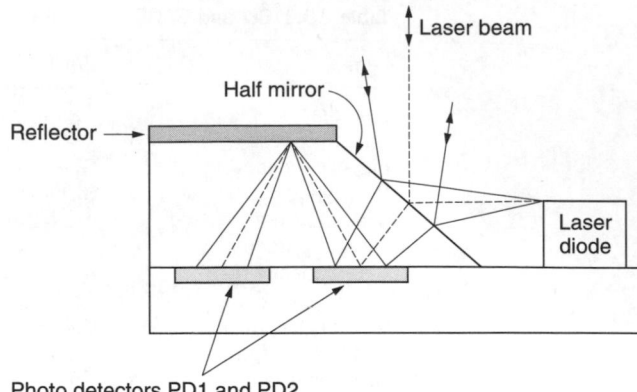

For focus and tracking, different additions and subtractions are made to define the error voltage.

The tracking is done with the TOP PUSH PULL principle. The tracking error signal is calculated as follows:

$$TE = (C + A' + B2 + B1') - (A + C' + B1 + B2')$$

In Figure 18.3, the 'on track' situation is presented. In case the laser beam shifts to the 'off track' position, the position of the reflection on the photo detector will also shift with a corresponding error signal as a result.

Figure 18.3 Tracking.

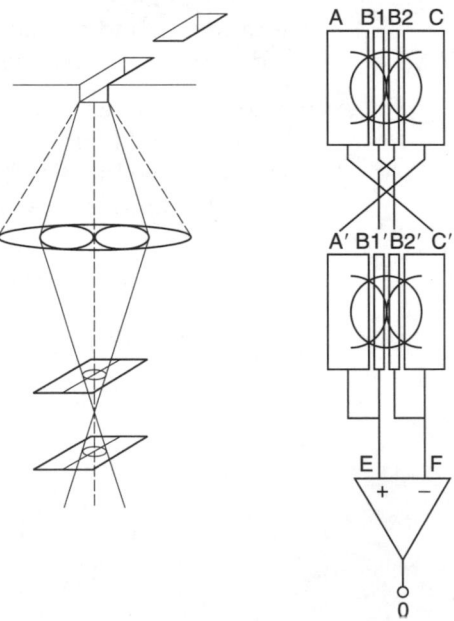

Figure 18.4 On focus.

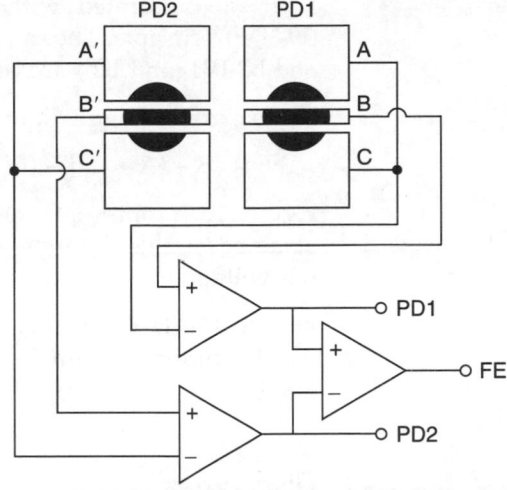

Figure 18.5 Off focus.

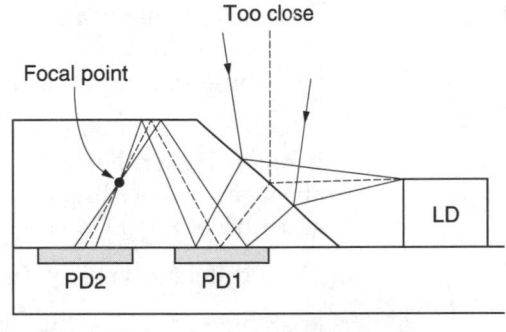

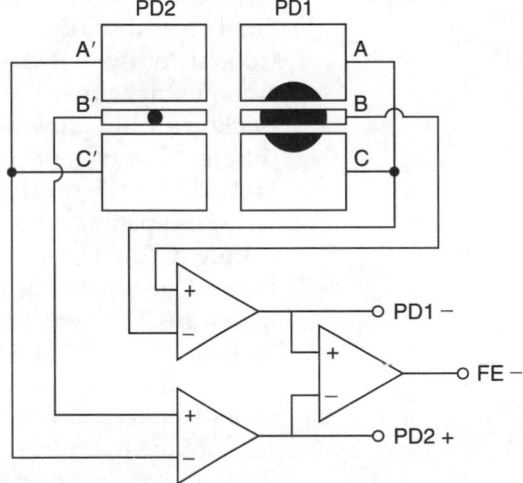

Focus is performed with the differential three-divided focus (D-3DF) principle. The same two photo detectors are used but B1 and B2 (B1′ and B2′) are connected as one detector field.

Focus error voltage will be:

$$FE = [B - (A + C)] - [B' - (A' + C')]$$

Figure 18.4 represents the situation where a good focus is obtained. In this situation, both reflections are equal and the output voltage will be zero.

In case the laser beam is out of focus, the reflection on PD1 will be different from PD2 and an error voltage is generated (see Figure 18.5).

Disc types

The same disc diameter is used for SACD as for CD. The maximum dimensions of the pits are different from CD. These are much smaller (0.4 μm) for SACD than for CD (0.83 μm). Also, the track width is about half of the track width of CD. The layer with these specifications is called the HD layer.

It becomes clear that the data storage capacity on this HD layer is much higher. For a single-layer disc, the storage capacity is about seven times higher than the conventional CD. The physical structure of SACD is very similar to DVD.

Three different types of disc are used.

(a) Single-layer disc (SL disc) (Figure 18.6)
     Consists of one HD layer and has a storage capacity of 4.7 GB
(b) Dual-layer disc (DL disc) (Figure 18.7)
     Consists of two HD layers for extended playback time. The storage capacity of this disc is about 8.5 GB. The disc thickness will be the same as the SL disc (1.2 mm). The two different layers are 0.6 mm apart.
(c) Hybrid disc (Figure 18.8)
     To be compatible with the conventional CD, this is a unique feature of SACD, as the disc consists of two totally different layers. One layer is the HD layer and can be played back only by the SACD player. The second layer, a CD layer, can be read by both the SACD and the conventional CD player is present on this disc. This possibility of backward compatibility is a major feature which sets SACD apart from other formats.
     Also for the hybrid disc, where the total thickness is 1.2 mm, the high-density layer is situated in the middle of the disc or at 0.6 mm from the surface.

The HD layer of the hybrid disc is semi-transparent and will be invisible for the CD laser beam with a numerical aperture of 0.45 and a wavelength of 780 nm.

The layer will become reflective for the SACD laser beam only.

Figure 18.6 Single-layer disc.

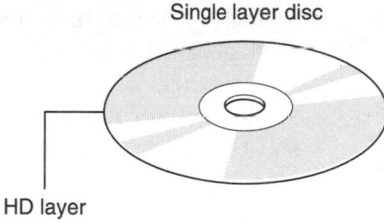

Single layer disc

HD layer

Figure 18.7 Dual-layer disc.

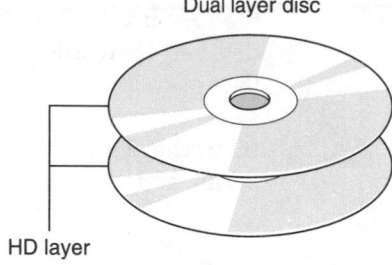

Dual layer disc

HD layer

Figure 18.8 Hybrid disc.

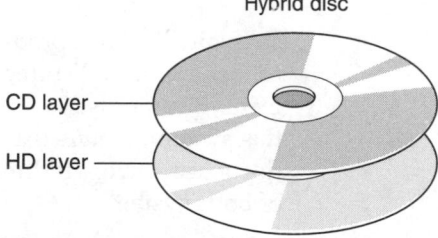

Hybrid disc

CD layer

HD layer

Disc type detection

In order to discriminate the type of disc that is inserted into the set, the difference in reflectivity and distance between the two

Table 18.2

	CD red book – compatible layer	Super Audio HD layer
Reflectivity	Reflective	Semi-reflective
Minimum pit length (μm)	0.83	0.4
Track pitch (μm)	1.6	0.74
Laser wavelength (nm)	780	650
Lens numerical aperture	0.45	0.6

layers is detected. While reading out the RF signal, the objective lens will be moved from the lower position to the upper. During this movement, the focus point will scan the thickness of the disc. At the place where the focus point reaches the surface of the disc, a small reflection will be returned to the detector, with a pulse on the RF output resulting. This pulse is the trigger of the timer into the microprocessor. Depending on the time between the first and second pulses (where the focus point reaches the second reflective layer), the set can discriminate the presence of a high-density layer in the middle of the disc. Remember that the disc thickness is 1.2 mm with an HD layer of 0.6 mm in the case of a DL or Hybrid disc.

Next, a discrimination of the upper layer has to be made. In the case of a dual-layer high-density disc, the reflection will be much lower compared to a CD layer on a Hybrid disc, that has to meet the specifications of the 'red book'. The difference in reflection is detected by reading out the tracking error voltage while moving the optical block from the centre to the outer circumference (focus servo is on). Depending on the level of the signal that is output from the TE servo, the judgement of CD/HD layer can be performed.

Watermarking

To avoid illegal reproduction, another request from the ISC was to include an anti-piracy system. SACD also meets this requirement. Both a visible and an invisible watermark are included in this system. Detailed information is not open to the public. For this reason, we will limit this explanation to a short description of both systems.

Visible watermarking

As the name itself implies, a visible watermark is a faint image printed on the signal side of the disc. This printing is hard to copy and hard to apply to the disc, since it must not disturb the

Table 18.3

Disc type	Reflection (%)
Single layer	60–85
Dual layer	18–30
Hybrid disc	15–30 (HD layer)
	>70 (CD layer)

reflection of the reading laser beam. A visible watermarking cannot be detected by the set. It gives the user the possibility of checking the disc by just looking at it. Visible watermarking is recommended by the ISC but is not obligatory.

Invisible watermarking

Invisible watermarking is a copyright protection system that has to be included in the disc. This type of protection is written on the disc substrate itself and makes it impossible for the set to read the disc in the case of an illegal copy. The idea behind invisible watermarking is to include the copyright protection information into the main signal but to remain 'invisible' to the original audio data.

To encode the copyright protection information into the signal, a new technology called Pit Signal Processing (PSP) has been developed. The basic feature in PSP is that the laser power of the beam on the recording side is modulated. A higher laser power results in a larger laser beam, changing the length and width of the pits.

Encoding and sector format

The analog input signal is converted to a DSD digital signal and divided into blocks of 2016 bytes of audio information per block.

Besides the audio information, also a great deal of supplementary data and synchronization has to be recorded on the SACD. Starting with a sampling frequency of 2.8224 MHz in a 1-bit system, the total amount of music data for 1 second of stereo music is $2 \times 2\,822\,400/8$ or 705.6 kbyte. Compared to the conventional CD (44.1 kHz, 16-bit) with 176.4 kbyte per second of music, this is four times the amount of data.

The main data are divided into blocks of 2048 bytes, which consists of 2016 bytes of audio information, header and supplementary information. Together with 4 bytes of Identification Data (ID), 2 bytes of ID Error Detection (IED) and 4 bytes of Error Detection code (EDC), a data sector is formed. Also, 6 bytes are reserved for expansion, resulting in 2064 bytes per data sector (Figure 18.9).

Thanks to the evolution to the more powerful Reed–Solomon Product Code (RSPC), error correction can be done over a much larger amount of data compared to the CIRC error correction for CD.

Figure 18.9 Data sectors.

Identification data (ID)	4 bytes
ID error detection (IED)	2 bytes
Reserved	6 bytes
Main data	2048 bytes
Error detection code (EDC)	4 bytes

The error correction information is applied to 16 data sectors, creating an ECC block.

The 2064 × 16 data bytes are scrambled into a matrix of 192 rows by 172 columns. Ten PI parity bytes are added to each row and 16 PO parity bytes are added to each column. With RSPC, at least 5 byte errors per row and 8 byte errors per column can be corrected. Several calculations after each other can even correct larger errors.

After this, the 192 data rows of the ECC block are divided into 16 blocks of 12 rows and the 16 PO rows are interleaved between the 16 blocks. In this way, a recording sector is created.

In fact, a recording sector consists of the information of one data sector (12 rows) or 2064 bytes + 12 × 10 PI bytes + 1 row of 182 PO bytes, or 2366 bytes.

Figure 18.10 ECC block.

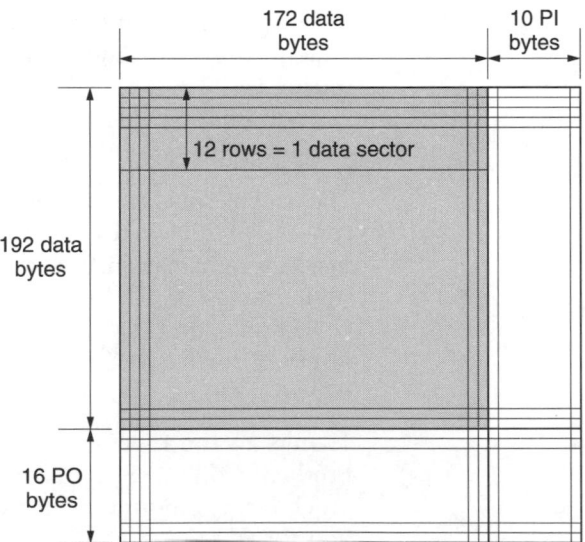

Now, each row of the recording sector is divided into two parts of 91 bytes. EFM+ is applied to each byte and a sync word of 32 bits is added to each part. The sync words are alternatively added to the pattern in a logical way, allowing the decoders to synchronize their timing.

Each sync word has a specific 'identification' pattern and a specific place in the physical sector:

SYNC0:	Start of the physical sector
SYNC1...4:	Start of an odd sync frame
SYNC5...7:	Start of an even sync frame

The pattern of the sync words are chosen from the additional list of words that meet the EFM recommendations (refer to EFM+), so no supplementary EFM conversion has to be applied to these sync words.

Using eight different sync words to the 13 rows of the recording sector results in a physical sector that consists of 13×2 (sync words per row) $\times$ (32 (sync bits) + 91 bytes $\times$ 2 (EFM+) $\times$ 8 (bits)) = 38 688 bits.

If we now return to the calculation of the data bit rate. One data sector has 2048 bytes of main data. Now, two different structures have been defined for the SACD format. A three-frame/ 14-sector and a three-frame/16-sector system.

In the case of a three-frame/14-sector system, the total amount of audio data in 14 sectors is 28 224 bytes. The rest is 84 bytes

Figure 18.11 Recording sector.

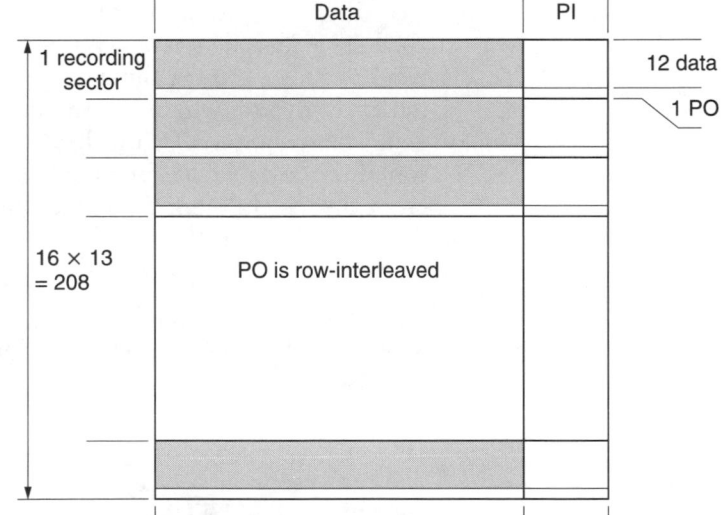

header information and 364 bytes supplementary data, giving a total of 14×2048 or 28 672 bytes.

One second of reading has 75 frames, meaning that the total audio data stream is 75/3 (three-frame system) $\times$ 28 224 (audio data per sector) or 705 600 bps.

A similar calculation can be performed for the three-frame/16-sector system, where the amount of supplementary data is different.

It becomes clear that the same conversion cannot be adopted for multi-channel purposes. The final bit rate would be too high, exceeding the maximum reading speed of the optical system. Therefore, the Direct Stream Transfer (DST) compression technique is used to reduce the amount of data and as a consequence the transfer speed. DST will be explained later.

The final definition for the Maximum Byte rate in the SACD version 1.1 format book for the audio and supplementary data is:

2CH (three frames/14 sectors): 716 800 B s^{-1} or 5.7344 Mbps
2CH (three frames/16 sectors): 819 200 B s^{-1} or 6.5536 Mbps
Multi-channel (DST): 1 873 920 B s^{-1} or 14.99136 Mbps

Disc structure

Three main areas are defined in the structure of the SACD disc. The innermost area or lead-in zone, where the information of the disc types (factory information, number of disc layers, etc.) is recorded, the program zone or data zone, and the lead-out zone. This last zone is used as a buffer zone in case the optical block is moved to the outer zone due to a shock. In the case of the Hybrid disc, where the upper layer has the same structure as a conventional CD, the lead-in zone of this CD layer also contains the table of contents. The middle zone for the dual-layer disc is different from the single-layer and Hybrid discs. Basically, the middle zone has the same function as the lead-out zone.

The Data zone of the SACD contains the TOC information together with the file system area, the Audio area and the Extra data area.

Figure 18.16 shows the complete structure recorded on the disc.

The file system area is an optional space where the information defined in the ISO-9660 or Unique Disc File (UDF) can be recorded.

Figure 18.12 Physical sector.

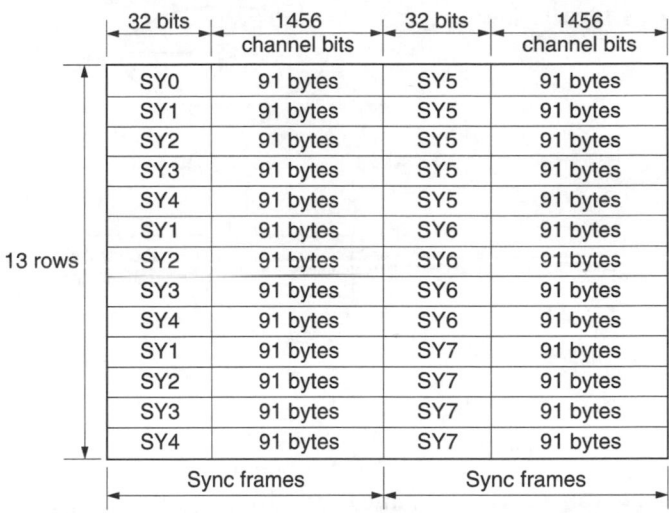

Figure 18.13 Hybrid disc.

Figure 18.14 Single-layer disc.

Figure 18.15 Dual-layer disc.

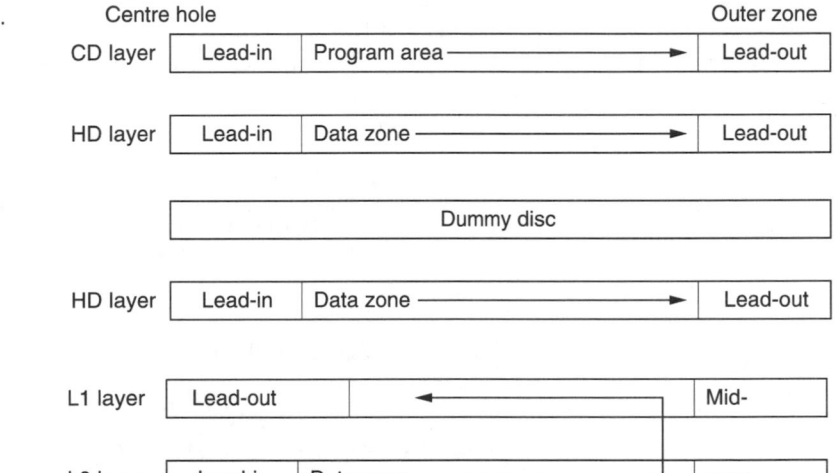

Different from CD, where the TOC holds the information about the number of tracks and the time recorded on the disc, the master TOC area used for SACD has information about the album and disc.

The main items in this master TOC are:

* Position and size of the area TOC.
* Disc and Album information (all information about disc type, disc catalogue number, genre, date and time, etc.).
* Text information (amount of text channels on the disc, language, etc.).
* Text (disc name, name of the artist, etc.).

Figure 18.16 Data structure.

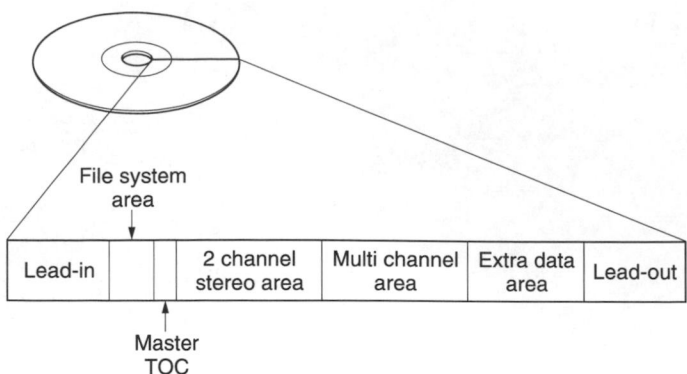

Taking a closer view of the Audio area, consisting of a Two-channel Stereo area and a Multi-channel area, shows that these areas are divided into three main blocks:

Both TOC1 and TOC2 contain the same information about the numbers of tracks recorded on the disc, total playing time, position of the different tracks, etc. The purpose of recording the same TOC information twice is for safety reasons in case the TOC1 area is damaged.

The actual audio data are recorded in the track area.

The extra data area is an optional part on the disc to record the data of the UDF or ISO9660 file system.

Direct stream transfer

As explained before, it is recommended by the ICS to store 74 minutes of two-channel high-quality sound or six-channel high-quality surround sound on the 4.7 GB HD layer of the optical disc. If we make a rough calculation of the memory needed for a stereo sound of 74 minutes, we can see that it would be impossible to record six-channel sound with the same specifications.

$$74 \times 60 \times 2 \times 2.8224 \text{ MHz}/8 \text{ or about 3.2 GB (for two channels)}$$

In order to make it possible to record the six-channel information, data reduction has to be adopted. Two different options were possible. A lossy compression method based on psychoacoustics, like Dolby Digital (AC3), or a lossless coding method.

The SACD format has chosen to use the new lossless coding: Direct Stream Transfer (DTS). Lossless coding methods were originally developed to reduce the amount of data for PC applications. After decoding, the original signal can be reconstructed bit for bit.

DTS is a very complex coding method, using framing, adaptive prediction and entropy encoding stages.

Entropy is based on the appearance probability of the symbols. A symbol that appears frequently has the shortest code, a symbol that does not appear frequently has the longest code. In this way, a substitution table of the most frequently appearing symbols can be formed.

A second basic method of reducing the amount of data is to shorten the code words. Instead of sending eight consecutive zero bits, a simple 8×0 will be encoded.

However, for DSD signals, the basic entropy method is not very efficient, since amplitude is converted into a bit stream and not into an absolute value as for PCM.

Therefore, a new adaptive prediction method has been developed. With this new method, DTS is capable of reducing the total amount of data by about 50% without losing any information.

The structure of a lossless coding system is shown in Figure 18.18. The input signal is framed to select an appropriate part. On this

Figure 18.17 Track/channel structure.

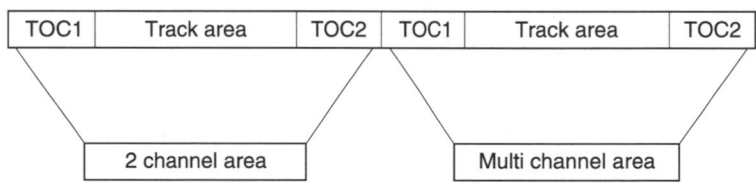

Figure 18.18 Structure of a lossless coding system.

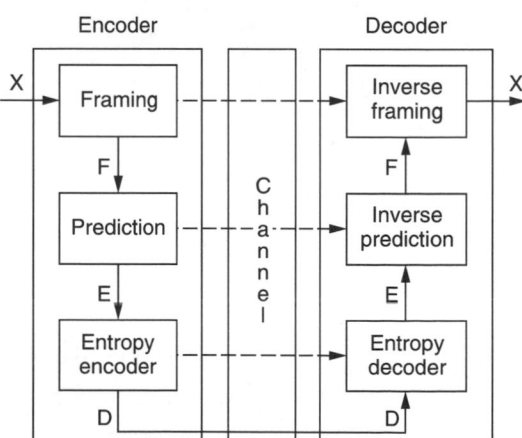

selection, prediction is done to remove redundancy. The entropy encoder will make the most efficient coding of the input signal.

Due to the adoption of the techniques of framing, prediction and entropy encoding, the final bit rate of a DTS encoder is not constant. A constant bit rate is desired in systems working with an optical disc system, since this corresponds with a constant linear disc speed. By adding a buffer control, the variable bit rate can be converted to a constant bit rate.

In conclusion, adaptive prediction used for DTS reduces the high bit rate of multi-channel DSD to store the entire 74 minutes of information on a single-layer disc.

Conclusion

With the SACD, Sony and Philips created a new standard that fulfils all the high-quality needs for the next generation of sound carriers. The SACD can serve not only for sound reproduction for home use, but also as a high-quality archiving system to replace the analog tape in recording studios.

Thanks to the sophisticated watermarking technology, the new format will also defeat piracy.

An additional advantage is that, with the hybrid disc, the SACD guarantees 100% compatibility with existing CD players.

19 DVD-Audio

Introduction

The latest development in the digital versatile disc (DVD) family is DVD-Audio. As shown in Figure 19.1, part 1 and part 2 of the 'DVD Specifications for Read-Only Disc' form the foundation for DVD-ROM. Together with the physical specifications (part 1) and the file system specifications (part 2), part 4 was also recently published to define the specifications for audio applications. Of course, according to part 3 (Video specifications) published before, video information can also be recorded on DVD-Audio.

Inspired by the music industry to create a high-quality recording and archiving system, the DVD Forum Audio Working Group

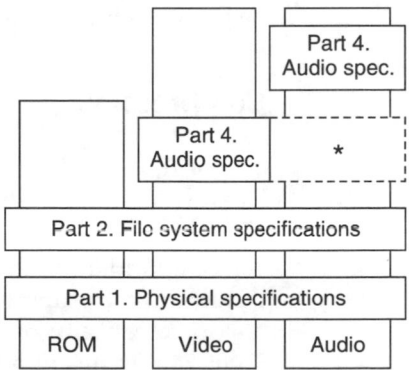

Figure 19.1 DVD specifications for DVD-Audio.

4, chaired by JVC and composed largely, but not exclusively, of consumer electronics hardware manufacturers, developed the DVD-Audio format.

Basic concept

Based on the recommendations of the International Steering Committee (ISC; also see Chapter 18) and making use of the storage capacity of 4.7 GB on the high-density layer of the DVD-ROM, DVD-Audio is a high-quality digital audio medium. Building on the knowledge of linear PCM encoding for the conventional CD (limited to a maximum bandwidth of 20 kHz), DVD-Audio uses a sampling frequency of up to 192 kHz. According to the Nyquist theorem explained in Chapter 2 of this book, this sampling frequency allows a bandwidth from DC to 96 kHz. In addition to the high sampling frequency, a 24-bit quantization is used giving a high signal-to-noise ratio.

The calculation of the theoretical signal-to-noise level (S/N) will thus be (refer to Chapter 3):

$$S/N \ (dB) = 6.02 \times n + 1.76$$

where n is the number of quanization bits.

In the case of a 24-bit system, the theoretical S/N ratio will be about 146 dB.

Besides these audio specifications it was also required to include multi-channel possibilities. DVD-Audio is capable of reproducing six-channel digital surround sound.

DVD versus CD

In the comparison given in Table 19.1, the main differences between both formats can be observed.

Compression

Taking into account that at a maximum sampling frequency and 24-bit resolution, the transfer rate becomes 9.6 Mbps (two channels, 192 000 samples per second at 24 bits per sample + error correction, identification and synchronization). At this rate, the maximum recording time on a single-layer HD disc (which has a storage capacity of 4.7 GB) will become less than 74 minutes. To fit 74 minutes of music onto one disc, a lossless compression

Table 19.1 Comparison between DVD-Audio and CD

Item	DVD-Audio	CD
Capacity	4.7 GB	640 MB
Size (cm)	8/12	8/12
Channels	6 max	2
Frequency response (kHz)	DC ~96	5–20
Dynamic range (dB)	144	96
Recording time (minutes)	74 or more in all modes	74
Max. transfer rate (Mbps)	9.6	1.4
Audio		
Audio signal format	PCM	PCM
Audio options	Dolby digital, DTS, MPEG, etc.	–
Sampling rate	Two-channel: 44.1, 88.2 or	
	48, 96, 192 kHz	44.1 kHz
	Multi-channel: 44.1, 88.2 or	
	48, 96 kHz	–
Quantization	12-, 20-, 24-bit	16-bit
Functions		
Still image	Yes	No
Real-time text	Yes	No

method, Meridian Lossless Packing (MLP), has been developed. Within MLP, three methods are used to reduce the data rate. Lossless waveform prediction is used to reduce inter-sample correlations, with a very large variety of special filters. In simple terms, this means that comparisons are made between a number of subsequent samples, and where possible, only the differences are encoded. Lossless processing and lossless matrixing are also used to reduce correlations between channels; similar to the inter-sample correlation method, here also a comparison is made, but now between channels. Hufman coding is used to reduce the data rate by efficiently encoding the most likely occurring successive values in the serial stream.

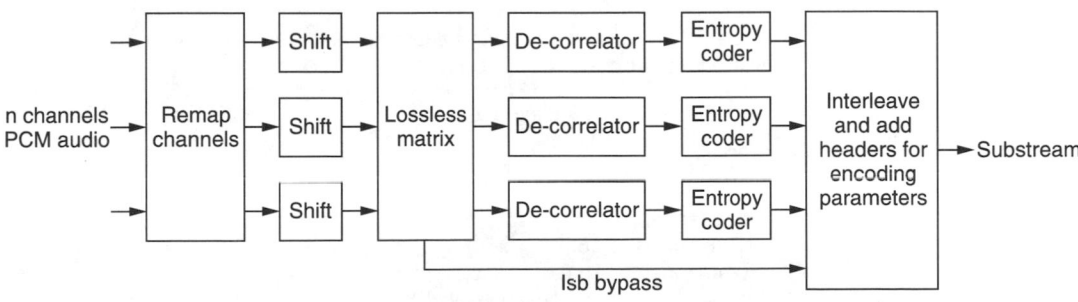

Figure 19.2 MLP encoder.

MLP reduces the total data amount to about 50% of the original without losing 1 bit after decoding. In practice, MLP is capable of handling up to 63 channels with sampling frequencies between 32 and 192 kHz, so it will perfectly match the requirements of DVD-Audio with maximum six channels and a maximum sampling frequency of 192 kHz (in two-channel mode).

It becomes clear that in the case of six-channel surround sound, the same high-quality standards cannot be achieved for all channels. If six-channel sound were encoded at 192 kHz/24 bits, an audio stream of 27.648 Mbit s^{-1} would be necessary to transfer the data, surpassing the limit of 9.6 Mbit s^{-1} defined for DVD even with MLP. In order to keep the playing time and bit rate to an acceptable level, a mixture of different sampling frequencies and resolutions can be used together with MLP compression.

For example, for a 5.1 surround sound, the front L, front R and centre channel can be encoded in 96 kHz/24 bits, where the left surround, right surround and LFE can be encoded in 48 kHz/16 bits.

Table 19.2 gives an overview of different sampling frequencies and the corresponding time on a single-layer disc at 24-bit resolution.

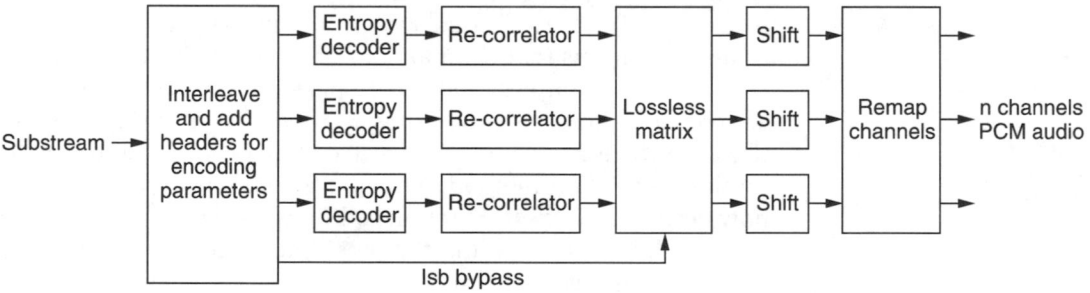

Figure 19.3 MLP decoder.

Table 19.2 Different sampling frequencies and corresponding times on a single-layer disc at 24-bit resolution

System	Sampling frequency (kHz)	No packing (min)	With MLP (min)
Two-channel	48	258	344
Two-channel	192	64	86
Six-channel	96	43	74
Five-channel (two groups)	Three channels at 96+ two channels at 48	64	111

Disc structure

The structure of a DVD-Audio is comparable to conventional CD except that, besides the tracks and indexes (both maximum 99), the album is also divided into a maximum of nine groups. Each group is a collection of a number of tracks. Each group or track is directly accessible by the user, making it very easy to navigate.

Optical read-out system

The optical block used for DVD-Audio is the same as for SACD. Tracking and focus servo are the same as explained in Chapter 18.

Options

Other requirements from the ISC were the possibility of including video or text and to create a reliable anti-piracy system.

Based on the DVD specifications for read-only disc parts 1–3, video and text can easily be included.

To meet the anti-piracy requirement, reserved fields to set control flags for copy-protection management are provided. It should be noted that some controversy on this subject has indeed delayed the final opening and use of DVD-Audio. For this reason, no details are given at this point.

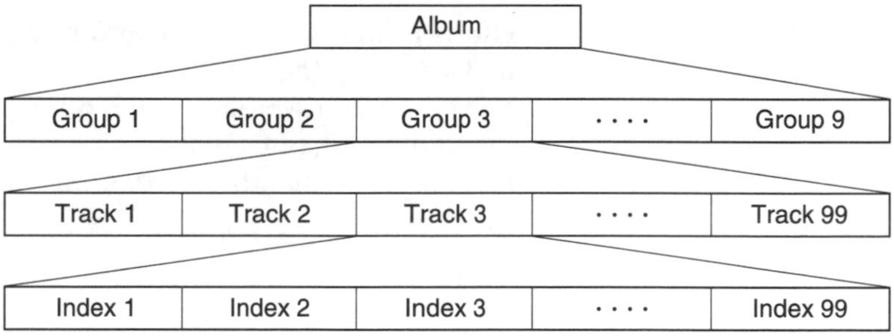

Figure 19.4 Visual structure of DVD-Audio.

Album				
Group 1	Group 2	Group 3	· · · ·	Group 9
Track 1	Track 2	Track 3	· · · ·	Track 99
Index 1	Index 2	Index 3	· · · ·	Index 99

20 Audio compression

Introduction

As mentioned and proven in a number of previous chapters, digital audio creates large amounts of data to be stored; the need for data compression became clear almost from the start of digital audio, but it also became clear that digital audio compression and data compression as used in PCs was a totally different game. Much research and development needed to be performed before some usable designs were available. One of the most noticeable efforts in this domain was performed by the Fraunhofer Institute in Germany, together with the University of Erlangen, who started to work around 1987 on a perceptual audio coding in the framework of Digital Audio Broadcast. A powerful algorithm to compress the digital data used for audio purposes was standardized.

MPEG

Based on the compression techniques developed by the Motion Picture Expert Group (MPEG) to compress the amount of data for video purposes, the MPEG audio coding family became reality.

Note that the MPEG theory, which is explained here, closely resembles the theory behind ATRAC, as used in MiniDisc and explained in Chapter 17. For the sake of completeness, and to demonstrate that both methods are indeed derived from the same line of thought, readers should be aware that some feeling of 'déja-vu' when reading the psychoacoustics behind MPEG might occur.

Depending on the MPEG layer, the original data are reduced up to 12 times. The basic technique to realize this is to make use of the masking effect and the psychoacoustics of the human ear, and to remove the redundant signals by making use of complex algorithms. The psychoacoustic curve has been set up by means of a group of test persons. By exploring several test tones to this group of persons, a detailed curve of the average human hearing sensitivity could be established. Out of these tests it was found that the human ear is most sensitive for frequencies between 2 and 4 kHz. Figure 20.1 is a representation of the threshold level of human hearing in a quiet room. The average person does not hear sounds from below that line.

The second part of the psychoacoustics is the masking effect or the covering of other signals. Due to a loud sound it can happen that the human ear does not notice weaker sounds. In Figure 20.2, it can be seen that when a tone of 1 kHz is emitted, the nearby frequencies with a lower level are covered. The normal threshold line will change momentarily due to this short burst.

Figure 20.1 Human hearing sensitivity.

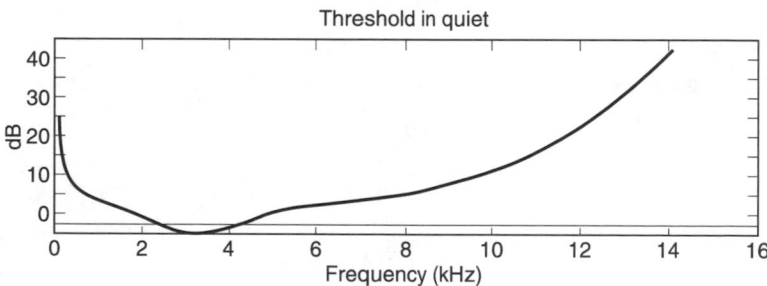

Figure 20.2 Masking effect.

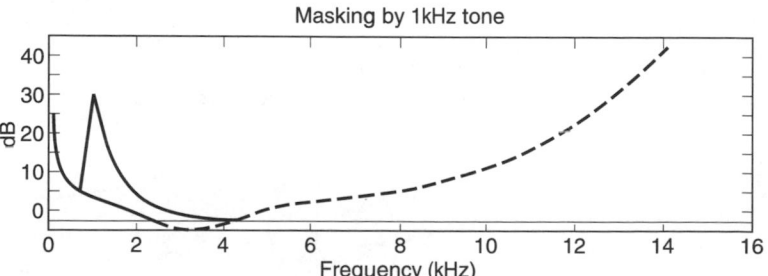

In the same way, a similar threshold line can be made for different test tones (Figure 20.3).

The third part is temporal masking. Right after playing a loud tone of a certain frequency, adjacent frequencies are masked even when the test tone has already stopped.

These items were set up by experiment and are used as a perceptual model to encode the digital information.

The basic idea of MPEG is the division of the incoming audio signal into 32 critical bands. Within these bands, the threshold level will be defined and all information below this threshold will be omitted.

The encoder part analyses the incoming signal and creates a filterbank. This filterbank is compared to the psychoacoustic model to estimate the noise level that can be covered (masked) by the signal. Subsequently, the encoder tries to allocate the available amount of bits (depending on the compression rate) to meet the bit rate and the masking requirements.

Suppose we have the next situation: an audio signal is analysed in the critical bands at the level presented in Figure 20.6. The level in band 8 is 60 dB. This will result in a masking effect of 12 dB in band 7. The level in band 7 is 10 dB, so it will be completely masked and there is no need to encode band 7. The noise level of band 8 is 12 dB, so encoding can be performed with 2 bits (refer to Chapter 3, section on 'Calculation of theoretical signal-to-noise ratio' pages 49–50).

Figure 20.3 Masking effect for different frequencies.

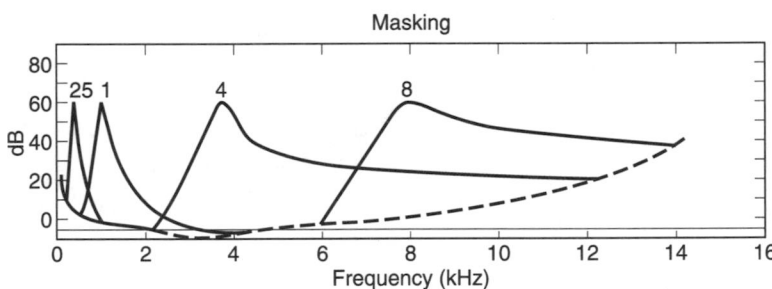

Figure 20.4 Temporal masking.

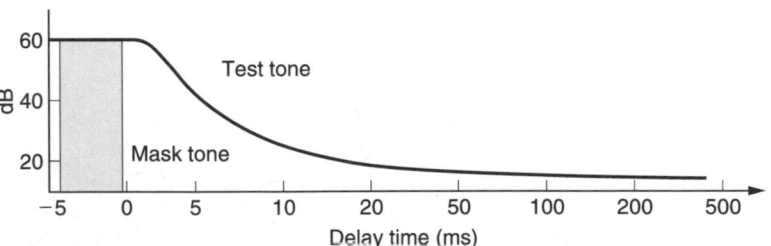

Figure 20.5 Encoder.

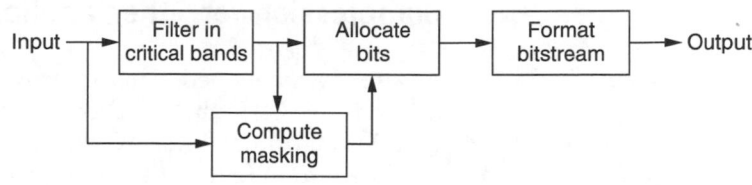

Figure 20.6 Band analysis.

Band	1	2	3	4	5	6	7	8	9	10	11	12	13	14	15	16
level (dB)	0	8	12	10	6	2	10	60	35	20	15	2	3	5	3	1

MPEG layer 1 will reduce the amount of data four times, which corresponds with a transfer rate of 384 kbps starting from a 16-bit PCM signal sampled at 44.1 kHz. Layer 2 reduces the data six to eight times. The very popular layer 3 or MP3 reduces the amount of data 10–12 times (corresponds with 128…112 kbps for a stereo signal), without losing the majority of the sound quality.

Obviously, the audio quality will also vary with the compression level; some tradeoff between the amount of compression and audio quality has to be foreseen, although design improvement continuously pushes the limits forward.

Based upon these psychoacoustic theories, and some basic specifications, MPEG audio layers become a kind of applied computer programming, which can be designed appropriately.

Figure 20.7 MP3 encoder.

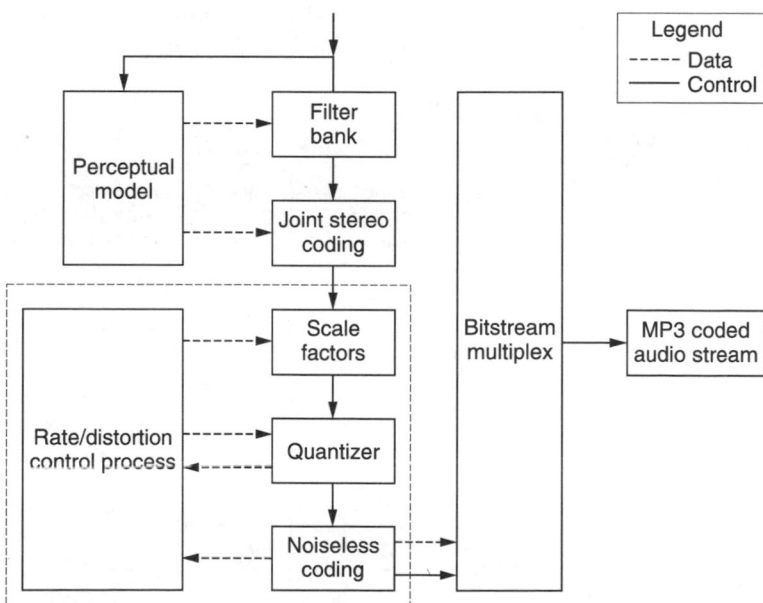

Compression for other applications

As already mentioned, Sony improved the existing compression method 'ATRAC' that is used in the MiniDisc. By means of complex algorithms, ATRAC version 3 is capable of compressing the data two to four more times. Depending on the compression chosen (in the case of MiniDisc Long Play or MDLP, this is called LP2 or LP4), a bit rate of 132 or 66 kbps can be achieved without losing most of the high-quality sound information.

Basically, ATRAC3 uses a more efficient method of allocating the bits than previous ATRAC versions. More frequency bands are used and more efficient use of the available bits is achieved by several recalculations of the original signal.

Further bit reduction for ATRAC3 and also for MP3 is realized by the adoption of a joint stereo coding scheme, which takes advantage of the similarity of the left and right channels in a stereo signal. In general, the difference between left and right is not very much, so the amount of data needed to encode these differences will be much smaller than the complete signal.

Multi-channel compression

With the further development of multi-channel audio, more powerful compressions were needed for this purpose. Starting with the Dolby AC1 encoding, where the surround channels are multiplexed in the stereo signals, sound quality and coding gain improved over the years. Already in 1992, the first cinema movies

Figure 20.8 ATRAC encoder–decoder.

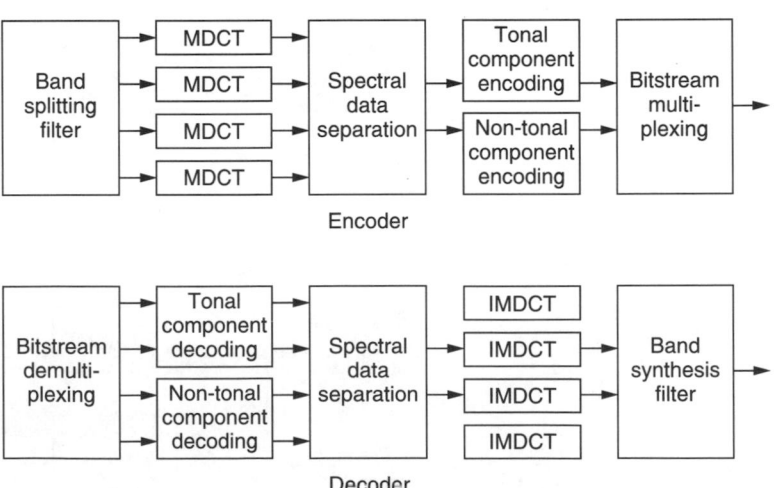

308

were presented with AC3 sound and in 1995 the **Dolby AC3** digital compression algorithm was widely accepted to encode 1 to 5.1 channel PCM converted signals into a serial bit stream.

A block diagram of the AC3 process is shown in Figure 20.9.

Under impulse of the manufacturers that came up with different carriers, different compression techniques were also developed to compress the multi-channel sound information. Most of these compression techniques are based on MPEG-2 Advanced Audio Coding (MPEG-2AAC), which was created by an international cooperation of the Fraunhofer Institute and companies like AT&T, Sony and Dolby. MPEG-2AAC became an international standard at the end of April 1999. For high-quality sound reproduction, compression without loss (DST for SACD and MLP for DVD-Audio) were developed.

In fact, MPEG-2AAC is the further development of the successful ISO/MPEG Audio layer 3. The main difference is that

Figure 20.9 AC3 encoder.

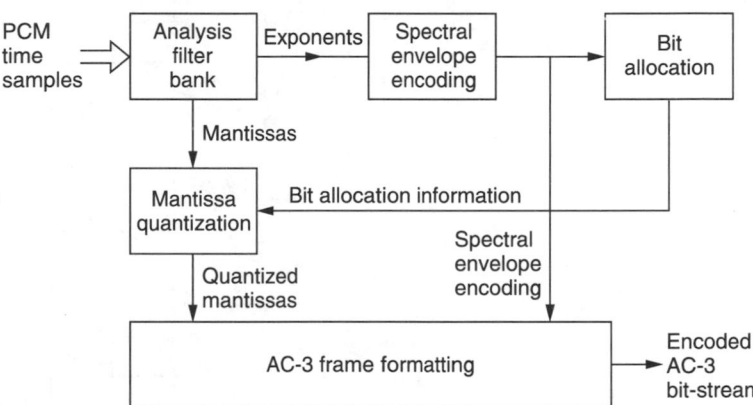

Figure 20.10 AC3 decoder.

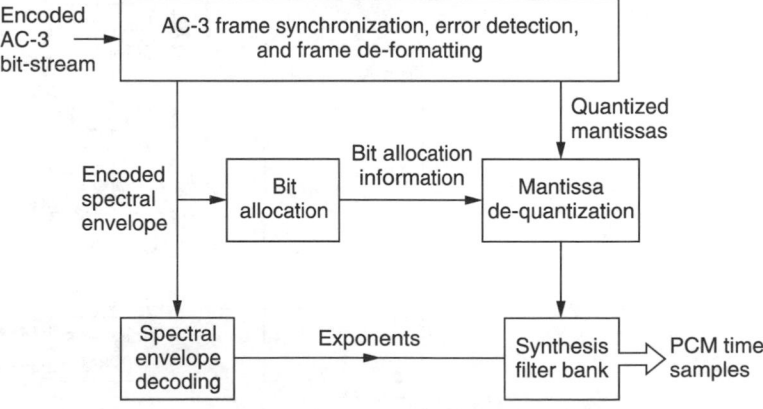

MPEG-2AAC can compress data between one and 48 channels with a sampling frequency between 8 and 96 kHz. The high-quality compression is achieved by some crucial differences to MP3:

- The filter bank used for MP3 is replaced by a plain Modified Discrete Cosine Transform (MDCT). The principle of MDCT is explained in Chapter 17.
- Temporal Noise Shaping (TNS) shapes the distribution of quantization noise in time by prediction in the frequency domain. In particular, voice signals are improved through this method.
- Prediction became a very common technique in the coding of audio signals. Certain sounds are easy to predict.

Figure 20.11 MPEG2-2AAC block.

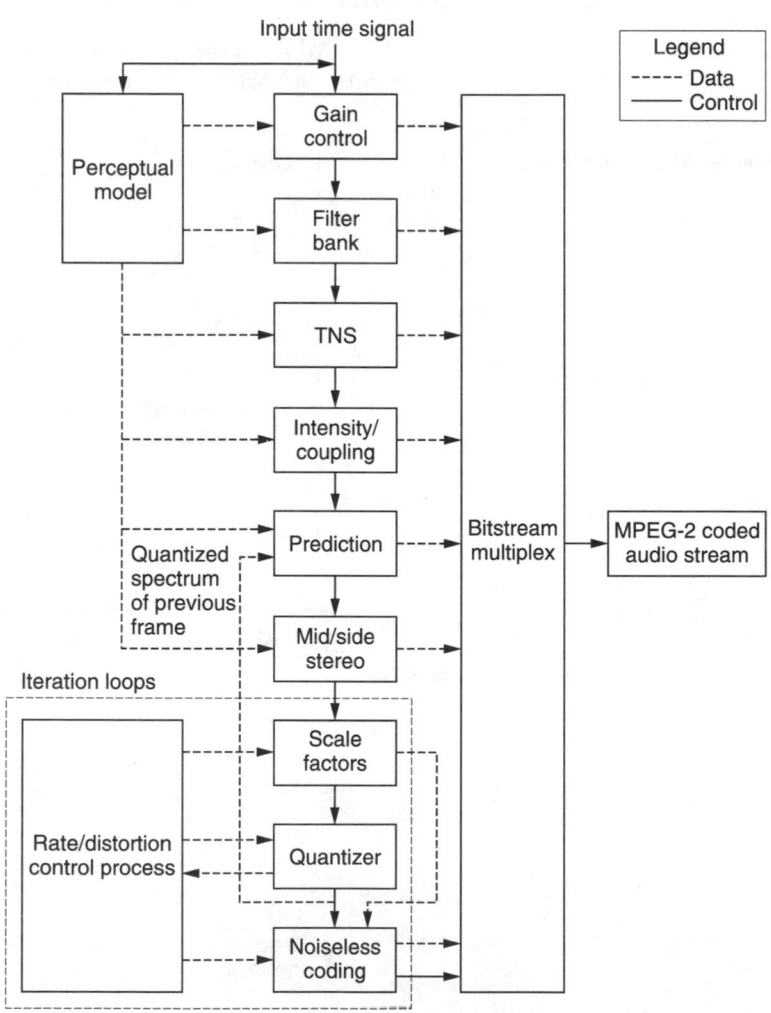

- The quantization resolution is made finer so that the available bits can be used more efficiently.

At present, MPEG-2AAC is the most advanced coding technology used as a basis to compress high-quality sound sources and is selected to be used in the Digital Radio Mondiale (DRM). DRM is a world consortium dedicated to form a single world standard for digital broadcasting in AM radio bands below 30 MHz. Its members are broadcasters, network operators, receiver and transmitter manufacturers, research institutes and standardization bodies.

For transportation of high-quality music via the internet, MPEG-2AAC will also play an important role.

Appendix 1: Error correction

Error correction is one of the most advanced areas in the entire field of digital audio. It is purely because of error-correction techniques that reliable digital recordings can be made, despite the frequent occurrence of tape dropouts.

In the next two sections, we will discuss the theory behind the EIAJ format P, Q and CRC codes. Because of the special nature and the complexity involved, a mathematical treatment of this topic is unavoidable. However, every attempt has been made to pool all the material together in a concise and systematic manner. Examples are included to make the situations more vivid. The only prerequisite in reading them is some knowledge of matrix algebra.

P, *Q* and the cyclic redundancy check code

In Figure A1.1, the P codes are generated by feeding the input data L_0–R_2 to an exclusive-or gate. Hence, we have:

$$P_0 = L_0 \oplus R_0 \oplus L_1 \oplus R_1 \oplus L_2 \oplus R_2 ...$$
$$P_1 = L_3 \oplus R_3 \oplus L_4 \oplus R_4 \oplus L_5 \oplus R_5$$
$$P_n = L_{3n} \oplus R_{3n} \oplus L_{3n+1} \oplus R_{3n+1} \oplus L_{3n+2} \oplus R_{3n+2} \tag{A1.1}$$

The symbol ⊕ (ring-sum) indicates modulo-2 summation, and obeys the following rules:

$$0 \oplus 0 = 0$$
$$0 \oplus 1 = 1$$
$$1 \oplus 0 = 1$$
$$1 \oplus 1 = 0$$

which is an exclusive-or operation. When considering expression A1.1 and applying the rules, the parity bit P_0 will be 0 for even numbers of logic levels, and 1 for odd numbers.

Q codes are generated by a matrix operation circuit, the principle of which is shown in Figure A1.2.

Figure A1.1

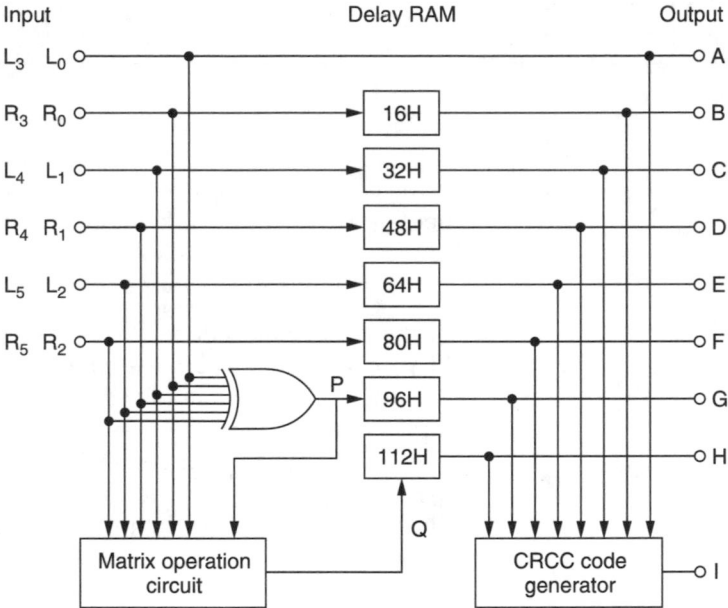

B-adjacent code generator

$$Q_0 = T^6L_0 + T^5R_0 + T^4L_1 + T^3R_1 + T^2L_2 + TR_2$$

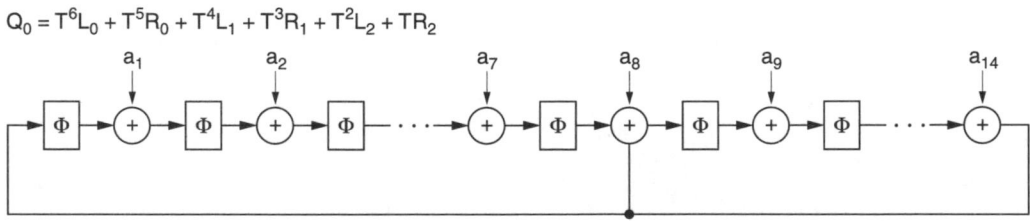

Figure A1.2

Initially, L_0, a 14-bit data word comprising bits a_1-a_{14}, is applied to the summing nodes and the shift register contents are all zeros. After one shift to the right, the shift register contents change to those shown in Figure A1.3. The single shift operation on word L_0 can be defined as TL_0.

Now, a second data word R_0, comprising bits b_1-b_{14}, is applied to the summing nodes. When the next shift occurs, the shift register contents change to those shown in Figure A1.4. This time, L_0 undergoes two shifts while R_0 is subjected to only one; the combined effect can be defined as $TTL_0 \oplus TR_0$, or $T^2L_0 \oplus TR_0$.

After six shift operations, the shift register contents constitute the code word Q_0. Mathematically, we can write:

$$Q_0 = T^6L_0 \oplus T^5R_0 \oplus T^4L_1 \oplus T^3R_1 \oplus T^2L_2 \oplus TR_2$$

Similarly, we have

$$Q_1 = T^6L_3 \oplus T^5R_3 \oplus T^4L_4 \oplus T^3R_4 \oplus T^2L_5 \oplus TR_5$$

and

$$Q_n = T^6L_{3n} \oplus T^5R_{3n} \oplus \ ... \ \oplus T^2L_{3n+2} \oplus TR_{3n+2}$$

The 1-bit shift and 2-bit modulo-3 summation are functions of T having the following forms:

1 shift

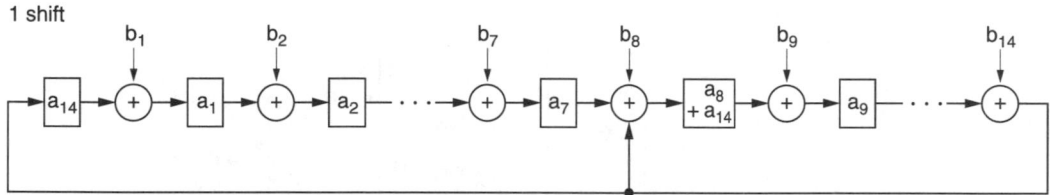

Figure A1.3

2 shift

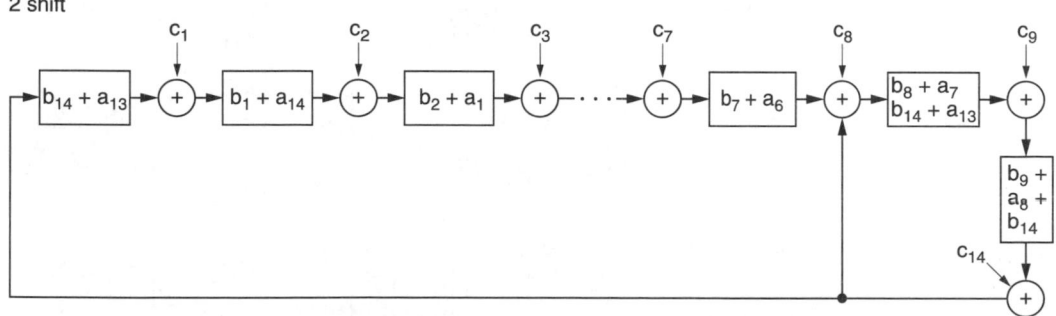

Figure A1.4

$$T = \begin{bmatrix} 00000000000001 \\ 10000000000000 \\ 01000000000000 \\ 00100000000000 \\ 00010000000000 \\ 00001000000000 \\ 00000100000000 \\ 00000010000000 \\ 00000001000000 \\ 00000000100000 \\ 00000000010000 \\ 00000000001000 \\ 00000000000100 \\ 00000000000010 \end{bmatrix} \quad \text{and} \quad TL_0 = T\begin{bmatrix} a_1 \\ a_2 \\ a_3 \\ a_4 \\ a_5 \\ a_6 \\ a_7 \\ a_8 \\ a_9 \\ a_{10} \\ a_{11} \\ a_{12} \\ a_{13} \\ a_{14} \end{bmatrix} = \begin{bmatrix} a_{14} \\ a_1 \\ a_2 \\ a_3 \\ a_4 \\ a_5 \\ a_6 \\ a_7 \\ a_8 \\ a_9 \\ a_{10} \\ a_{11} \\ a_{12} \\ a_{13} \end{bmatrix}$$

If we let

$$\left[L_0 = a_1 a_2 \ldots a_{14} \right]$$

$$\left[R_0 = b_1 b_2 \ldots b_{14} \right]$$

$$\left[L_1 = c_1 c_2 \ldots c_{14} \right]$$

$$\vdots$$

$$\left[R_2 = f_1 f_2 \ldots f_{14} \right]$$

then we have the following:

$$TR_2 = \begin{bmatrix} f_{14} \\ f_1 \\ f_2 \\ f_3 \\ f_4 \\ f_5 \\ f_6 \\ f_7 \\ f_8 \oplus f_{14} \\ f_9 \\ f_{10} \\ f_{11} \\ f_{12} \\ f_{13} \end{bmatrix} \quad T^2 L_2 = \begin{bmatrix} e_{13} \\ e_{14} \\ e_1 \\ e_2 \\ e_3 \\ e_4 \\ e_5 \\ e_6 \\ e_7 \oplus e_{13} \\ e_8 \oplus e_{14} \\ e_9 \\ e_{10} \\ e_{11} \\ e_{12} \end{bmatrix} \quad T^3 R_1 = \begin{bmatrix} d_{12} \\ d_{13} \\ d_{14} \\ d_1 \\ d_2 \\ d_3 \\ d_4 \\ d_5 \\ d_6 \oplus d_{12} \\ d_7 \oplus d_{13} \\ d_8 \oplus d_{14} \\ d_9 \\ d_{10} \\ d_{11} \end{bmatrix}$$

$$T^4 L_2 = \begin{bmatrix} c_{11} \\ c_{12} \\ c_{13} \\ c_{14} \\ c_1 \\ c_2 \\ c_3 \\ c_4 \\ c_5 \oplus c_{11} \\ c_6 \oplus c_{12} \\ c_7 \oplus c_{13} \\ c_8 \oplus c_{14} \\ c_9 \\ c_{10} \end{bmatrix} \quad T^5 R_0 = \begin{bmatrix} b_{10} \\ b_{11} \\ b_{12} \\ b_{13} \\ b_{14} \\ b_1 \\ b_2 \\ b_3 \\ b_4 \oplus b_{10} \\ b_5 \oplus b_{11} \\ b_6 \oplus b_{12} \\ b_7 \oplus b_{13} \\ b_8 \oplus b_{14} \\ b_9 \end{bmatrix} \quad T^6 L_0 = \begin{bmatrix} a_9 \\ a_{10} \\ a_{11} \\ a_{12} \\ a_{13} \\ a_{14} \\ a_1 \\ a_2 \\ a_3 \oplus a_9 \\ a_4 \oplus a_{10} \\ a_5 \oplus a_{11} \\ a_6 \oplus a_{12} \\ a_7 \oplus a_{13} \\ a_8 \oplus a_{14} \end{bmatrix}$$

Now, let us see how a Q code can be generated if data are input serially. Consider the following scheme:

1   One of the feedback loops is controlled by a switch that turns on whenever 15 shifts are made.
2   The exclusive-or operation is only active during a shift register (SR) shift.

The configuration shown in Figure A1.5 applies.

Then, after 14 shifts the contents of the two SRs are as shown in Figure A1.6a. For an additional shift we have the situation shown in Figure A1.6b, which we recognize as TL_0. As we continue, we have the situation shown in Figure A1.6c. After the 30th shift, the contents of the SRs are as shown in Figure A1.6d which is T^2L_0 + TR_3 (when compared with the two column matrices TR_2 and T^2L_2). Note that for this scheme to work, the serial data have to be sent in the following format:

$$a_{14}a_{13}a_{12}a_{11}...a_1 \qquad bb_{14}b_{13}b_1 \qquad bc_{14}c_{13}$$

where $b = 0$. Finally, the reader can easily verify that, after 90 shifts, the contents of the two SRs constitute the word Q.

The PCM-F1 employs cyclic redundancy check code (CRCC) for detecting code errors. The encoding can be easily implemented by using shift registers, while the decoding scheme becomes simple because of the inherent well-defined mathematical structure. In mathematical terms, a code word is a code vector because it is an n-tuple from the vector space of all n-tuples. For a linear code C with length n and containing r information digits, if an n-tuple

$$V^{(1)} = (V_{n-1}, V_1, ..., V_{n-2})$$

obtained by shifting the code vector of C

$$V = (V_0, V_1, V_2, ..., V_{n-1})$$

cyclically one place to the right is also a code vector of C, then linear code C is called a cyclic code. From this definition, it follows that no matter how many times the digits in V are shifted

Figure A1.5

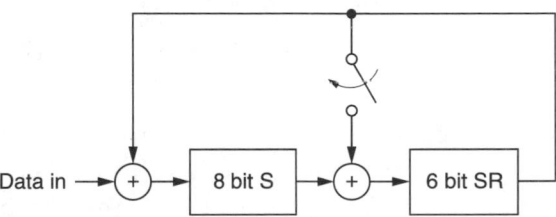

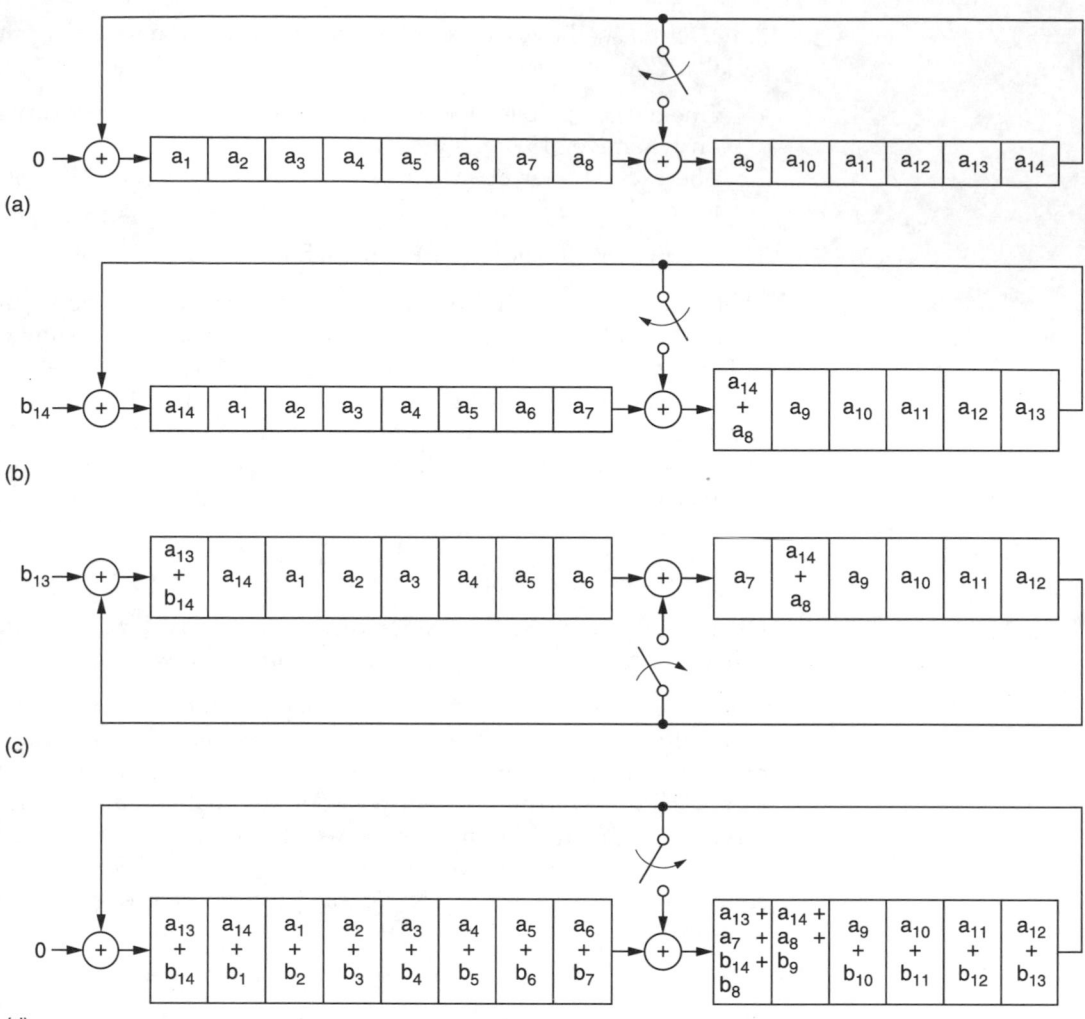

Figure A1.6

cyclically to the right, the resulting *n*-tuple is also a code vector. The components of a code vector define the coefficients of a polynomial. Therefore, if we denote $V(X)$ as the code polynomial of V, we then have:

$$V(X) = V_0 + V_1 X + V_2 X^2 + \ldots + V_{n-1} X_{n-1}$$

or, in a particular case, where we have the code word shown in Figure A1.7, the code polynomial is written as

$$V(X) = X^{11} + X^{10} + X^8 + X^5 + X^3 + 1$$

where + stands for modulo-2 summation. Moreover, every code polynomial $V(X)$ can be expressed in the following form:

$$V(X) = M(X)g(X)...$$

$$= \left(M_0 + M_1 X + M_2 X^2 + ... + M_{r-1} X^{r-1}\right)$$

$$\times \left(1 + g_1 X + g_2 X^2 + ... + g_{n-r-1} X^{n-r-1} + X^{n-r}\right) \qquad \text{(A1.2)}$$

where M_0, M_1, ..., M_{r-1} are the information digits to be encoded and $g(X)$ is defined as a generator polynomial. Hence, the encoding of a message $M(X)$ is equivalent to multiplying it by $g(X)$. Note that in any particular cyclic code, there only exists one generator polynomial $g(X)$. However, if we encode the information digits according to Equation A1.2, the orders of the message digits are altered. Hence, a different scheme must be used. The code can be put into a systematic form (see Example 3 below) by first multiplying $M(X)$ by $xn - r$ and then dividing the result by $g(X)$, hence,

$$X^{n-r}M(X) = Q(X)g(X) + R(X)$$

Adding $R(X)$ to both sides, we have

$$X^{n-r}M(X) + R(X) = Q(X)g(X) \qquad \text{(A1.3)}$$

where $Q(X)$ is the quotient and $R(X)$ is the remainder.

In modulo-2 summation, $R(X) + R(X) = 0$.

If we define the left-hand side of Equation A1.3 as our transmission polynomial $T(X)$, then

$$T(X) = Q(X)g(X)$$

which means that $T(X)$ can be divided exactly by $g(X)$. In this case, the encoding is equivalent to converting the information polynomial $M(X)$ into $T(X)$.

If, in the recording process, $T(X)$ changes to $E(X)$ then errors are detected. In the PCM-F1, when the CRCC indicates that one word in a horizontal scanning period H is incorrect, the remaining words in that period will also be considered erroneous.

At this point, it is natural to ask what kind of properties $g(X)$ must have in order to generate an (n, r) cyclic code, i.e., generating an n-bit code from a message of r bits. We have the following theorem.

Theorem. If $g(X)$ is a polynomial of degree $n - r$ and is a factor of $X^n + 1$, then $g(X)$ generates an (n, r) cyclic code.

Figure A1.7

14	13	12	11	10	9	8	7	6	5	4	3	2	1
0	0	1	1	0	1	0	0	1	0	1	0	0	1

X^{11} X^{10} $\quad$ X^8 $\quad$ X^5 $\quad$ X^3 $\quad$ $X^0 = 1$

Now, we will show how a 4-bit cyclic code can be generated from a message of 3 bits. We hope that the whole situation will be clarified by the following simple examples.

Example 1. Since $n = 4$ and $r = 3$, we are searching for a polynomial of degree 1. Because:

$$X^4 + 1 = (X + 1)(X^3 + X^2 + X + 1)$$

our $g(X)$ will be $X + 1$. Note that $g(X)$ is primitive (irreducible).

Example 2. Consider the messages 000, 100, 010, ..., 111 where the LSB is at the extreme left of each word. From Equation A1.2, we have

for 100, $V(X) = 1(1 + X) = 1 + X$
for 101, $V(X) = (1 + X^2)(1 + X) = 1 + X + X^2 + X^3$

For these two cases, the coded words are 1100 and 1111 respectively. Proceeding as before, we have the following:

MESSAGES			CODE			
LSB		MSB	LSB			MSB
0	0	0	0	0	0	0
1	0	0	1	1	0	0
0	1	0	0	1	1	0
1	1	0	1	0	1	0
0	0	1	0	0	1	1
1	0	1	1	1	1	1
0	1	1	0	1	0	1
1	1	1	1	0	0	1

This code has a minimum distance of 2, hence it can detect a single error in the message; it is not systematic, i.e., we do not have a code where the first three digits are the unaltered message digits and the last one is a check bit.

It is not difficult to see that this code can be implemented by the circuit shown in Figure A1.8, which employs two exclusive-or gates and two SRs. For this scheme to work, the message should be input as:

$$M_1M_2M_3M_1 0 M_1'M_2'M_3'M_1'0... \text{ etc.}$$

The SRs reset whenever five shifts are made.

Figure A1.8

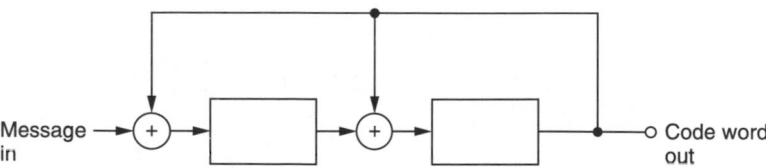

Message in → (+) → [ ] → (+) → [ ] → Code word out

The decoder has the configuration shown in Figure A1.9.

After four data bits come out of the AND gate, we look at the fourth bit. If this bit is zero, the three message bits are considered correct; otherwise, an error is made. Hence, the SRs reset after every four shifts. Also, the message input sequence is:

$$C_1C_2C_3C_4C_1'C_2'C_3'C_4'\dots \text{ etc.}$$

The circuit works because when we multiply $a_1a_2a_3$ by 11 we get the following:

$$
\begin{array}{ccc}
a_1 & a_2 & a_3 \\
 & 1 & 1 \\
\hline
a_1 & a_2 & a_3 \\
 & a_1 & a_2 & a_3 \\
\hline
a_1 & a_1+a_2 & a_2+a_3
\end{array}
$$

Hence, with the circuit shown in Figure A1.10, in the second shift, a_1 is out.

From these five shifts, we get $0, a_1, a_1 + a_2, a_2 + a_3$, which matches the elements generated by the multiplication.

However, for this to work, the message input must have the following format:

$$M_1M_2M_3M_10M_1'M_3'M_1'0\dots$$

and the SRs reset whenever five shifts are made.

Hence, for the decoded message $D_1D_2D_3$, we have:

Figure A1.9

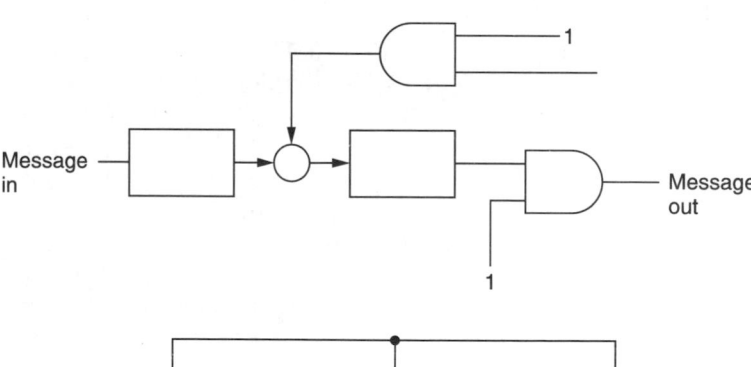

Figure A1.10

$$D_1 = a_1 \cdot 1$$
$$D_2 = ((a_1 \cdot 1) + a_2) \cdot 1$$
$$D_3 = (((a_1 \cdot 1) + a_2) \cdot 1 + a_3) \cdot 1$$

Also,

$$R = D_3 + a_4$$

For zero error detection, $D_3 = a_4$. This is true when we look at the LSB columns of the message and code table. It is also clear that the decoder circuit implements the above division.

Example 3. We can form a systematic code by using Equation A1.3. Take the last message word (1110 in the above example), then the message polynomial is $M(X) = 1 + X + X^2$. Since $X^{n-r} = X$ in this case, we have $XM(X) = X + X^2 + X^3$.

Dividing $XM(X)$ by the generator polynomial $g(X) = 1 + X$, we have

$$
\begin{array}{r}
X^2 \qquad +1 \\ \hline
X+1 \,\big|\, X^3 + X^2 + X \\
X^3 + X^2 \\ \hline
X \\
X + 1 \\ \hline
1
\end{array}
$$

The remainder $R(X) = 1$. Thus, the code polynomial is

$$T(X) = X^3 + X^2 + X + 1$$

which is 1111.

Proceeding as before, we get the following systematic code:

MESSAGES			CODE			
LSB		MSB	LSB			MSB
0	0	0	0	0	0	0
1	0	0	1	1	0	0
0	1	0	1	0	1	0
1	1	0	0	1	1	0
0	0	1	1	0	0	1
1	0	1	0	1	0	1
0	1	1	0	0	1	1
1	1	1	1	1	1	1

Assuming that message bits are input as

$$M_1 M_2 0 M_1' M_2' M_3' M_0' \ldots$$

The code can be implemented by the circuit shown in Figure A1.11.

Figure A1.11

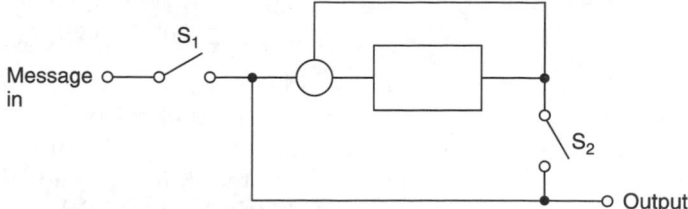

Initially, S_2 is off and S_1 is on. After M_3 is fed into the circuit, S_1 switches off, S_2 turns on and an additional shift by the SR generates the parity check code. Then, S_2 switches off and S_1 turns on to repeat the process (Figure A1.12).

The decoder operates in the same manner as the encoder. However, an AND gate is inserted to check the parity received and the one generated by the SR. If they do not match, an error is detected.

Erasure and b-adjacent decoding

After the CRCC decoder detects a code error in each data block, error correction is implemented in the parity decoder. First, the syndrome S_{P0} (corrector) is calculated as follows:

$$S_{P0} = L_0 + R_0 + L_1 + L_2 + P_0$$

where

$$P_0 = L_0 + R_0 + L_1 + R_1 + L_2 + R_2$$

If there is no error, $S_{P0} = 0$, because in modulo-2 summation, adding any two identical quantities is zero.

On the other hand, suppose a single error R_1' (i.e., all 16 bits are incorrect) in a single message block is made. In this case, $S_{P0} = 1$.

If we add this to R_1', correction can be made because:

Figure A1.12

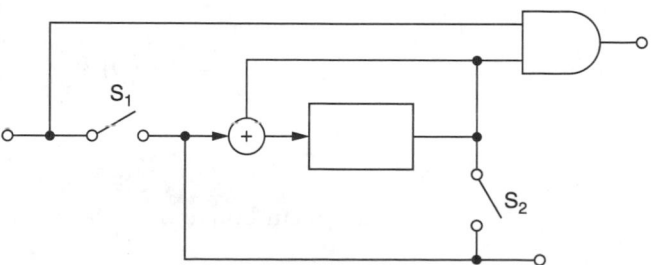

If $R_{1X} = 0$, $R_{1,X} = 1$ where R_{1X} is any bit in R_1
$R_{1X} = 1$, $R_{1,X} = 0$ where R_{1X} is any bit in R_1

and, in either case, we have $R_1' + S_{P0} = R_1$. However, if two errors occur, this method of correction – called the erase method – fails to recover the original information. Since interleaving is employed in the PCM-F1, original bits in a block are shuffled to the other blocks by a 16H delay. Effectively, this method is capable of correcting errors occurring successively in 2048 bits.

In b-adjacent decoding, the syndromes S_{P0} and S_{Q0} are calculated as follows:

$$S_{P0} = L_{0r} + R_{0r} + L_{1r} + L_{2r} + R_{2r} + P_{0r} \cdots \tag{A1.4}$$

$$S_{00} = T^6L_{0r} + T^5R_{0r} + R^4L_{1r} + T^3R_{1r} + T^2L_{2r} + Q_{0r} \cdots \tag{A1.5}$$

where

$$P_0 = L_0 + R_0 + L_1 + R_1 + L_2 + R_2{}^*$$

$$Q_0 = T^6L_0 + T^5R_0 + T^4L_1 + T^3R_1 + T^2L_2 + TR_{2r}{}^*$$

$^*L_{0r}, \ldots, R_{2r}, P_{0r}, Q_{0r}$ are the information received.

Suppose the received information R_{1r} and R_{2r} contain errors, then:

$$R_1' = R_1 + E_{R1} \cdots \tag{A1.6}$$

$$R_2' = R_2 + E_{R2} \cdots \tag{A1.7}$$

where R_1 and R_2 are the original messages transmitted, and E_{R1} and E_{R2} are the errors added.

Substituting R_1' and R_2' in place of R_{1r} and R_{2r} in Equations A1.4 and A1.5, we can easily get

$$S_{P0} = E_{R1} + E_{R2} \cdots \tag{A1.8}$$

$$S_{Q0} = T^3E_{R1} + TE_{R2} \cdots \tag{A1.9}$$

Adding T^{-1} to both sides of Equation A1.9, then

$$T^{-1}S_{Q0} = T^2E_{R1} + E_{R2} \cdots \tag{A1.10}$$

Adding S_{P0} to both sides of Equation A1.10, we get

$$S_{P0} + T^{-1}S_{Q0} = (T^2E_{R1} + E_{R2}) + S_{P0}$$
$$= (T^2E_{R1} + E_{R2}) + (E_{R1} + E_{R2})$$
$$= (T^2 + I)\, E_{R1} \cdots \tag{A1.11}$$

Rearranging Equation A1.11, we get:

$$E_{R1} = (T^2 + I) - 1(S_{P0} + T^{-1}S_{Q0}) \cdots \tag{A1.12}$$

And from Equation A1.8:

$$E_{R2} = E_{R1} + S_{P0} \ldots \tag{A1.13}$$

Before we go on, let us summarize the situation as follows: we first calculate the syndromes S_{P0} and S_{Q0} according to Equations A1.4 and A1.5. After some algebraic manipulations (if only two or less errors are made during the transmission of data), the error pattern can be calculated using Equations A1.12 and A1.13. Once these are known, we can correct the error using Equations A1.6 and A1.7, where:

$$R_1 = R_1{}' + E_{R1}$$
$$R_2 = R_2{}' + E_{R2}$$

Note that in modulo-2 summation, since $1 + 1 = 0$, $1 = -1$. By this method, 4096 successive bit errors can be corrected.

In what follows we shall demonstrate how the decoding is done by going through an example.

Unfortunately, the matrix T consists of 14×14 elements. Even for a simple example such as that shown below, a lot of algebraic manipulations are involved. Therefore, in order to avoid the many steps of purely mechanical computation from clouding the main issue, we shall show only the important steps or results. Anyway, what is important here is to get a 'feel' of how decoding is carried out.

Example. We have the following data:

	LSB													MSB
L_0	0	0	0	0	0	0	0	0	0	0	0	0	0	0
R_0	1	0	0	0	0	0	0	0	0	0	0	0	0	0
L_1	0	1	0	0	0	0	0	0	0	0	0	0	0	0
R_1	0	0	1	0	0	0	0	0	0	0	0	0	0	0
L_2	0	0	0	1	0	0	0	0	0	0	0	0	0	0
R_2	0	0	0	0	1	0	0	0	0	0	0	0	0	0
P_0	1	1	1	1	1	0	0	0	0	0	0	0	0	0
Q_0	0	0	0	0	0	0	1	0	0	0	0	0	0	0

Suppose that, after the data are transmitted, the received data become:

L_0	0	0	0	0	0	0	0	0	0	0	0	0	0	0
R_0	1	0	0	0	0	0	0	0	0	0	0	0	0	0
L_1	0	1	0	0	0	0	0	0	0	0	0	0	0	0
$R_1{}'$	0	0	1	1	1	0	0	0	0	0	0	0	0	0
L_2	0	0	0	1	0	0	0	0	0	0	0	0	0	0
$R_2{}'$	0	0	0	0	0	0	0	0	0	1	1	1	0	0
P_0	1	1	1	1	1	0	0	0	0	0	0	0	0	0
Q_0	0	0	0	0	0	0	1	0	0	0	0	0	0	0

We can find E_{R1} and E_{R2} as follows.

First, we have to find S_{P0} and S_{Q0}. Using Equation A1.4 it is easy to see that

$$S_{P0} = \begin{bmatrix} 0 \\ 0 \\ 0 \\ 1 \\ 0 \\ 0 \\ 0 \\ 0 \\ 0 \\ 1 \\ 1 \\ 1 \\ 0 \\ 0 \end{bmatrix}$$

Similarly, from Equation A1.5 and the column matrices TR_2, T^2L_2, ..., T^6L_0 listed in the previous section, we find

$$S_{Q0} = \begin{bmatrix} 0 \\ 0 \\ 0 \\ 0 \\ 0 \\ 1 \\ 1 \\ 1 \\ 0 \\ 0 \\ 1 \\ 1 \\ 1 \\ 0 \end{bmatrix}$$

After going through some algebraic manipulations, we find $T - 1$, $(T^2 + 1)$ and finally $(T^2 + 1)^{-1}$. We have

$$S_{P0} + T^{-1}S_{Q0} = \begin{bmatrix} 0 \\ 0 \\ 0 \\ 1 \\ 1 \\ 1 \\ 1 \\ 0 \\ 0 \\ 0 \\ 0 \\ 0 \\ 0 \\ 0 \end{bmatrix}$$

Moreover, using Equation A1.12, we obtain:

$$E_{R1} = \begin{bmatrix} 0 \\ 0 \\ 0 \\ 1 \\ 1 \\ 0 \\ 0 \\ 0 \\ 0 \\ 0 \\ 0 \\ 0 \\ 0 \\ 0 \end{bmatrix}$$

and finally from Equation A1.13,

$$E_{R2} = \begin{bmatrix} 0 \\ 0 \\ 0 \\ 0 \\ 1 \\ 0 \\ 0 \\ 0 \\ 0 \\ 1 \\ 1 \\ 1 \\ 0 \\ 0 \end{bmatrix}$$

Now, it can be verified that by taking the modulo-2 summation of R_1' and E_{R1}, R_2 and R_{R2}, the correct transmitted data are obtained.

Notes

1. $(T^2 + 1)^{-1}$ can be constructed as follows:

We have

$$\left(T^2 + I\right)L_0 = \begin{bmatrix} a_1 + a_{13} \\ a_2 + a_{14} \\ a_1 + a_3 \\ a_2 + a_4 \\ a_3 + a_5 \\ a_4 + a_6 \\ a_5 + a_7 \\ a_6 + a_8 \\ a_7 + a_9 + a_{13} \\ a_8 + a_{10} + a_{14} \\ a_9 + a_{11} \\ a_{10} + a_{12} \\ a_{11} + a_{13} \\ a_{12} + a_{14} \end{bmatrix} = M$$

Because $(T^2 = 1)^1$, $M = L_0$, then for the first row of $(T^2 + 1)^{-1}$ we must have

[0 0 1 0 1 0 1 0 1 0 1 0 1 0]

so that

[1 0 0 1 0 1 0 1 0 1 0 1 0 1 0] $M = a_1$

2. We can find T^{-1} and $(T^2 + 1)$ in a manner similar to the above.

3. It can be verified easily that $(T^2 + 1)^{-1}(T^2 + 1) = 1$.

When we multiply the first row and the first matrix by the first column of the second one, we have

(0.1) + (0.0) + (1.1) + (0.0) + (1.0) + (0.0) + (1.0) +
(0.0) + (1.0) + (0.0) + (1.0) + (0.0) + (1.0) + (0.0) = 1

Thus, the first element of the matrix is 1. Similarly, we can verify that all the other elements, except the ones at the diagonal, are zero.

$$Q_0 = T^6 L_0 + T^5 R_0 + T^4 L_1 + T^3 R_1 + T^2 L_2 + TR_2$$

$$T^{-1} = \begin{bmatrix} 0&1&0&0&0&0&0&0&0&0&0&0&0&0 \\ 0&0&1&0&0&0&0&0&0&0&0&0&0&0 \\ 0&0&0&1&0&0&0&0&0&0&0&0&0&0 \\ 0&0&0&0&1&0&0&0&0&0&0&0&0&0 \\ 0&0&0&0&0&1&0&0&0&0&0&0&0&0 \\ 0&0&0&0&0&0&1&0&0&0&0&0&0&0 \\ 0&0&0&0&0&0&0&1&0&0&0&0&0&0 \\ 0&0&0&0&0&0&0&0&1&0&0&0&0&0 \\ 0&0&0&0&0&0&0&0&0&1&0&0&0&0 \\ 0&0&0&0&0&0&0&0&0&0&1&0&0&0 \\ 0&0&0&0&0&0&0&0&0&0&0&1&0&0 \\ 0&0&0&0&0&0&0&0&0&0&0&0&1&0 \\ 0&0&0&0&0&0&0&0&0&0&0&0&0&1 \\ 1&0&0&0&0&0&0&0&0&0&0&0&0&0 \end{bmatrix}$$

$$T^{-2} + I = \begin{bmatrix} 1&0&0&0&0&0&0&0&0&0&0&0&1&0 \\ 0&1&0&0&0&0&0&0&0&0&0&0&0&1 \\ 1&0&1&0&0&0&0&0&0&0&0&0&0&0 \\ 0&1&0&1&0&0&0&0&0&0&0&0&0&0 \\ 0&0&1&0&1&0&0&0&0&0&0&0&0&0 \\ 0&0&0&1&0&1&0&0&0&0&0&0&0&0 \\ 0&0&0&0&1&0&1&0&0&0&0&0&0&0 \\ 0&0&0&0&0&1&0&1&0&0&0&0&0&0 \\ 0&0&0&0&0&0&1&0&1&0&0&0&0&0 \\ 0&0&0&0&0&0&0&1&0&1&0&0&0&0 \\ 0&0&0&0&0&0&0&0&1&0&1&0&0&0 \\ 0&0&0&0&0&0&0&0&0&1&0&1&0&0 \\ 0&0&0&0&0&0&0&0&0&0&1&0&1&0 \\ 0&0&0&0&0&0&0&0&0&0&0&1&0&1 \end{bmatrix}$$

$$\left(T^{-2}+I\right)^{-1}=\begin{bmatrix}0&0&1&0&1&0&1&0&1&0&1&0&1&0\\0&0&0&1&0&1&0&1&0&1&0&1&0&1\\0&0&0&0&1&0&1&0&1&0&1&0&1&0\\0&0&0&0&0&1&0&1&0&1&0&1&0&1\\0&0&0&0&0&0&1&0&1&0&1&0&1&0\\0&0&0&0&0&0&0&1&0&1&0&1&0&1\\0&0&0&0&0&0&0&0&1&0&1&0&1&0\\0&0&0&0&0&0&0&0&0&1&0&1&0&1\\1&0&1&0&1&0&1&0&1&0&0&0&0&0\\0&1&0&1&0&1&0&1&0&1&0&0&0&0\\1&0&1&0&1&0&1&0&1&0&1&0&0&0\\0&1&0&1&0&1&0&1&0&1&0&1&0&0\\1&0&1&0&1&0&1&0&1&0&1&0&1&0\\0&1&0&1&0&1&0&1&0&1&0&1&0&1\end{bmatrix}$$

Appendix 2: Sampling theorem

Sampling picks out values $f(nT)$ from a signal $f(t)$, at regular intervals. This is equivalent to the multiplication of $f(t)$ with the signal $s(t)$, given by the expression:

$$f_s(t) = \sum_{n=-\infty}^{\infty} Tf(nT)\delta(t-nT)$$

where (t) is a delta function.

The Fourier transform $F_s(\omega)$ of $f_s(t)$ is given by the expression:

$$F_s(\omega) = \int_{-\infty}^{\infty} \sum_{n=-\infty}^{\infty} Tf(nT)\delta(t-nT)\, e^{-j\omega t} \mathrm{d}t$$

where $\omega_s = 2\pi/T$. When two functions are multiplied in time their transforms in the frequency domain are convoluted. For this reason, the spectrum $F(\omega)$ is repeated at multiples of the sampling frequency. Function $f(t)$ may be recovered from $F_s(\omega)$ by first multiplying by a gating function $G(\omega)$, illustrated in Figure A2.1. This results in the expression:

$$F(\omega) = F_s(\omega)G(\omega)$$

Now, the inverse transform of $G(\omega)$ is given by the expression:

$$\frac{\sin \omega_0 t}{t}$$

Figure A2.1 Recovery of a sampled signal.

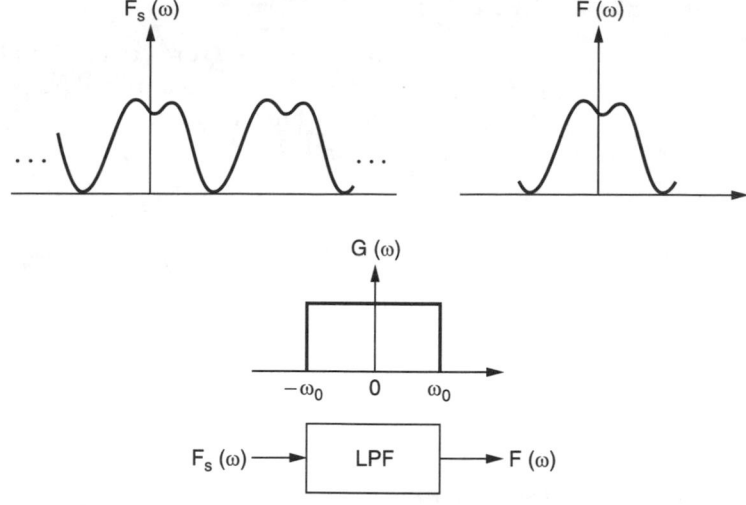

and if:

$$F_s(\omega) \leftrightarrow f_s(t)$$

then:

$$G(\omega) \leftrightarrow g(t)$$

so that:

$$F_s(\omega)G(\omega) \leftrightarrow f_s(t) * g(t)$$

where the symbol * denotes convolution. This gives:

$$f(t) = \sum_{n=-\infty}^{\infty} Tf(nT)\delta(t-nT) * \frac{\sin \omega_0 t}{T_0 \omega_0 t / 2}$$

$$= \sum_{n=-\infty}^{\infty} Tf(nT)\frac{\sin \omega_0(t-nT)}{T_0 \omega_0(t-nT)/2}$$

And, when $T_0 = 2T$:

$$f(t) = \sum_{n=-\infty}^{\infty} f(nT)\frac{\sin \omega_0(t-nT)}{\omega_0(t-nT)}$$

This result, the sampling theorem, relates the samples $f(nT)$ taken at regular intervals to the function $f(t)$.

The sampling theorem is important when considering the bandwidth of a sampled signal. For example, a function $f(t)$ can only be properly reconstructed when samples have been taken at the Nyquist rate, $1/T = 2/T_0$. In practical terms, this means that the sampling frequency should be twice that of the highest signal component frequency, i.e.:

$$F_s = 2f_0$$

and so, to make sure that signal component frequencies greater than half the sampling frequency are not sampled, an anti-aliasing filter is used.

Appendix 3: Serial copy management system (SCMS)

The introduction of new digital recording systems like DAT, MiniDisc and DCC made it perfectly possible to produce high-quality copies of existing materials. When using digital inter-connection, there is virtually no sound quality degradation. This raises the question for protection of copyrighted material. The SCMS limits the ability of consumer digital audio recording equipment to make digital transfers. SCMS allows a single generation copy of a copyrighted digital source. For example, a copyrighted CD can be digitally copied to a DAT recorder once (and only once). The SCMS system prevents further digital copies of that tape. Figure A3.1 shows some examples of possible situations. Note that, regardless of the source, it is always possible to make copies by using analog connections.

Three possible cases exist when digitally copying a digital source.

- The source is not copyrighted: digital copy permitted.
- The source is copyrighted: a first generation copy is permitted.
- The source is a digital copy of copyrighted material: digital copy prohibited.

Each of the digital recording systems contains some auxiliary data, needed for correct recovery of the digital audio signal. The SCMS copy bits can be found in the auxiliary data, e.g., in case of DAT they are contained in ID6 bit 6 and bit 7 (see Table 15.4).

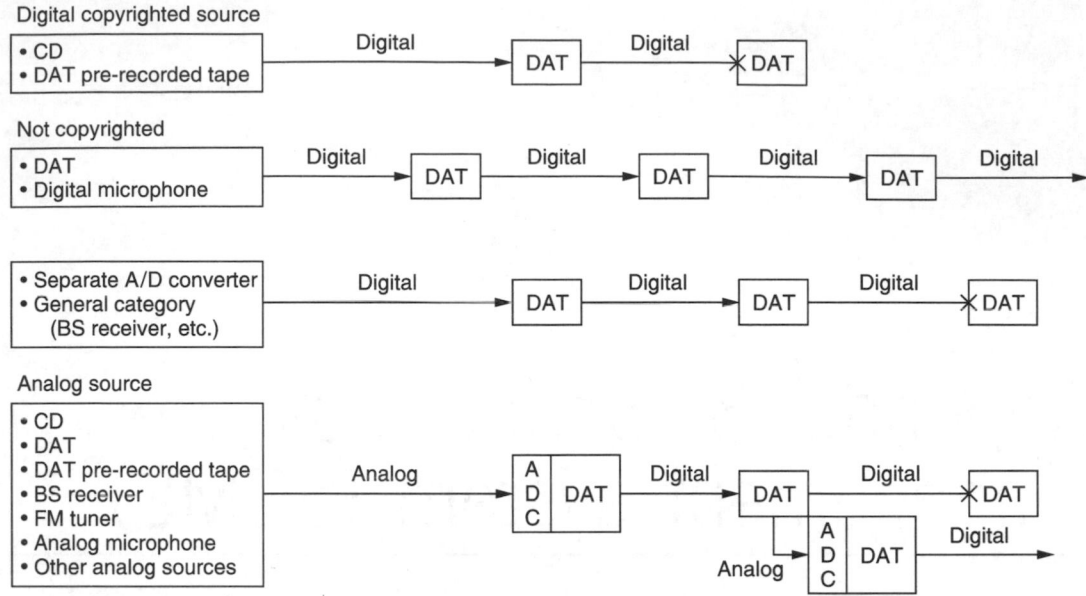

Figure A3.1 Different situations when copying via the SCMS system.

In the Digital Audio Interface Format (S/PDIF), the copy bits are included in the channel status bits.

The SCMS copy management routine is shown in Figure A3.2. The SCMS software checks the channel status bit of the digital audio signal. Several requirements must be fulfilled on the recording side before copying can start. The recording machine will alter the copy bits according to the new situation.

Note the following.

- A digital copy is never possible when the digital audio signal is not for consumer use.
- When the category code is unidentified, the source is considered copyrighted. A single digital copy is allowed, but new copy bits will be set to prevent further copies.
- When the category code is 'general' or 'A/D converter', the recording is considered to be a (copyrighted) master recording. The copy bits will be set to allow one further digital copy.
- If the channel status bits do not carry a copy prohibited code, the source is considered not copyrighted. Digital copies will always be possible.
- Even when the copy prohibited code is set, certain categories (e.g., CD) allow for a single digital copy with the new copy bits set to copy prohibited.
- In all other cases a digital copy is prohibited.

Figure A3.2 SCMS
management routine.

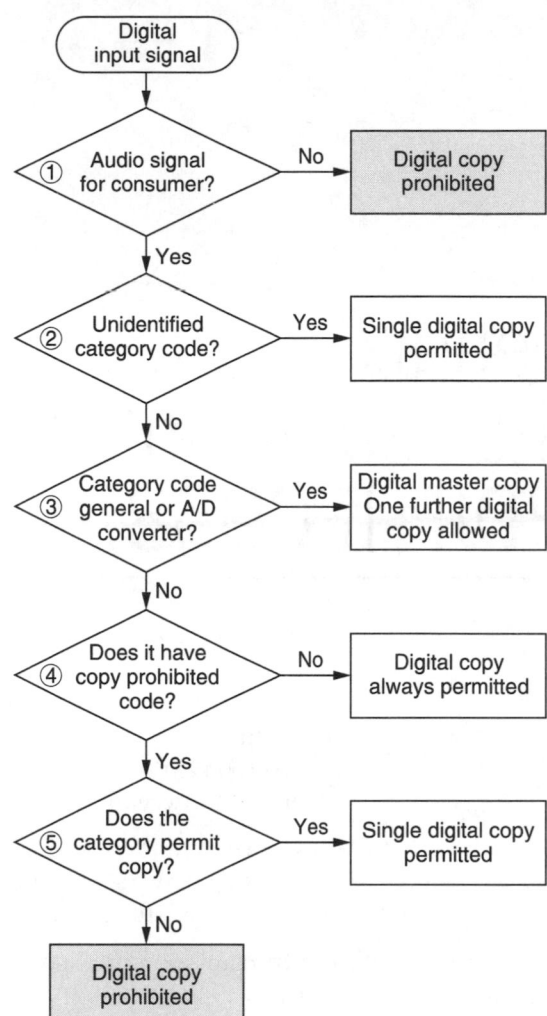

Appendix 4:
Digital audio interface format (S/PDIF-IEC958)

An interface format is designed for use between digital audio sets. This format was originally defined jointly by Sony and Philips (Sony/Philips Digital Interface Format – S/PDIF), but has become an international standard.

This format allows digital connection of digital audio sets, such as CD players, DAT players and others, thus excluding any quality loss related to multiple D/A–A/D conversion.

The transmission method can be either standard (coax cable) or optical.

In the S/PDIF format, both left and right channels are transmitted on one line, they are multiplexed.

Note that the SCMS is included in this S/PDIF format in order to avoid illegal copying. Figure A4.1 shows the S/PDIF data format.

Each subframe will give full information on one audio sample. If the audio sample is less than 20 bits, the remaining bits are meaningless. Two subframes make one frame, giving a left and right channel sample.

192 frames make one block, the channel status bits of one block, i.e., 192 bits taken from each channel bit in each subframe give more information about type of audio, sampling rate, etc.

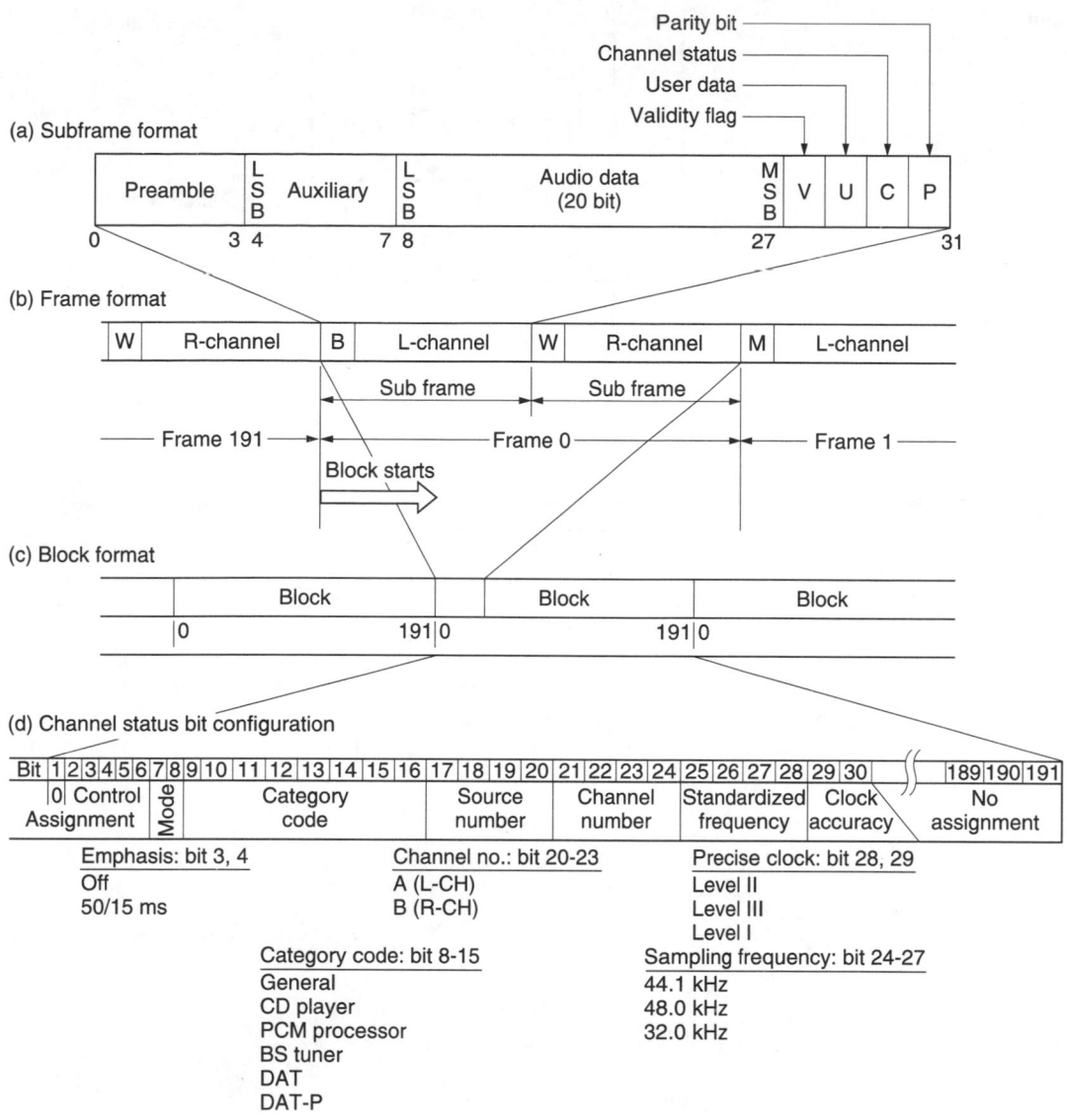

Figure A4.1 S/PDIF data format.

Channel coding

For transmission of the S/PDIF format, a channel coding/ bi-phase modulation type was chosen to ensure proper clocking. If the input data is '1', there is a transient in the middle of the bit; if the input data is '0', there is no transient in the middle of the bit; at the end of each bit there will also be a transient. Figure A4.2 shows the channel coding.

Figure A4.2 Channel coding.

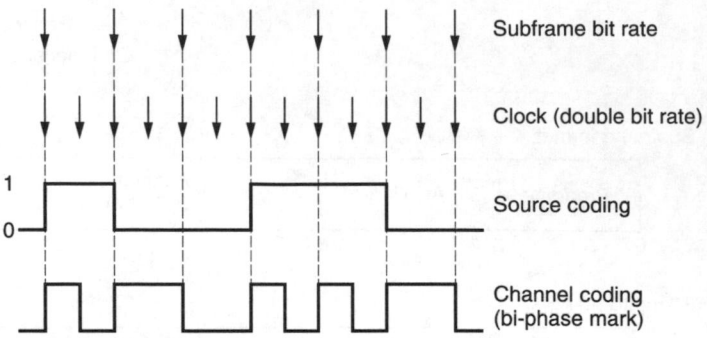

The preambles of each subframe are encoded in a unique way, as shown in Figure A4.3.

Figure A4.3 Preamble coding.

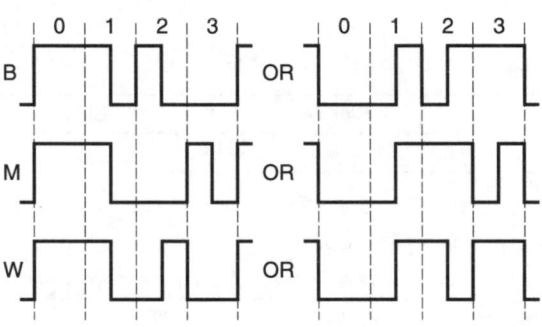

Index

 **Focal Press**

www.focalpress.com

Join Focal Press on-line

As a member you will enjoy the following benefits:

- an email bulletin with **information on new books**
- a regular **Focal Press Newsletter**:
    - featuring a selection of new titles
    - keeps you informed of **special offers, discounts and freebies**
    - alerts you to **Focal Press news and events** such as author signings and seminars
- complete access to **free content** and reference material on the focalpress site, such as the focalXtra articles and commentary from our authors
- a **Sneak Preview** of selected titles (sample chapters) *before* they publish
- a chance to have your say on our **discussion boards** and **review books** for other Focal readers

Focal Club Members are invited to give us feedback on our products and services.
Email: worldmarketing@focalpress.com – we want to hear your views!

Membership is **FREE**. To join, visit our website and register. If you require any further information regarding the on-line club please contact:

Emma Hales, Marketing Manager
Email: emma.hales@repp.co.uk
Tel: +44 (0) 1865 314556
Fax: +44 (0)1865 314572
Address: Focal Press, Linacre House,
Jordan Hill, Oxford, UK, OX2 8DP

Catalogue

For information on all Focal Press titles, our full catalogue is available online at www.focalpress.com and all titles can be purchased here via secure online ordering, or contact us for a free printed version:

USA
Email: christine.degon@bhusa.com
Tel: +1 781 904 2607

Europe and rest of world
Email: jo.coleman@repp.co.uk
Tel: +44 (0)1865 314220

Potential authors

If you have an idea for a book, please get in touch:

USA
editors@focalpress.com

Europe and rest of world
focal.press@repp.co.uk